Cellulose and Its Derivatives: Applications, and Future Perspectives

Cellulose and Its Derivatives: Applications, and Future Perspectives

Editor

Shuangquan Yao

MDPI

Basel • Beijing • Wuhan • Barcelona • Belgrade • Novi Sad • Cluj • Manchester

Editor
Shuangquan Yao
Guangxi University
Nanning, China

Editorial Office
MDPI
St. Alban-Anlage 66
4052 Basel, Switzerland

This is a reprint of articles from the Special Issue published online in the open access journal *Polymers* (ISSN 2073-4360) (available at: https://www.mdpi.com/journal/polymers/special_issues/Cellulose_and_Its_Derivatives_Applications_and_Future_Perspectives).

For citation purposes, cite each article independently as indicated on the article page online and as indicated below:

Lastname, A.A.; Lastname, B.B. Article Title. *Journal Name* **Year**, *Volume Number*, Page Range.

ISBN 978-3-0365-9498-9 (Hbk)
ISBN 978-3-0365-9499-6 (PDF)
doi.org/10.3390/books978-3-0365-9499-6

Contents

About the Editor . . . vii

Sherif S. Hindi, Jamal S. M. Sabir, Uthman M. Dawoud, Iqbal M. Ismail, Khalid A. Asiry, Zohair M. Mirdad, et al.
Nanocellulose-Based Passivated-Carbon Quantum Dots (P-CQDs) for Antimicrobial Applications: A Practical Review
Reprinted from: *Polymers* **2023**, *15*, 2660, doi:10.3390/polym15122660 . . . 1

Enas Hassan, Shaimaa Fadel, Wafaa Abou-Elseoud, Marwa Mahmoud and Mohammad Hassan
Cellulose Nanofibers/Pectin/Pomegranate Extract Nanocomposite as Antibacterial and Antioxidant Films and Coating for Paper
Reprinted from: *Polymers* **2022**, *14*, 4605, doi:10.3390/polym14214605 . . . 33

Xingyu Liang, Yan Yao, Xiao Xiao, Xiaorong Liu, Xinzhou Wang and Yanjun Li
Pressure-Steam Heat Treatment-Enhanced Anti-Mildew Property of Arc-Shaped Bamboo Sheets
Reprinted from: *Polymers* **2022**, *14*, 3644, doi:10.3390/polym14173644 . . . 57

Feitian Bai, Tengteng Dong, Zheng Zhou, Wei Chen, Chenchen Cai and Xusheng Li
Enhancing for Bagasse Enzymolysis via Intercrystalline Swelling of Cellulose Combined with Hydrolysis and Oxidation
Reprinted from: *Polymers* **2022**, *14*, 3587, doi:10.3390/polym14173587 . . . 69

Qiuxiao Zhu, Tingting Wang, Xiaoping Sun, Yuhe Wei, Sheng Zhang, Xuchong Wang and Lianxin Luo
Effects of Fluorine-Based Modification on Triboelectric Properties of Cellulose
Reprinted from: *Polymers* **2022**, *14*, 3536, doi:10.3390/polym14173536 . . . 83

Chunxia Zhu, Shuyu Pang, Zhaoxia Chen, Lehua Bi, Shuangfei Wang, Chen Liang and Chengrong Qin
Synthesis of Covalent Organic Frameworks (COFs)-Nanocellulose Composite and Its Thermal Degradation Studied by TGA/FTIR
Reprinted from: *Polymers* **2022**, *14*, 3158, doi:10.3390/polym14153158 . . . 97

Na Wang, Baoming Xu, Xinhui Wang, Jinyan Lang and Heng Zhang
Chemical and Structural Elucidation of Lignin and Cellulose Isolated Using DES from Bagasse Based on Alkaline and Hydrothermal Pretreatment
Reprinted from: *Polymers* **2022**, *14*, 2756, doi:10.3390/polym14142756 . . . 113

Jinwei Zhao, Zhiqiang Gong, Can Chen, Chen Liang, Lin Huang, Meijiao Huang, et al.
Adsorption Mechanism of Chloropropanol by Crystalline Nanocellulose
Reprinted from: *Polymers* **2022**, *14*, 1746, doi:10.3390/polym14091746 . . . 127

Jiatian Zhu, Yuqi Bao, Luxiong Lv, Fanyan Zeng, Dasong Du, Chen Liang, et al.
Optimization of Demineralization and Pyrolysis Performance of Eucalyptus Hydrothermal Pretreatment
Reprinted from: *Polymers* **2022**, *14*, 1333, doi:10.3390/polym14071333 . . . 139

Yan Li, Mingzhu Yao, Chen Liang, Hui Zhao, Yang Liu and Yifeng Zong
Hemicellulose and Nano/Microfibrils Improving the Pliability and Hydrophobic Properties of Cellulose Film by Interstitial Filling and Forming Micro/Nanostructure
Reprinted from: *Polymers* **2022**, *14*, 1297, doi:10.3390/polym14071297 . . . 153

About the Editor

Shuangquan Yao

Shuangquan Yao obtained his Ph.D. degree majoring in Light Chemical Engineering in 2017 and was a visiting student at the State University of New York, USA, between 2016 and 2017. Since 2019, he has worked at Guangxi Key Laboratory of Clean Pulp and Papermaking and Pollution Control, Guangxi University, in the research group of Prof. Shuangfei Wang, an academic at the Chinese Academy of Engineering. His research focuses on the efficient separation and high-value utilization of lignocellulosic biomass. He has currently published more than 89 papers in Applied Catalysis B: Environmental, Bioresource Technology, Carbohydrate Polymers, ACS Sustainable Chemistry & Engineering, ACS Applied Materials, etc.

polymers

MDPI

Review

Nanocellulose-Based Passivated-Carbon Quantum Dots (P-CQDs) for Antimicrobial Applications: A Practical Review

Sherif S. Hindi [1,*], Jamal S. M. Sabir [2], Uthman M. Dawoud [3], Iqbal M. Ismail [4], Khalid A. Asiry [1], Zohair M. Mirdad [1], Kamal A. Abo-Elyousr [1,5], Mohamed H. Shiboob [6], Mohamed A. Gabal [7], Mona Othman I. Albureikan [2], Rakan A. Alanazi [1] and Omer H. M. Ibrahim [1]

1 Department of Agriculture, Faculty of Environmental Sciences, King Abdullaziz University (KAU), P.O. Box 80208, Jeddah 21589, Saudi Arabia; ralanazi0010@stu.kau.edu.sa (R.A.A.); omerhoooo@gmail.com (O.H.M.I.)
2 Department of Biological Sciences, Faculty of Sciences, King Abdullaziz University (KAU), P.O. Box 80208, Jeddah 21589, Saudi Arabia
3 Department of Chemical and Materials Engineering, King Abdullaziz University (KAU), P.O. Box 80208, Jeddah 21589, Saudi Arabia
4 Department of Chemistry, Faculty of Science, Center of Excellence in Environmental Studies, King Abdullaziz University (KAU), P.O. Box 80208, Jeddah 21589, Saudi Arabia
5 Plant Pathology Department, Faculty of Agriculture, Assiut University, Assiut 71526, Egypt
6 Department of Environment, Faculty of Environmental Sciences, King Abdullaziz University (KAU), P.O. Box 80208, Jeddah 21589, Saudi Arabia
7 Department of Chemistry, Faculty of Science, King Abdullaziz University (KAU), P.O. Box 80208, Jeddah 21589, Saudi Arabia
* Correspondence: shindi@kau.edu.sa; Tel.: +96-656-676-0086

Abstract: Passivated-carbon quantum dots (P-CQDs) have been attracting great interest as an antimicrobial therapy tool due to their bright fluorescence, lack of toxicity, eco-friendly nature, simple synthetic schemes, and possession of photocatalytic functions comparable to those present in traditional nanometric semiconductors. Besides synthetic precursors, CQDs can be synthesized from a plethora of natural resources including microcrystalline cellulose (MCC) and nanocrystalline cellulose (NCC). Converting MCC into NCC is performed chemically via the top-down route, while synthesizing CQDs from NCC can be performed via the bottom-up route. Due to the good surface charge status with the NCC precursor, we focused in this review on synthesizing CQDs from nanocelluloses (MCC and NCC) since they could become a potential source for fabricating carbon quantum dots that are affected by pyrolysis temperature. There are several P-CQDs synthesized with a wide spectrum of featured properties, namely functionalized carbon quantum dots (F-CQDs) and passivated carbon quantum dots (P-CQDs). There are two different important P-CQDs, namely 2,2′-ethylenedioxy-bis-ethylamine (EDA-CQDs) and 3-ethoxypropylamine (EPA-CQDs), that have achieved desirable results in the antiviral therapy field. Since NoV is the most common dangerous cause of nonbacterial, acute gastroenteritis outbreaks worldwide, this review deals with NoV in detail. The surficial charge status (SCS) of the P-CQDs plays an important role in their interactions with NoVs. The EDA-CQDs were found to be more effective than EPA-CQDs in inhibiting the NoV binding. This difference may be attributed to their SCS as well as the virus surface. EDA-CQDs with surficial terminal amino (-NH_2) groups are positively charged at physiological pH (-NH^{3+}), whereas EPA-CQDs with surficial terminal methyl groups (-CH_3) are not charged. Since the NoV particles are negatively charged, they are attracted to the positively charged EDA-CQDs, resulting in enhancing the P-CQDs concentration around the virus particles. The carbon nanotubes (CNTs) were found to be comparable to the P-CQDs in the non-specific binding with NoV capsid proteins, through complementary charges, π-π stacking, and/or hydrophobic interactions.

Keywords: microcrystalline cellulose; nanocrystalline cellulose; carbon quantum dots; functionalization; passivation; antiviral therapy; norovirus

Citation: Hindi, S.S.; Sabir, J.S.M.; Dawoud, U.M.; Ismail, I.M.; Asiry, K.A.; Mirdad, Z.M.; Abo-Elyousr, K.A.; Shiboob, M.H.; Gabal, M.A.; Albureikan, M.O.I.; et al. Nanocellulose-Based Passivated-Carbon Quantum Dots (P-CQDs) for Antimicrobial Applications: A Practical Review. *Polymers* **2023**, *15*, 2660. https://doi.org/10.3390/polym15122660

Academic Editor: Shuangquan Yao

Received: 3 March 2023
Revised: 8 May 2023
Accepted: 16 May 2023
Published: 12 June 2023

1. Introduction

1.1. Nanocelluloses (NCs)

The most prevalent renewable organic substance on Earth is cellulose [1–4]. It can be extracted from plants, algae, and bacteria. Higher plants have primary and secondary cell walls that are made up of cellulose, hemicelluloses, lignin, and pectin. The distinctions between primary and secondary cell walls in terms of chemical make-up and structure are what give rise to the plant kingdom's variety [1].

Several distinct types of nanoscale cellulosic fillers are possible due to the hierarchical and multilevel structure of cellulose. In addition to its nanocrystalline forms, cellulose also exists in an amorphous state that is randomly arranged in a spaghetti-like configuration, giving it a lower density. On the other hand, because they are vulnerable to intense acid attack, amorphous parts can be eliminated while leaving crystalline regions intact under certain circumstances [1,3,5].

As shown in Figure 1, the anatomical structure of a typical wood tissue is clear (Figure 1a), besides showing some macerated fibers (MFs), as presented at Figure 1b, which are the famous natural resource of the cellulose precursor for the MCCs and NCCs products. Furthermore, cellulosic microfibrils are confirmed to be a consequence of crystalline and amorphous regions.

Cellulose-rich sources such as wood contain amorphous regions (Figure 1c) of cellulosic microfibrils that are degraded by acid hydrolysis to produce highly crystalline nanoparticles. Self-organization into a chiral nematic (cholesteric) liquid crystal phase with a helical configuration is a remarkable feature of NCCs. With the help of this remarkable property, dried NCC film can be utilized for security documents, mirrorless lasing, and liquid crystal displays (LCDs and LEDs). Size, dimensions, and other NCCs' geometrical properties are also influenced by the composition of the cellulose precursors [2–11].

The amorphous regions are less dense than the crystalline domains and are constructed in a random manner like a spaghetti pattern (Figure 1c). As a result, the crystalline regions may remain unharmed while the amorphous regions are vulnerable to acid attack. Depending on their precursors, the majority of cellulosic materials contain crystalline and amorphous areas in varying proportions. The way that the cellulose molecules are organized has a significant impact on the physicochemical characteristics of the material. The majority of chemical reagents can only enter amorphous regions and can interact with crystallite surfaces [3,5] to create MCC and/or NCCs (Figure 1d–k).

The MCC is a partially hydrolyzed cellulose [2,4]. It can be obtained industrially from wood or lignocellulosic residues including linters, flosses, stalks, straw, rags, or shells of agricultural crops. The MCC is favorable in pharmaceutical, food, and cosmetic industries due to its high content of crystalline domains of the cellulosic microfibrils [2]. The MCC is one of the most important tableting excipients due to its outstanding dry binding properties of tablets for direct compression.

The nanometer range encompasses sizes larger than a few atoms and smaller than the visible light spectrum [4,11]. Due to their distinct mechanical characteristics, chirality, sustainability, and accessibility, colloidal NCCs rods with high aspect ratio (100–250 in length and 4–10 nm in width) have gained significant popularity in international markets [3,5,11].

Illustrating the large scale of the NCCs noticed in Figure 1, it is arisen from a so-called novel crystallographic phenomenon termed as crystal growth (Figure 1a–j). When NCC particles are approaching each other in an acidic aqueous atmosphere at a relatively warm temperature condition, they are susceptible to agglomerating electrostatically up to microscale particles, termed as pseudo-microcrystalline cellulose (PMCCs), which differ from ordinary MCCs in terms of their origin. For more illustration, the PMCC is agglomerated directly from NCC upon its crystal growth, while the ordinary MCCs are ingrained directly from cellulosic microfibrils harvested from plant's cell wall. Despite both PMCCs and MCCs being situated within the microscale zone, they differ in their internal construction, especially crystallographic properties, namely crystallinity index (CI), crystallite size (CS), and lattice spacing (LS). It is worth mentioning that the NCCs have higher CI and CS, and lower LS than the MCCs, as examined by XRD.

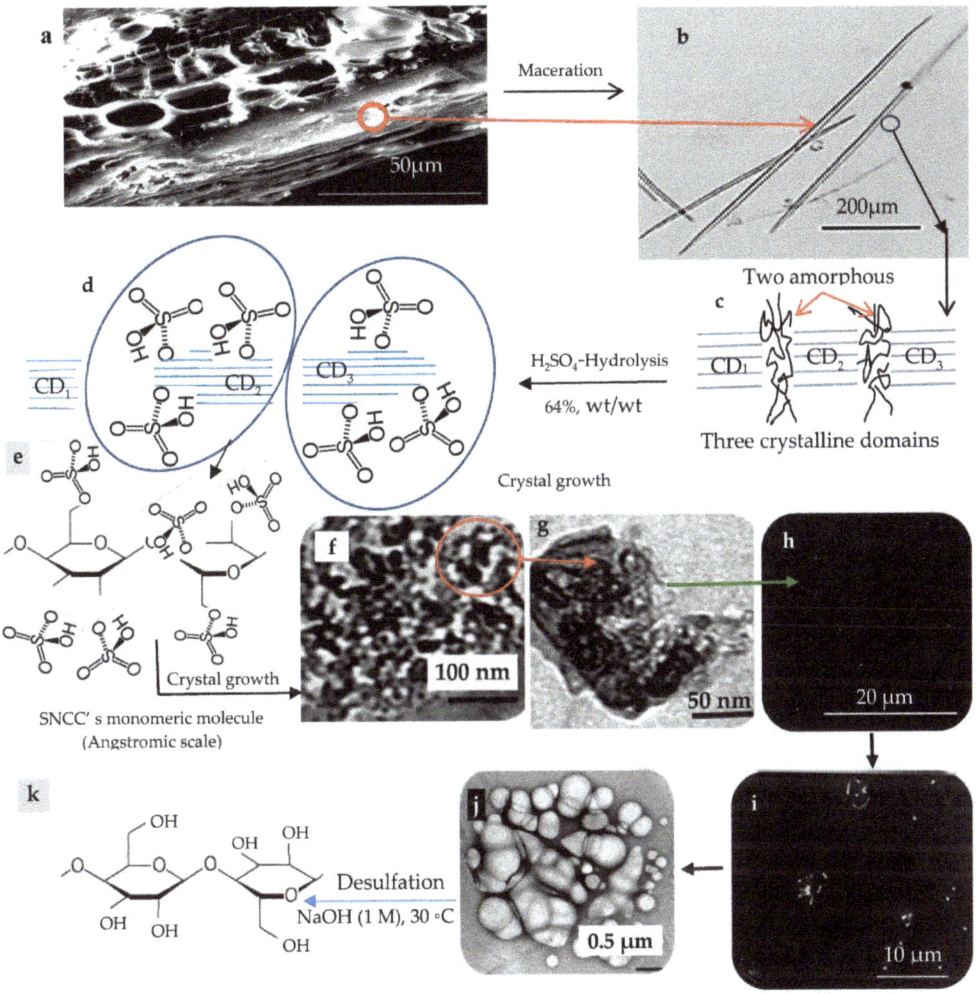

Figure 1. Formation of sulphated nanocrystalline cellulose (SNCCs): (**a**) SEM micrograph of anatomical structure of a typical wood tissue. (**b**) An optical image of macerated fibers. (**c**) The crystalline and amorphous domains within a microfibril. (**d**) SNCCs crystallite grafted by sulphated groups. (**e**) A monomeric molecule of SNCC. (**f**) TEM micrographs of SNCCs colony, and (**g**) Close-up image the SNCCs colony. (**h**) SEM micrographs of spreading and converging of the SCMCs. (**i**) A single colony with wider particles due to agglomeration. (**j**) SCMCs aggregation of single and multiball-shaped microcrystalline cellulose (SMCCs). (**k**) Desulphated cellobiose unit.

For the sulphate groups (Figure 1e), grafted as a result of the acid hydrolysis of cellulosic microfibrils or MCC using sulfuric acid, we think that these functional groups may play an essential role in the agglomeration (upon crystal growth) and dissociation of micrometric particles (upon ingraining the NCCs from MCCs). Before synthesizing the CQDs from the SNCCs, they are desulphated using sodium hydroxide, as is seen in Figure 1k [12].

The nature of the cellulose precursors as well as the hydrolysis circumstances, such as duration, temperature, ultrasound treatment, and material purity, affect the geometrical properties of the NCCs, such as size, dimensions, and form [13–15]. The rod-like structure of the charged NCCs creates an anisotropic liquid crystalline phase above a critical concentration [4].

For the medicinal applications of the NCs, cellulose nanocrystals have the potential to be cutting-edge nanomaterials, according to Marpongahtun, et al. [11]. Due to their exceptional qualities, including good mechanical capabilities, low density, and an inherent renewable nature, nanocelluloses have gained a lot of attention in recent years [7]. These qualities make them ideal candidates for use as reinforcing nanofillers for various polymers. Additionally, CNCs have a number of benefits as starting materials for the creation of carbon structures, including a high fixed carbon content, low cost, and the exceptional ability to assemble into various morphologies (such as single nanoparticles, films, filaments, or aggregates). Then, specific carbon structures can be created by thermally decomposing these various CNC assemblies [16–21].

1.2. CQDs

The CQDs are small carbon nanoparticles (less than 10 nm in size) with some form of surficial passivation [22–24]. They possess the following properties: brightly fluorescent, non-toxic, ecofriendly, made with simple synthetic techniques, and have photocatalytic skills comparable to those of nanoscale semiconductors [11,25–27]. They have also attracted a lot of attention because of their stable photoluminescence properties, wide ranges of excitation and emission spectra, excellent biocompatibility, and little cytotoxicity effects on biological components. C-Dots are crucial in a number of applications [11]. The chemical modification of CQDs by adding organic molecules to their surfaces has created a novel class of materials with unique characteristics [23,28]. Valuable applications cover chemical and biological sensing, bioimaging, nanomedicine, photocatalysis, and electrocatalysis [25]. Among their unique properties is also their photo-catalytic antimicrobial function [27,29]. The CQDs with visible light illumination were found to be highly effective in inhibiting *Escherichia coli* cells, which can be attributed to their photodynamic effect [30].

1.3. P-CQDs

The surface modification of CQDs is an important target for selective application such as bioimaging and can be performed by either passivation (Figure 2a–d) or functionalization (Figure 2e) processes. The passivation process is the infliction of an outer layer of a shield material over a core material via a chemical reaction. This process is performed by constructing a core-shell model combined from passivation agents (such as EDA and EPA) that surround the hard fluorescent core of the CQDs and improve fluorescence emissions [31]. The process of surface functionalization (Figure 2e) involves adding functional groups to the surface, such as carboxyl, carbonyl, and amine groups, which can act as surface energy traps and change the fluorescence emission of CQDs. Surface chemistry or interactions such as coordination, interactions, covalent bonding, etc., can result in surface functionalization. The oxygenous characteristic of carbon quantum dots makes covalent bonding with functionalizing chemicals possible.

Functionalized carbon quantum dots have superior photoreversibility, high stability, strong biocompatibility, and minimal toxicity when compared to naked carbon quantum dots. Occasionally, a small number of molecules can serve as both passivating and functionalizing agents, requiring no extra post-synthesis modifications [29,32,33]. To compete with their rivals, such as organic dye molecules and inorganic semiconductor quantum dots, carbon quantum dots must have a high emission quantum yield. In addition to surface passivation and functionalization, one can use the heteroatom and nitrogen doping of carbon quantum dots to increase the quantum yield by up to 83%.

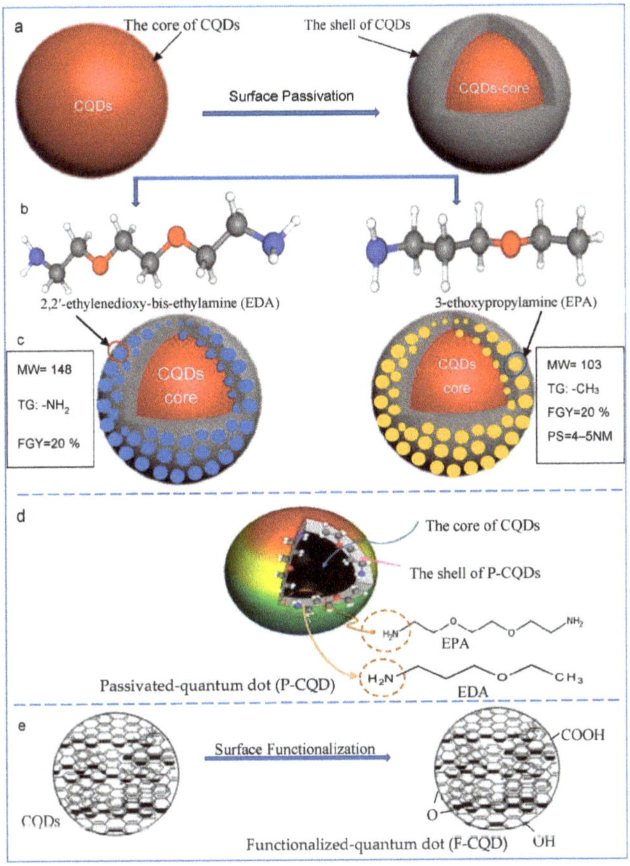

Figure 2. Schematic construction of surface modification of carbon quantum dots (CQDs): a surface passivation. (**a**) The spherical core and the thin layer shell of CQDs, (**b**) chemical structure of 2,2′-ethylenedioxy-bis-ethylamine (EDA) and 3-thoxypropylamine (EPA) which will be grafted on the CQD surface, (**c**) passivated CQDs, where MW is molecular weight of the surface molecule, TG is the terminal group of the surface molecule, FGY is fluorescence quantum yield, and PS is particle size. (**d**) 3D-ilustration schematic model for the grafted EDA and EPA, and (**e**) surface functionalization.

1.4. Applications of CQDs

1.4.1. Industrial Field

CQDs have numerous applications in industrial fields [34] due to their enormous surface area, high electric conductivity, and quick electric charge transfer, as well as high physiochemical properties including crystallization, dispersibility in different liquids, and photoluminescence. In particular, the small size, superconductivity, and rapid electron transfer of CQDs endow the CQDs-based composites with improved electric conductivity and catalytic activity. In addition, CQDs have huge surficial functional groups that could facilitate the preparation of electrical active catalysts, which plays an important role in electrochemistry due to promoting charge transfer within and/or between molecules of these composites. By adjusting the size, shape, surface functional groups, and heteroatom doping of CQDs, it is possible to tailor their distinctive electrical and chemical structures. Rich organic groups that have been grafted onto the surface of CQDs make it possible for water molecules to easily adsorb there while also providing active coordinating sites for metal ions to produce CQD hybridized catalysts. The engineering of the electronic structures of the nearby carbon atoms within CQDs is greatly aided by the heteroatoms (such as N, S, and P) doped in CQDs [35].

Moreover, CQDs have been utilized to fabricate thin-film composite membranes for forward osmosis derived from oil palm biomass into polysulfone, which increased water flux and improved antibacterial performance [36] and nanofiller [37], packaging sheets [38,39], and lubricant additives [40].

Furthermore, there are many applications of CQDs in the field of electrocatalysis such as the reduction and/or evolution of oxygen, hydrogen, or CO_2, as well as bifunctional catalysts, drug delivery, bioimaging, biosensing, optronic, solar cells, light-emitting diodes (LEDs), and fingerprint recovery [35].

1.4.2. Medicinal Field

The CQDs were reported to have medicinal therapeutic effects [15,16,24,40–55]. It was indicated that all these biomass-derived CQDs contain the nitrogen element, which might be from the proteins, amino acids, and nucleic acids in the biomass [34]. Furthermore, metal-containing CQDs (Figure 3) are divided into four types that can be used as antimicrobial agents: metal ion-doped CQDs, metal nanoparticle-decorated CQDs, CD/metal oxide nanocomposites, and CQD/metal sulfide nanocomposites [34]. For photoresponsive CQD, photosensitive agents (photosensitizers) are sensitized by light in the presence of oxygen to generate ROS, such as free radicals and singlet oxygen [56,57]

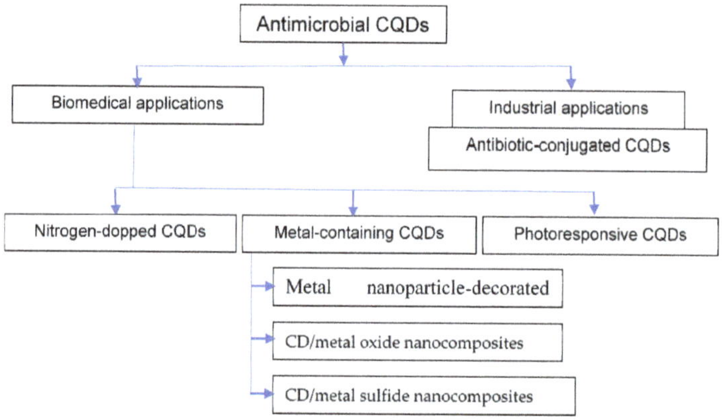

Figure 3. Scheme illustrating the different types of antimicrobial CQDs for biomedical and industrial applications.

Bacterial Field

Several mechanisms were proposed to illustrate the effects of CQDs on typical bacterial cells [38,41,58–62]. The antimicrobial CQDs have been leveraged for coating the surface of orthopedic implant materials [58].

Positively charged CQDs (p-CQDs) effectively combat multidrug resistant (MDR) bacteria and can prevent the formation of biofilms, whereas n-CQDs significantly enhanced bone regeneration [41].

Incorporating water-dispersible and photoluminescent CQDs into bacterial nanocellulose (BNC) film was found to have protective activities against microbes, oxidants, and ultraviolet, making it suitable for food packaging [38]. The behavior of this biocomposite can be revealed by the hydrogen bonding interaction between CQDs and the surficial carboxyl, hydroxyl, and carbonyl groups of BNC, leading to the formation of the CQD–BNC film.

Bacterial biofilm (BB) is a key issue in the medical industry. The BBs were found to be colonized and to damage a wide range of medical implants and devices [59].

In addition, biofilms have major efficacy in many industries including oil, gas, and water production [60] due to causing metal corrosion in engineered systems.

In the complex process of biofilm formation, microorganisms grow and attach to surfaces in an irreversible manner. They also secrete extracellular polymeric substances (EPS) that help the formation of an extracellular matrix (ECM) and alter the phenotype of the organisms in terms of growth rate and gene transcription [61].

Although numerous conventional antimicrobial treatments have been employed to stop the development of mature biofilms or to remove them, these agents frequently require high dosages and are toxic, which poses serious risks to ecological and environmental systems as well as public health. Recent research on the newly created CQDs has had a substantial impact on efforts aimed at both prevention and eradication [11].

There are three general mechanisms illustrating the effects of CQDs on bacterial cells, namely electrostatic interaction, the disruption of the cytoplasm in which the internalization and intercalation occur in the bacterial membrane of the cytoplasm as a result of the charge alteration on the cell surface, and photodynamic inactivation with reactive oxygen species (ROS) production and DNA damage [62].

Viral Field

The semiconductor quantum dots can be used in labeling enveloped viruses for single virus trafficking [63]. Due to the importance of human noroviruses (NoVs), this review was focused on novel technical therapy using CQDs. NoVs are known for acute gastroenteritis outbreaks [64–66]. Great considerations were directed towards chemical and physical disinfection methods of human pathogens, especially norovirus (NoV) known as virus-like particles (VLPs) GI.1 and GII.4 [67,68]. This is due to the fact that there are currently no licensed vaccines or therapeutics for the prevention or treatment of human noroviruses. Moreover, a lack of well-defined infection models for such viruses, either in vitro or in vivo, has limited the development of their countermeasures [69]. Finally, these viruses are known for their resistance against traditional sanitizers and disinfectants [70]. However, most of these methods have been used for antibacterial applications and have been extended to be antiviral agents.

In the last couple of years, the use of nanoparticles as an antiviral strategy has gained much attention [14,15], which includes, but is not limited to, silver nanoparticles [71], gold-copper core-shell [72], TiO_2 coupled with the illumination of low-pressure UV light [73], and passivated-carbon quantum dots (P-CQDs) which should be pithily considered [74].

A group of viruses known as NoVs (family: Calicivirdae) is distinguished by their single-stranded RNA and lack of an envelope. They consist of six genogroups (GI-GIV), which can be further divided into various genetic genotypes based on the sequencing of their capsids [64]. Examples of these are GI, which has nine genotypes, and GII, which has 22 genotypes [64]. It is worth mentioning that human infection is caused by the genogroups GI, GII, and GIV [75].

Gastroenteritis is a common cause of morbidity and mortality among all ages of individuals, and it results from a large variety of bacteria, parasites, and viruses [66]. Serovar is a distinct variation that may occur within a species of bacteria, virus, or immune cells, which can be used for classifying them according to their cell surface antigens.

It was reported by Patel et al. [66] that developing protocols for direct serovar purposes will be an important area of studying NoVs due to these viruses having not yet been cultivated. Expressed VLPs from different NoV strains were found to be useful as immunogens to produce hyperimmune animal sera, and as antigens to assess serum antibody responses to infection. Identifying a cellular NoV receptor and researching potential host–cell interactions have both been performed using VLPs. Human histo-blood group antigens (HBGAs) have been shown to function as NoV infection receptors.

It is known that histo-blood group antigens (HBGAs) determine the host's susceptibility to NoV infection. Protection from viral infection is provided by antibodies that prevent NoVs–HBGAs binding [76]. The NoVs engage in strain-specific infection interactions with

HBGAs in intestinal tissues as receptors or attachment factors [77,78]. It is important to note that HBGAs are terminal assemblies of glycan chains that are complex and highly polymorphic carbohydrates. They mostly consist of the ABO, secretor, and Lewis groups. Moreover, HBGAs are widely distributed on the mucosal epithelia of the gastrointestinal tract, where they serve as anchors for NoVs to begin infection [79]. According to earlier research, intestinal bacteria that express HBGA or synthetic HBGAs may promote NoV infection in B cells [80].

2. Material and Methods

2.1. Synthesis of CQDs

For the synthesis of the CQDs, their precursor differs according to the synthesis route (Figure 4), either a top-down [25] or bottom-up route [81], and whether natural materials (Tables S1 and S2), especially nanocelluloses (MCC and NCC), are used, as shown in Table S2, or synthetic based precursors (Table S3). As shown in Figure 4, the 'top-down' synthetic route breaks down larger carbon assemblies such as graphite, carbon nanotubes, nano-diamonds, or carbon nano-powders [25] into CQDs below 10 nm. On the other hand, the 'bottom-up' synthetic route is a building process that begins from small precursors such as glucose, carbohydrates, citric acid, and polymer–silica nanocomposites [81].

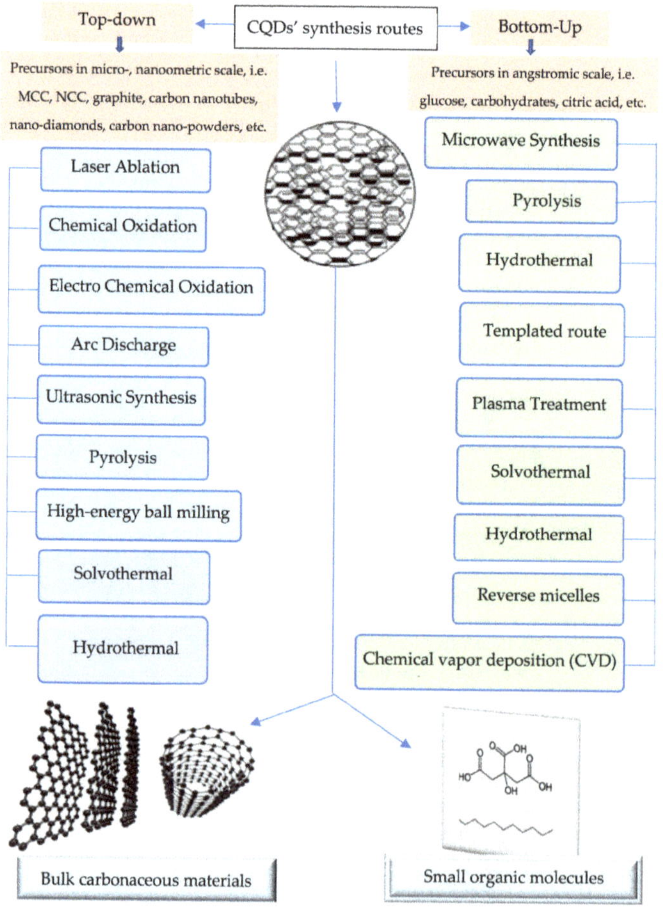

Figure 4. The synthesis routes of carbon quantum dots (CQDs).

The 'top-down' synthetic route breaks down larger carbon structures such as graphite, carbon nanotubes, nano-diamonds, or carbon nano-powders [25] into CQDs below 10 nm using laser ablation [59,74], arc discharge [82], high energy ball milling [83], and electrochemical techniques [84]. In addition, chemical oxidation with acid reinforces quick CODs with good characteristics [85–87].

On the other hand, the 'bottom-up' synthetic route is a building process that begins from small precursors such as glucose, carbohydrates, citric acid, and polymer–silica nanocomposites [81]. There are several synthesis methods via the bottom-up route, namely combustion/thermal/hydrothermal [88–93], plasma treatment [94], supported synthesis [95–97], solution chemistry approaches [91,92,98–100], and the cage-opening of fullerenes [101]. Regardless of their synthesis procedure, the resulting CQDs have different particle sizes, and thereby require complex separation processes to obtain mono-dispersed CQDs. Some of the explored post-synthesis separation techniques include dialysis [89], chromatography [84,102], gel electrophoresis [103], and ultra-filtration [104].

Synthesis of CQDs from Natural Resources

CQDs can be ingrained from plethora of macro-natural resources, as shown in Table S1 [11,16,105–140], as well as nano-natural resources of MCC and NCC (Tables S1 and S2) [3,6–9,141–174]. It was reported by Marpongahtun, et al. [11] that due to the fragmentation of the cellulose structure into tiny bits that carbonize to produce the CDs, CQDs were probably created during the thermal decomposition of the NCCs. Through a straightforward thermal pyrolysis method without any surface passivation, this work has successfully demonstrated the conversion of cellulose nanocrystals from oil palm empty fruit into fluorescing CQDs. The materials produced by pyrolysis at various temperatures exhibit various fluorescence and morphological characteristics.

Through a straightforward thermal pyrolysis method without any surface passivation, this work has successfully demonstrated the conversion of cellulose nanocrystals from oil palm empty fruit into fluorescing CQDs. The materials produced by pyrolysis at various temperatures exhibit various fluorescence and morphological characteristics.

Synthesis of Microcrystalline Cellulose (MCC)

The MCC can be synthesized by different processes such as reactive extrusion, enzyme mediated, steam explosion, and acid hydrolysis. The latter process is performed using mineral acids such as H_2SO_4, HCl, and HBr as well as ionic liquids (Table S2) in order to dissolve the amorphous regions, and, subsequently, the remaining the crystalline domains [6–9]. The degree of polymerization (DP) of the MCC is typically less than 400, while that for NCC is more than 400 extending to several thousands of (1→4)-β-d-glucopyranose units.

After synthesizing MCC (Figure 5), CQDs were ingrained from MCC (Figure 6) and prepared under hydrothermal conditions [16].

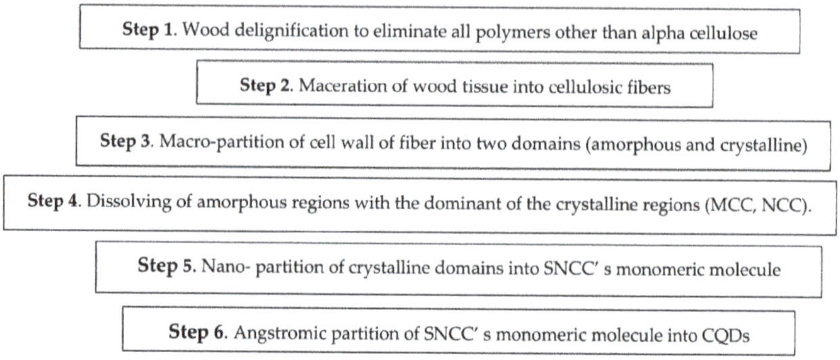

Figure 5. Schematic representation of deriving CQDs from macro-, nano-, and angstrom-structured cellulosic tissues.

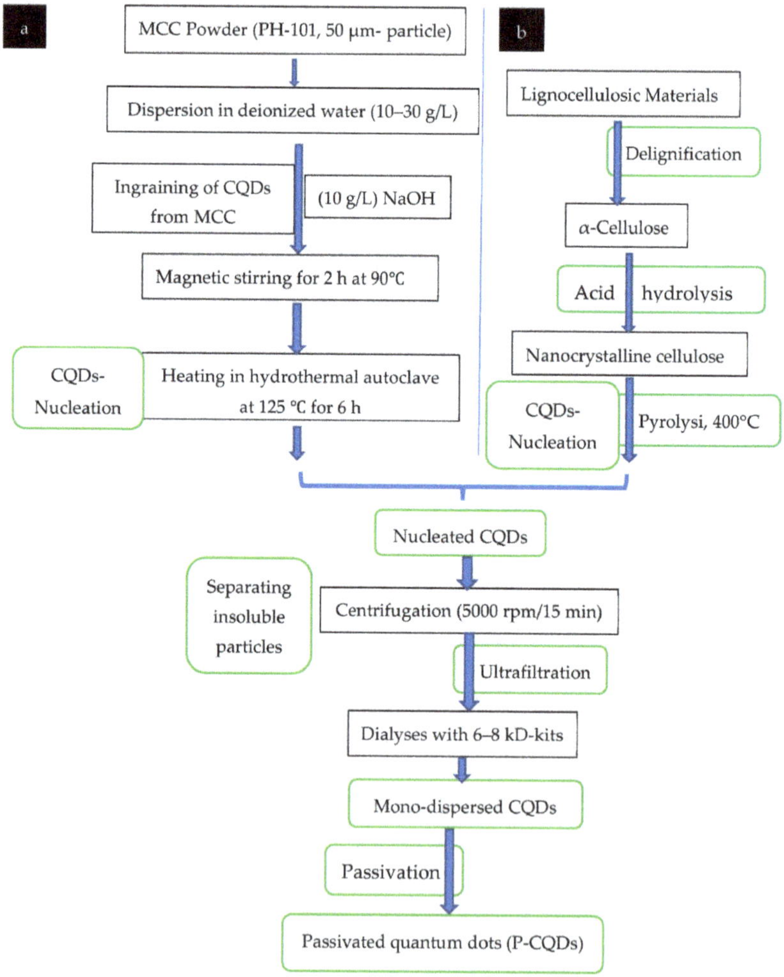

Figure 6. Synthesis of CQDs from (**a**) MCC, and (**b**) NCC [11,16].

Synthesis of Nanocrystalline Cellulose (NCC)

I. Ordinary Synthesis Methods

NCCs were synthesized by Hindi [3,5] by macerating cellulosic fibers with H_2SO_4, 64% *w*/*w* at 70 °C and stirring continuously for an hour. Deionized water was used to dilute the solution up to 20 times in order to stop the reaction. The unhydrolyzed fibers were removed from the suspension by centrifuging it at 1500 rpm, then for 20 min at 14,000 rpm to extract the NCCs. The precipitate was recovered, centrifuged again, and dialyzed until neutralized against deionized water. The NCC synthesis did not undergo any sonication exposure [3,5–10].

II. Cryogenic Synthesis Methods

A novel procedure for synthesizing NCCs, issued in December 2018, was invented by Hindi and Abohassan [6]. The patentability cornerstone of this patent is using liquid nitrogen vapors and/or its liquor for cooling the resultant NCCs to force them to be agglomerated, and, subsequently, precipitated. This cooling technique is termed as the lyophilizing process or cryogenic method. SEM and TEM analyses revealed that the obtained forced precipitates are nanoscale constructions (50–100 nm), although the agglomerated particles may reach up to several micrometers in diameter via the crystal growth phenomenon [6].

To obtain these NCCs, oven-dried MFs powder (10 g) is indirectly subjected to liquid nitrogen vapor. Then, once the frozen concentrated sulfuric acid (98.06%) is melted, it is allowed to saturate the lyophilized MFs powder in a ratio of 1:1 (wt/wt) by suction. A series of successive vacuums and releasing vacuums was performed as an alternative to the blending process to assist and accelerate the complete penetration of the acid into all interior pores of the MF structure. The acid-saturated MFs were re-lyophilized to maintain the synthesized crystalline particles from corrosion by the acid. Once the hydrolysis process had finished, a mixture of cold distilled water with tiny flakes in a ratio of 1:1 wt/wt was added to the NCCs synthesized. Then, two subsequent vacuum filtration steps were performed immediately after the dilution of the NCCs, namely primary filtration and secondary filtration. The primary filtration was performed using a textile (mesh), while the secondary filtration was applied using the Gooch crucible filter [6]. The simplicity of this patent is extended to cover the collection of the NCCs without needing to use the centrifugation process, obtaining nutrients' NCCs via an ordinary washing process without needing to use the dialysis process, and using simple machinery helpful for cheap mass production of the NCCs.

III. Removal of NCCs' sulfate groups

Sulfate groups were hydrolytically cleaved from CNCs following established procedures [12]. About 1% wt. dispersions of CNCs were treated with 1 M NaOH at 60 °C for 5 h. Then, the reaction was quenched by a 10-fold dilution with distilled water and centrifuged at 12,000 rpm at 4 °C for 20,121 min. Consequently, desulfated CNCs were re-dispersed and dialyzed against distilled water for one week to remove traces of NaOH.

Converting MCC into NCC

A simple, fast, economical, and ecofriendly method was invented for producing NCC from MCC using frozen concentrated H_2SO_4 and cooling with hair-shaped ice [10]. There are many benefits of using MCC as a starting material instead of cellulosic fibers for the synthesis of NCC, such as using less of the cellulosic precursor. As it consumes less concentrated acid, the MCC precursor can be easily handled within the synthesis apparatus because it is a powder, compared to the fibrous cellulose, and, finally, MCC is less susceptible to degradation by acid hydrolysis compared to other cellulosic precursors.

Other limitations of conventional NCC production processes include the requirement for the use of expensive machinery, such as sonication baths, sonication props, centrifuges, dryers, lyophilizers, and spray-driers or a complicated series of process steps such as requirements for centrifugation, sonication, neutralization, dialysis, and/or subsequent drying of an NCC product. Consequently, there is a need for a less complicated process that produces NCC in less time and at a lower cost.

After synthesizing NCCs, they are converted to CQDs in the manner illustrated in Figures 5 and 6 [11,16].

2.2. Synthesis of CQDs from Synthetic Resources

Besides the possibility of synthesizing CQDs from natural resources, they can be produced from synthetic precursors such as suitable organic acids, salts, or carbonaceous materials, as presented in Table S3 [14,74,75,83–86,173,175–192]. Moreover, the CQDs can be synthesized from carbon nanopowders by the top-down route using nitric acid (8 M) under reflux for 48 h (Figure 7). After cooling the reaction liquor and centrifugation at 1000× *g*, the supernatant is discarded, while the precipitate is dispersed in water. The new liquor is dialyzed and centrifuged at 1000× *g* to retain the supernatant. Upon the subsequent dehydration, nanometric CQDs can be collected and are used in the subsequent functionalization process [15,29,32,33].

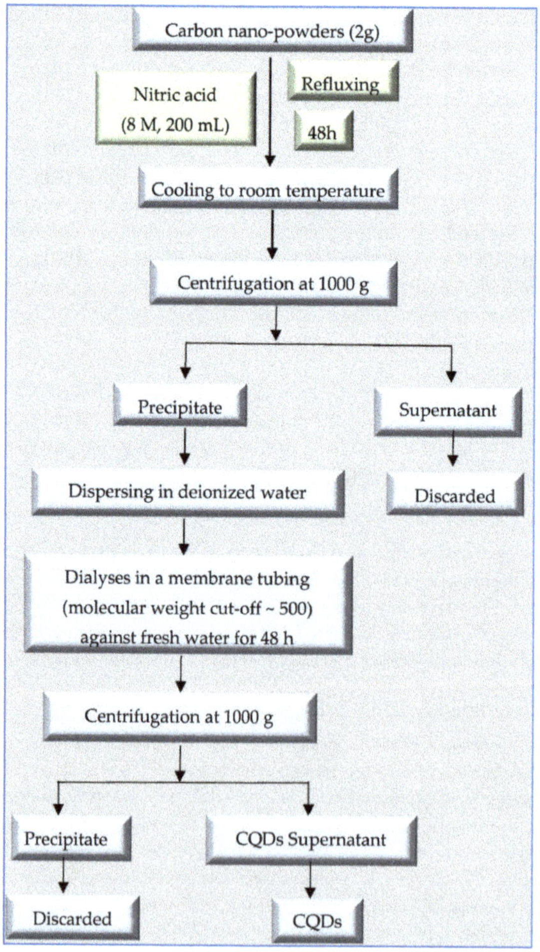

Figure 7. Synthesis of CQDs from carbon nanopowders.

2.3. Syntheses of P-CQDs

The difference between the functionalization and passivation processes of a CQD to produce P-CQDs is shown in Figure 2. For the passivation process, two different P-CQDs can be synthesized, namely EDA-CQDs using 2,2′-ethylenedioxy-bis-ethylamine [15,29], and EPA-CQDs using 3-ethoxypropylamine, as presented in Figure 2b–d [15,29,32,33]. Furthermore, the surface functionalization of CQDs can be achieved by gifting chemical groups such as carboxyl, hydroxyl, oxygen atom, etc., to the CQDs′ surface (Figure 2e).

As shown in Figures 2 and 8, the synthesized CQDs are chemically passivated to yield either EDA- or EPA-CQDs [40]. First, the CQDs are allowed to react with $SOCl_2$ in order to form the acid chloride intermediates which are more active than their carboxylic group precursors, and form amides by a reaction with the amine-terminated molecules [15].

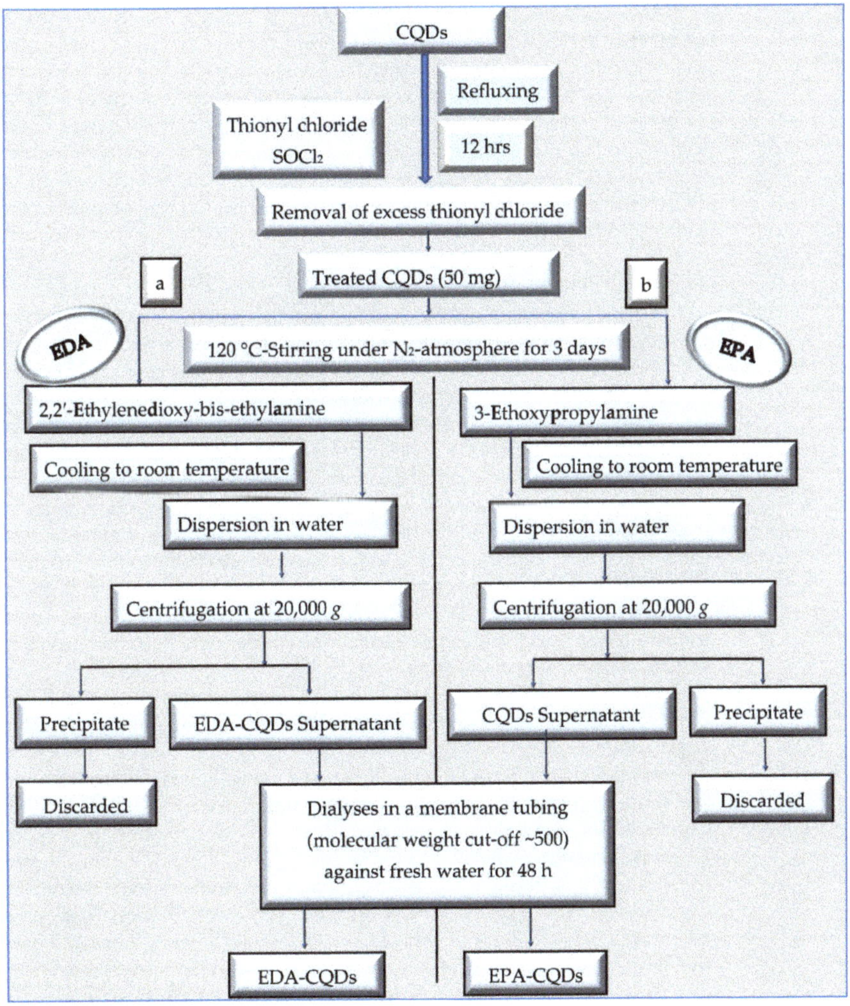

Figure 8. Synthesis of chemically passivated-CQDs (P-CQDs): (**a**) EDA-CQDs, and (**b**) EPA-CQDs [15,32,33].

2.3.1. Characterization of Nanocelluloses

There are huge studies that were conducted to characterize the suitability of different natural precursors for the synthesis of CQDs [190,191].

In order to evaluate the MCC quality, several characteristics were tested, including particle size, density, compressibility index, angle of repose, powder porosity, hydration swelling capacity, moisture sorption capacity, moisture content, crystallinity index, crystallite size, and mechanical properties such as hardness and tensile strength. Furthermore, thermogravimetric analysis (TGA) and differential thermal analysis (DTA) or differential scanning calorimetry (DSC) are also important properties to evaluate the thermal behavior of the MCC under thermal stresses.

2.3.2. Characterization of CQDs

Several techniques are used to characterize the CQDs such as nuclear magnetic resonance (NMR), X-ray diffraction (XRD), transmission electron microscope (TEM), Fourier transform infrared spectroscopy (FTIR), fluorescence spectrophotometer, ultraviolet (UV) spectroscopy, UV–vis absorption spectra, and atomic force microscopy (AFM), as reported by Xu et al. [192], Singh et al. [59,193], and Joo et al. [63]. Moreover, the interactions of various polypeptides with individual carbon nanotubes (CNTs), both multiwall (MW) and single wall (SW), were investigated by Li et al. [194] using atomic force microscopy (AFM). The characterization procedures were performed to test for bacteria [13,61,195–197] and viruses [15,16,24,197–209].

2.3.3. Evaluation of Viral Therapeutic Efficacy of P-CQDs

Briefly, saliva samples from healthy adult volunteers, including blood type A, B, and O, are collected [15]. The pretreatment process is performed using phosphate-buffered saline (PBS). The plates are blocked with Super-Block T20 (PBS) blocking buffer as shown in Figure S5. Collected saliva samples from individuals with blood types A, B, and O are collected and pretreated with phosphate-buffered saline (PBS). The samples are immediately boiled for 5 min and centrifuged at 10,000× g for 5 min. The collected supernatant is diluted to 1:2000 in PBS. For coating the plates with HBGAs, an aliquot of 50 μL saliva dilution was used to coat 96-well plates at 4 °C overnight. Unbound saliva was removed and the wells were rinsed three times with super-block T20 (PBS) buffer to yield a high signal-to-noise ratio in the detection system [15]. Noticeably, a similar difference in effectiveness between EDA-CQDs and EPA-CQDs was found in their antiviral function [27], where EDA-CQDs were more effective than EPA-CQDs in inhibiting norovirus virus-like particles binding to histo-blood group antigen receptors, due primarily to the difference in surface charge status between the two CQDs.

The final solution is treated with 3,3′,5,5′-tetramethylbenzidine (TMB) peroxidase developer, and the absorbance is measured at the wavelength of 450 nm using a microplate reader (Figure S2).

As presented at Figure S3 [15,77,79], the enzyme-linked immunosorbent assay (ELISA) test is used to evaluate the binding capacity between the EDA-CQDs and EPA-CQDs and human NoVs–VLPs antibody (GI.1 or GII.4) using two standard antibodies: (1) primary antibody, namely mAb 3901 for the strain 'GI.1', or mAb NS14 for the strain GII.4, and (2) secondary antibody such as horseradish peroxidase (HRP) having 44,173.9-dalton glycoprotein with 6 lysine residues for labeling goat anti-mouse IgG antibody [15]. It produces a colored, fluorometric, or luminescent derivative of the labeled molecule when incubated with a proper substrate, allowing it to be detected and quantified [194].

The sodium dodecyl sulphate polyacrylamide gel electrophoresis (SDS-PAGE) test and Western blotting protocol (Figure S4) are used for evaluating the effect of EDA- and EPA-CQDs on VLP capsid protein. Different concentrations of EDA- and EPA-CQDs (20 or 60 μg/mL) are applied to treat VLPs (GI.1 or GII.4) [15].

On medium-binding 96-well polystyrene plates, EDA- and EPA-CQDs at various doses ranging from 0 to 60 g/mL are employed to treat either GI.1 or GII.4 VLPs. The reaction solutions are removed and the wells are twice washed with phosphate-buffered saline (PBS) following the addition of a specific amount of PBS, agitation, and incubation for 30 min.

For one hour, PBS-blocking buffer is used to block the wells. Each well is twice washed with phosphate buffered saline with tween 20 (PBST) buffer after the blocking solution has been discarded. After that, 50 aliquots of 1 g/mL anti-GI are added. To interact with the bound GI.1 or GII.4 VLPs, 1 VLP antibody (mAb 3901) or anti-GII.4 VLP antibody (mAb NS14) is added to each well. Each well is put into a solution containing goat anti-mouse IgG that has been HRP-labeled before being incubated at 37 °C for 1 h. The wells are then twice rinsed with PBST. PBST is used to wash the plates. Tetramethylbenzidine (TMB) peroxidase is used to create the end product, and its absorbance is measured at 450 nm.

The gel containing the VLPs-treated P-CQDs (GI.1-VLP/EDA-CQDs and GI.1-VLP/EPA-CQDs) is used for staining (Figure S5) and Western blotting (Figure S6). The gel used for staining was previously prefixed with a 50% methanol and 7% acetic acid solution, stained by GelCode Blue stain, and imaged using an infrared imaging system as explained.

For 30 min, 1.5 mL centrifuge tubes were continuously shaken at the setting level of 2 at an ambient temperature. Following the CQDs treatments, 5 μL of 1 × NuPAGE LDS sample buffer, 2 μL of 1 M DTT, and 3 μL deionized water were added to each tube. After 10 min of incubation at 70 to 80 °C, all of the samples were placed onto 2 precast 1.0 mm × 10-well NuPAGE® 4–12% Bis-Tris gels (Life Technologies, Grand Island, NY, USA). For each well, the loading volume was adjusted to be 10 L.

The gels were run for one hour at 200 V in 1 MOPS SDS running buffer. One gel was used for Western blotting, and the other was used for staining. The gel for staining was pretreated for 15 min with a solution of 50% methanol and 7% acetic acid, and then washed three times for 5 min with deionized water. The GelCode Blue stain was applied to the gel and shaken continuously for 1 h before being washed with deionized water for 1 h to remove the stain. Infrared imaging equipment was then used to image the gel (Figure S5).

As presented at Figure S6, regarding Western blotting, the gel is treated with NuPAGE® Transfer Buffer and 10% MeOH packaged within the nitrocellulose membrane using Hoefer Semi-Dry Transfer Apparatus. The membrane is blocked with blocking buffer and PBS.

The gel was transferred to a nitrocellulose membrane for Western blotting (Figure S6), which is blocked using blocking buffer and PBS at room temperature for one hour. Both GI.1/antibody mAb 3901 and GII.4/antibody mAb NS14 underwent primary antibody treatment using PBST and blocking buffer. After incubating the antibody solution at 4 °C with gentle shaking for the entire night, it was discarded. The membrane was treated with 0.5 μg of goat anti-mouse IRDye® 800CW antibodies in PBST and blocking buffer at ambient temperature for 1 h after being rinsed 5 times with PBS plus 0.05% Tween 20 (PBST) for 5 min each time. The membrane was first washed with PBST five times for approximately five minutes each while being shaken, followed by a soak in deionized water, and then an IR imaging system was applied (Figure S6).

3. Results and Discussion

The properties of ordinary CQDs as well as P-CQDs (EDA- and EPA-CQDs) and their inhibitory rate on microbial defense are presented in Table 1 (for viruses) and Table 2 (for bacteria and fungi).

Table 1. Estimated mean values [1] of the P-CQDs for inhibition of the NoVs.

Property of the CQDs	[2] HBGA's Type	Concentration μg.mL^{-1}	VLS Strains	P-CQDs	
				EDA	EPA
Inhibitory rate of CQDs on NoVs, %		8		90	8
		16		92	24
		32		88	26
Inhibition of HBGA binding, %	A	2	GI.1	93.6	53.3
			GII.2	88	61.2
		5	GI.1	100	93
			GII.2	100	100
	B	2	GI.1	74.5	36.4
			GII.2	78.2	38.2
		5	GI.1	99.1	75.5
			GII.2	100	81.8
	O	2	GI.1	79.1	54.4
			GII.2	59.5	38.9
		5	GI.1	100	77.2
			GII.2	100	79.3

[1] Mathematically-estimated from Dong et al. [15]; [2] Histo-blood group antigens.

Table 2. Mean values of the P-CQDs for inhibition of bacteria.

Property of the CQDs		[1] AT Hour	[2] Conc. μg.mL^{-1}	P-CQDs		CQDs		Reference
				EDA	EPA	[3] HT	[4] IR	
Particle size, nm				4–5	4–5			[200]
Molecular weight				148	103			
Surficial terminal group				$-NH_2$	$-CH_3$			
Fluorescence quantum yield, %				~20	~20			
Viable cell number of bacteria, CFU/mL			0	11×10^6	11×10^6			
			0.1	9×10^3	5×10^6			
			0.2	0.4×10^3	1.5×10^6			
Inhibitory effect OF ADE-CQDs on a bacterial biofilm formation, %		1	10	95.86				[13]
			20	100				
			30	100				
		2	10	72.2				
			20	96				
			30	100				
		3	10	34.25				
			20	41				
			30	50				
Minimum inhibitory concentration, μg/mL	[5] Gram$^+$-bacterium					250	350	[16]
	[6] Gram$^-$-bacterium					100	300	
	[7] Unicellular fungi					350	400	

[1] Addition time, [2] Concentration, [3] Hydrothermally-synthesized, [4] Infrared-assisted synthesized, [5] *Staphylococcus aureus*, [6] *Escherichia coli*, [7] *Candida albicans*.

3.1. *Viral Therapy of P-CQDs on NoVs*

3.1.1. Absolute Efficacy

P-CQDs have an inhibitory effect on the binding of VLPs to HBGA receptors. Human HBGAs are recognized by NoVs as attachment factors or receptors having a significant impact on the host's susceptibility to NoV infection [200,201]. It has been discovered that norovirus binding to HBGAs is extremely varied but strain-specific.

Based on the binding of different norovirus strains to HBGAs, several binding patterns have been discovered and divided into two primary binding groups [206], and a model of norovirus/HBGA binding has also been put forth [78]. Other studies revealed that Norwalk VLPs lacked the binding to saliva samples obtained from nonsecretors, and that saliva from type B individuals did not bind or only weakly bound to the Norwalk virus [199]. A retrospective study revealed that type O individuals had a significantly higher infection rate than those with other blood types [198].

The approximate mean values obtained from studying the impact of EDA- and EPA-CQDs on the binding of GI.1 and GII.4 VLPs to salivary HBGAs from blood type A, B, and O are reported in Table 1. The binding to type 'A' salivary HBGA receptors was entirely blocked for GI.1 VLPs treated with EDA-CQDs at 5 g/mL (100% inhibition), demonstrating a highly effective inhibition impact of EDA-CQDs on GI.1 VLP's binding to HBGA receptors.

When GII.4 VLP bound to type A HBGA receptors was treated with 5 g/mL EDA-CQDs, the same quantitative inhibition (100%) was seen (Table 1). Even at lower CQD concentrations, the inhibitory effect persisted, as seen by the more than 80% inhibition in GI.1 and GII.4 VLP bindings after treatment with 2 g/mL EDA-CQDs (Table 1).

The findings revealed that the diverse strains of VLPs had a similar inhibitory effect to EDA-CQDs on HBGA receptor binding. Although slightly less potent on a rising concentration basis, EPA-CQDs were still quite effective in the same inhibition.

As seen in Table 1, treatment with EPA-CQDs at concentrations of 5 g/mL and 2 g/mL inhibited the binding of GI.1 VLPs to type A HBGA receptors by 91% and 51%, respectively. A similar suppression of GII.4 VLPs was seen after treatment with EPA-CQDs (Table 1). These results demonstrate that EDA- and EPA-CQDs had equally potent inhibitory effects on the two strains of VLPs' ability to bind to type B and type O HBGA receptors. Investigating the inhibition to type A HBGA receptors, shown in Table 1, also exhibited the dot concentration dependence and difference between the two types of CQDs (EDA- and EPA-CQDs). The findings for the two different strains of VLPs indicated that EDA-CQDs were more efficient than those for EPA-CQDs in preventing VLP binding to all three types of HBGA receptors.

The differing surface charge status and hydrophobicity characteristics between the two types of CQDs may be to blame for the different effectiveness. While EPA-CQDs with surface methyl (-CH_3) terminal groups are not charged, EDA-CQDs with surface amino (-NH_2) terminal groups tend to be altered positively at physiological pH (-NH_3^+).

Even though the mechanistic details of the interactions of the CQDs with the VLPs and the resulting inhibition effects are probably very complex [78,80], the negatively charged VLPs should be more attractive to the positively charged EDA-CQDs, leading to a higher "local concentration" of the dots around the VLP particles. One of the potential explanations for the CQDs' observed strong inhibitory effects is that they bind to the surface of the VLPs and physically block the binding sites for the HBGA receptors.

According to the X-ray crystal structure of the NoVs prototype GI.1 [207], it has two domains, namely the shell (S) domain and the protrusion (P) domain. The HBGA receptor binding interfaces are found at the top of the 'P' domain and contain pockets for binding carbohydrates. The binding of HBGAs to the viral capsid protein is stabilized by these pockets, which include many dispersed amino acid residues that form large hydrogen bond networks with individual saccharides [208,209]. However, some of the complexities in the HBGA binding interactions have been reported [11], including capsid P domain loop movements, alternative HBGA conformations, and HBGA rotations. This is because the binding of norovirus to human HBGA is a typical protein–carbohydrate interaction in which the protruding domain of the viral capsid protein serves as an interface for the oligosaccharide side-chains of the HBGAs40.

In fact, employing sera from immunized animals or sick humans aids the blockage of NoV HBGA binding sites and has been employed as a proxy for a NoV neutralization experiment [202,203]. It was discovered that protection against infection in NoV-vaccinated chimpanzees and against sickness in infected human volunteers could be connected with the serum's capacity to prevent VLP–HBGA interactions [65,76]. These investigations suggest that a promising method for avoiding HuNoV infection is to inhibit the HuNoV capsid from recognizing its binding sites on host cells. As a result, the CQDs reported efficient inhibition of the NoV VLPs, as shown in Table 1; it may be viewed as an application of this tactic.

In addition, the antiviral activities of the P-CQDs are summarized in Figure 9. NoV infects the host cell via surficial cell receptors leading to the formation of syncytium and subsequent gradual degradation, which are responsible for spreading viral infection. With the recent advances in nanotechnology, NoV infection can be detected and inactivated specifically through different pathways such as targeted tagging or by blocking surficial viral proteins.

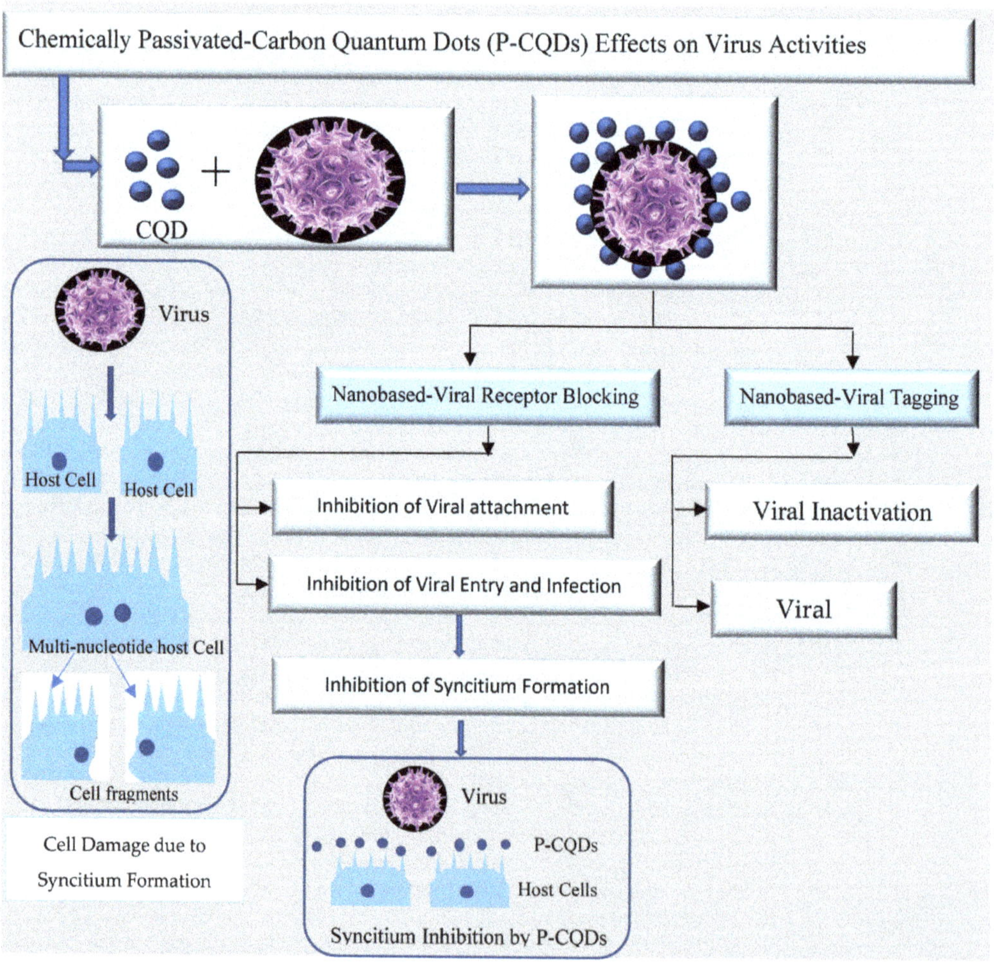

Figure 9. The potential role of the passivated-carbon quantum dots (P-CQDs) in detecting and inactivating human infection by noroviruses through targeted tagging and by blocking viral surface proteins.

It is well known that HBGAs, as receptors, play an essential role for host susceptibility to NoV infection [189,210]. Although the binding between NoVs and HBGAs is highly diverse, it is strain-specific. Numerous patterns of such binding have been identified and classified into two major groups, proposing a suitable model [73]. It was indicated by Hutson et al. [198] that type O individuals had a significantly higher infection susceptibility rate than those with other blood types. On the other hand, other studies showed that Norwalk VLPs did not bind to saliva samples collected from nonsecretors, especially for type B individuals [198].

Both EDA- and EPA-CQDs showed a strong inhibition effect in the binding between the two strains of VLPs and the HBGA samples collected from type B- and type O-individual receptors [15]. Furthermore, the EDA-CQDs were found to be more effective than EPA-CQDs in inhibiting VLPs' binding to HBGA receptors. This difference may be attributed to the quality (positive, negative, or neutral) and amount of surficial charge for the P-CQDs as well as the virus surface. EDA-CQDs with the surficial terminal amino (-NH_2) groups are positively charged at physiological pH (-NH^{3+}), whereas EPA-CQDs with surficial terminal methyl (-CH_3) groups are not charged. As VLPs are negatively charged, they are perhaps more attractive to the positively charged EDA-CQDs, leading to a higher accumulation of the P-CQDs around the NoV particles. However, the mechanism of the interactions between the P-CQDs and viruses is likely very complex [78,80].

The strong inhibitory effects of P-CQDs against the NoVs can be attributed to the physical blocking that occurred as a result of the binding between the P-CQDs and the surficial active sites on the virus. XRD investigations revealed that the NoV strain 'GI.1' contains two domains: (1) the shell domain (SD), and (2) the protruding domain (PD) which contains the HBGA–carbohydrate complex formed via a hydrogen bond network [198]. Accordingly, the binding between the human HBGA and P-domain is a carbohydrate–protein complex [211]. Some of these complexes include movements of the binding interaction as well as conformations and/or rotations of the HBGA. The blocking of NoVs' binding sites by using sera from immunized individuals could be classified under such a strategy [80,201,202]. Moreover, Dong et al. [15] showed that the EDA-CQDs are more effective in the inhibition of the NoVs (GI.1), binding to the first antibody (mAb 3901) compared to binding to the GII.4 mAb NS14. This difference between EDA- and EPA-CQDs may be attributed to their difference in surficial charge status.

In addition, for both EDA- and EPA-CQDs treatments using different concentrations, the NoV-strain 'GI.1' was inhibited in its binding to mAb3901 antibodies more effectively than the strain 'GII.4' in its binding to mAb NS14 [15]. This might be due to the capsid structure difference in the two strains of NoVs (GI.1 and GII.4), involving in NoVs–antibody interactions. Furthermore, no significant difference was detected in the P-CQDs inhibitory effect on the binding of both strains of NoVs to HBGA receptors [15].

After the P-CQDs treatments, the quantity of NoVs fragments found by Western blotting was found to be unchanged. It is known that mAb 3901-antibody may bind to both the full-lengths of 58 KDa and a 32 KDa of the protein fragments found in the P domain of the NoVs strain (GI.I) protein bands [203–205], identifying a continuous epitope on the C-terminal of the capsid protein [205]. Furthermore, the lowest band in the Western blot is probably a fragment that contains this sequence because the antibody mAb 3901 also identifies a domain between amino acids 453 and 495. The mAb NS 14-antibody binds to the capsid protein and additional protein fragments that contain the identified epitopes, just like it does for the other NoVs strain (GII.4). Obviously, the protein band patterns found in Western blotting for both strains of the NoVs (GI.1 and GII.4) were nearly identical to those found in SDS-PAGE found by GelCode Blue staining. Therefore, the findings portray that the viral proteins were not degraded by the P-CQDs, since these proteins still retained their virgin sequences of the amino acids and were able to react with their antibodies again [15].

3.1.2. Comparative Efficacy for Other Carbon Nanomaterials (CNMs)

The most crucial characteristics governing the behavior of CQDs and subsequent applications are absorption, photoluminescence (PL), and electroluminescence [28]. Generally, the optical absorption peaks of CQDs in the UV-visible region are usually estimated as the π-π^* transition of sp^2 conjugated carbon and n-π^* transition of hybridization with a heteroatom such as N, S, P, etc. Surface passivation or modification processes can be used to modify the absorption property [35].

The PL is one of the most wonderful features of CQDs. Generally, the distinct dependence of the emission wavelength and intensity is one of the uniform features of the PL for CQDs. The reason for this unique phenomenon may be the optical selection of nanoparticles with a different size, or CQDs with different emissive traps on the surface. The variation in particle size and PL emission can be reflected from the broad and excitation-dependent PL emission spectrum [73,74].

Zhang et al. [79] studied the emission behaviors of CQDs at 470 nm wavelength with various concentrations. It was found that the PL strength of the CQDs solution first increased and then decreased with the increase in their concentration [35].

Similar to semiconductor nanocrystals, CQDs can display electroluminescence (ECL), which can be used in electrochemical fields [35]. It was reported by Zhang et al. [79] that a CQDs-based light-emitting diodes (LED) device could be used, in which the emission color ranging from blue to white can be controlled by the driving current.

In order to comprehend the luminescence process of CQDs based on the band gap emission of the conjugated p domain and the edge effect generated by another surface defect, two models of CQDs were put forth by SK et al. [19]. The quantum confinement effect (QCE) of p-conjugated electrons in the sp^2 atomic framework is the source of the photoluminescence (PL) features of the fluorescence emission of CQDs from the conjugated p domain, which may be modified by altering their size, edge configuration, and shape. The sp^2 and sp^3 hybridized carbon and other surface defects of CQDs cause fluorescence emission, and even the fluorescence intensity and peak position are connected to this defect.

At low pHs, the interaction is dominated by adhesion forces resulting from electrostatic interactions between the protonated amine groups of polylysine and carboxylic groups on acid-oxidized multi-wall carbon nanotubes (Ox-MWCNTs), whereas at high pHs, adhesion forces via hydrogen bonding between the neutral $-NH_2$ groups of polylysine and the $-COO^-$ groups of the Ox-MWCNTs are detected [193].

Furthermore, it was discovered that the adhesion force for oxidized multiwalled carbon nanotubes (Ox-MWCNTs) increased with the oxidation time, while it was negligible for oxidized single-wall carbon nanotubes (Ox-SWCNTs) because the latter had carboxylate groups attached only to the nanotube tips as opposed to both the sidewall and the tips. Additionally, it was shown that proteins with aromatic moieties, such as poly-tryptophan, exhibited a stronger adhesion force with Ox-MWCNTs than polylysine because of the additional pi-pi stacking interaction between the polytryptophan chains and CNTs. [193].

The binding ability between various CNMs and viral capsid proteins has been reported [17,18,20,21,193]. The CNTs and P-CQDs can be non-specific binders to NoVs' capsid proteins through complementary charges, π-π stacking, and/or hydrophobic interactions [17,18,194].

It was reported that van der Waals forces are responsible for the binding between fullerene (C_{60}) and lysozyme, whereas polar solvation and entropy were reported to be detrimental to this binding [20]. Furthermore, C_{60} was reported to inactivate HIV-proteases by integrating with proteins to form hybrid functional assemblies [21]. Similarly, the inhibition of NoVs' capsid protein by the P-CQDs may occur due to the combination of several driving forces for blocking the active sites on NoVs with the HBGA receptors [15].

The van der Waals force was found to be the primary driving factor responsible for the binding between fullerene and lysozyme, whereas polar solvation and entropy are detrimental to such bindings [20]. It was shown that C60 might suppress the activity of HIV proteases by integrating with proteins to form hybrid functional assemblies [21], which is more pertinent to the blockage of receptor sites.

As a result, a conceptually similar explanation for the observed inhibition of NoVs-VLPs could be that the CQDs interact with the capsid protein of VLPs by combining several driving forces, which then prevents the active sites on NoVs-VLPs from binding to HBGA receptors.

In addition, a similar surface charge effect has been reported on silver nanoparticles' antimicrobial activity, where positively and negatively charged silver nanoparticles exhibited the highest and lowest bactericidal activities, respectively [29]. As such, there have been recent studies on inducing charges onto the surface of silver nanoparticles for higher antimicrobial efficacy [32,33,200]. The results reported here suggest that the same strategy may be exploited in the design and preparation of CQDs with higher antibacterial efficacy [15].

3.2. *Bacterial Therapy Efficacy of P-CQDs*

Biofilm formation is a complex process in which microorganisms irreversibly attach to and grow on a surface and produce extracellular polymeric substances (EPS) that facilitate the attachment and formation of an extracellular matrix (ECM), resulting in the altered phenotype of the organisms with respect to growth rate and gene transcription [61].

Important characteristics of EDA-CQDs and EPA-CQDs are listed in Table 2. A Gram-positive laboratory model bacteria, *Bacillus subtilis*, was used to evaluate the antimicrobial efficiencies of each of the CQDs with different surface passivation (EDA-CQDs and EPA-CQDs) for probing the surface charge effect. As is clear in Table 2, EDA- and EPA are small molecules, with molecular weights of 148 and 103 g/mol, respectively, and they are structurally similar but their corresponding CQDs differ in terms of terminal groups on the dot surface, $-NH_2$ in EDA-CQDs vs. $-CH_3$ in EPA-CQDs. The former can be positively charged at physiological pH as $-NH_3^+$, but not the latter. The observed fluorescent quantum yields of the EDA-CQDs and EPA-CQDs used in the study were both ~20% [196].

Additionally, it is evident from Table 2 that EPA-CQDs and EDA-CQDs at 0.1 and 0.2 mg/mL to *Bacillus subtilis* cells have antibacterial action in terms of a reduction in viable cell counts after treatments with light illumination for one hour. At 0.1 mg/mL, EPA-CQD treatment minimally reduced the number of viable *Bacillus subtilis* cells, but EDA-CQD therapy was significantly more successful, causing a 3.26 log drop in viable cells.

EPA-CQD treatment reduced the number of viable *Bacillus subtilis* cells by about 0.84 log at a CQD concentration of 0.2 mg/mL, whereas EDA-CQD treatment reduced the number of viable cells by around 5.8 log at the same concentration. EDA-CQDs consistently outperformed EPA-CQDs in terms of their antibacterial action toward *Bacillus subtilis* cells at both tested CQD doses, as was to be expected.

These findings demonstrated how crucial surface charge is for CQD interactions with bacteria and the performance of their antibacterial activity. Stronger binding-like interactions between EDA-CQDs and the bacterial cells will result in a higher "local concentration" of EDA-CQDs on the bacterial surface, making them more effective in antibacterial actions against the bacterial cells [196]. The negatively charged bacterial surface must favor the positively charged end groups ($-NH_3^+$) on EDA.

As is clear from Table 2, using 10 μg/mL of the EDA-CQDs is very effective in inhibiting the biofilm formation for all the addition times used (1, 2 and 3 h) compared to 20 and 30 μg/mL, as is also indicated by Dong et al. [13]. When the 10 μg/mL CQDs were added at 1, 2, and 3 h after the initiation of biofilm growth, the inhibitory effect on the final biofilm formation was decreased from 95.86% (at 1 h) to 72.2% (at 2 h) reaching to about 34.25% (3 h), as indicated by different researchers [13,61,197]. Furthermore, the time of CQDs' addition during biofilm growth had a significant effect on the process of biofilm formation up to the final product stage.

These results are logical when considering the interactions between EDA-CQDs and bacterial cells during biofilm formation. At the early stage during biofilm formation, no thick extracellular polymeric substances (EPS) are produced around the bacteria, and most of the bacterial cells are still planktonic so that the added EDA-CQDs can bind and interact with bacteria efficiently to inactivate the cells before they can form a biofilm; thus, this explains the observed high inhibitory effects on biofilm formation.

In addition, these findings demonstrated how crucial surface charge is for CQD interactions with bacteria and the performance of their antibacterial activity. Stronger binding-like interactions between EDA-CQDs and the bacterial cells will result in a higher "local concentration" of EDA-CQDs on the bacterial surface, making them more effective in antibacterial actions against the bacterial cells [196]. The negatively charged bacterial surface must favor the positively charged end groups ($-NH_3^+$) on EDA.

Bacterial cells multiply and the extracellular matrix (ECM) gradually becomes stronger with the development of biofilm if CQDs are added 4–5 h after the start of biofilm growth. As the bacteria expand, the development of an ECM network may make it more difficult for CQDs to enter the biofilm and for EDA-CQDs to interact directly with the bacterial cells. These contacts and interactions are especially important to the light-activated EDA-CQDs' antibacterial function.

The production of electrons and holes, which are trapped at various stabilized surface defect sites, requires quick charge transfers and separation for a better representation of photoexcitation of the EDA-CQDs. Due to the short half-lives of these redox species, these separated redox pairs are attributed with making significant contributions to the observed antibacterial activities [13,197], largely in the near-neighbor manner due to the short-lived nature of these redox species. Their radiative recombinations produce emissive excited states that are responsible for the fluorescence's noticeable brightness and color, as well as the production of traditional reactive oxygen species (ROS), which also aid in the antibacterial effect. Although the ROS are still transient, the poor diffusion circumstances caused by the ECM network during the biofilm formation may also interfere with their antibacterial properties.

Due to the restriction associated with the requirement for the CQDs to penetrate into the biofilm, EDA-CQDs with light activation are therefore more effective in preventing biofilm formation before the bacterial cells have the chance and time to form the network structure toward the biofilm, and are less effective when the biofilm formation is already well underway. This restriction was made clearer in an investigation of the removal of mature biofilms using EDA-CQDs and the same visible light exposure [13].

Moreover, based on the investigation performed by Mogharbel et al. [16] who examined the microbicide potency for the embedded CQDs against three distinct bacterial strains, including a Gram-positive bacterial strain (*Staph. aureus*), Gram-negative bacterial strain (*Escherichia coli*), and fungal strain (*C. albicans*), as shown in Table 2, the superior microbicide potency of CQDs against several bacterial strains has been confirmed [13,16,61,197]. This effect was attributed to the decorative hydroxyl groups: (i) decorative oxygen-containing groups are responsible for the mortal effects of the prepared CQDs against the tested microbial cells through the generation of reactive oxygen species (ROS); (ii) the liberated ROS act by killing the microbial cells, as ROS adhere to them and then penetrate the microbial cell wall to motivate the oxidative stress by deteriorating DNA and RNA.

Additionally, ROS contribute to mitochondrial dysfunction, lipid peroxidation, inhibition of intracellular protein synthesis, progressive deterioration of the cell wall, and, ultimately, apoptotic cell death. The efficiency of hydrothermal conditions in the formation of small and size-controllable CQDs that are easily able to penetrate the microbial cell wall for eventual cell demise is attributed to the fact that CQDs-HT demonstrated significantly higher microbicide potentiality [16].

4. Conclusions and Future Perspectives

Biomass has a carbon chain which is why it is considered as an excellent option for the production of carbon materials. Nanocrystalline cellulose could become a potential source for fabricating carbon quantum dots which are affected by pyrolysis temperature.

The large surface area, good electric conductivity, and fast electric charge transfer of carbon quantum dots endow them with a great potential for a wide spectrum of applications. Luminescent carbon quantum dots are interesting newcomers in the category of nanomaterials, emerging with more and more advanced applications in the fields of chemical sensors, bioimaging, nanomedicine, drug delivery, and electrocatalysis. The unique electronic and chemical structures of carbon quantum dots can be tuned by controlling their size, shape, surficial functional groups, and heteroatom doping.

The 2,2′-ethylenedioxy-bis-ethylamine-carbon quantum dot and 3-ethoxypropylamine-carbon quantum dot were found to be highly effective to inhibit noroviruses from binding to histo-blood group antigens receptors on human cells with inhibition efficiencies of 100% and 85–99%, respectively.

In the future, we hope to discover more precursors and invent more economic, simple, and innovative synthetic methods and novel promising applications to increase the potential of these valuable carbon materials. In addition, more efforts must be made to simplify the traditional machinery used for the synthesis process of carbon quantum dots, especially in the collection of the nanometric dots by centrifugation, the neutralization of the dot's supernatant by dialysis, and the standardization of the dot size. Furthermore, the good findings in regard to using chemically-passivated carbon quantum dots for the prevention and therapy of norovirus must be extended to cover other epidemic pathogens, especially coronaviruses (COVID-19).

Supplementary Materials: The following supporting information can be downloaded at https://www.mdpi.com/article/10.3390/polym15122660/s1, Table S1. Synthesis routes of CQDs from natural macro-precursors and their applications; Table S2. Technical procedures and chemical reagents used for nanocelluloses productions; Table S3. Synthesis of CQDs from synthetic precursors and their applications; Figure S1. Pretreatment of well-plates with histo-blood group antigens (HBGA) present in saliva; Figure S2. Testing binding capacity between viruses like particles (VLPs) and standard antibodies; Figure S3. Testing binding capacity between viruses like particles (VLPs) and passivated-CQDs (EDA-CQDs and EPA-CQDs) using enzyme-linked immunosorbent assay (ELISA) test; Figure S4. Effect of the passivated-quantum dot (P-CQDs): (EDA-CQDs and EPA-CQDs) on viruses like particles (VLPs)-capsid protein using sodium dodecyl sulphate polyacrylamide gel electrophoresis (SDS-PAGE) test and western blotting protocol; Figure S5. Preparation of the gel containing the viruses like particles (VLPs)-passivated-CQDs (GI.1-VLP/EDA-CQDs and GI.1-VLP/EPA-CQDs) used for staining; Figure S6. Preparation of the gel containing the viruses like particles (VLPs)-passivated-CQDs (GI.1-VLP/EDA-CQDs and GI.1-VLP/EPA-CQDs) used for western blotting.

Author Contributions: Planning, supervision, conceptualization, and methodology, S.S.H.; designing the scientific plan of this study, revision, modification, and validation, J.S.M.S.; validation, formal analysis, and writing—review and editing, U.M.D.; enhancing environmental concepts of the CQDs as well as natural resources, validation, and review, I.M.I.; software and assistance with writing—original draft preparation, K.A.A.; preparation and assistance with the modification of the article, Z.M.M.; preparation and assistance with microbial knowledge, K.A.A.-E.; practical participation in reviewing the reported examinations and collecting the chemical information related to the article's topic, M.H.S.; revising the collected chemical information, especially that concerned with the nanoscience field, M.A.G.; planning, revision, validation of all bacterial topics, M.O.I.A.; design and drawing of the schematic representations, figures, and tables, R.A.A.; conceptualization, and methodology belonging to nanocelluloses and carbon quantum dots. O.H.M.I. All authors have read and agreed to the published version of the manuscript.

Funding: This work was funded by the Deanship of Scientific Research (DSR), KAU, Jeddah, under grant no. G-115/155/1432.

Institutional Review Board Statement: Not applicable.

Data Availability Statement: The supporting data for the reported results, including a link to the publicly archived datasets analyzed or generated during the study, can be found under the above-mentioned patents: Justia Patents Search https://patents.justia.com/patent/11060208 (accessed on 17 October 2021).

Acknowledgments: The corresponding author are deeply thankful to the DSR, KAU, Jeddah, for funding this research work.

Conflicts of Interest: The authors declare no conflict of interest.

Nomenclature

Symbol	Definition	Symbol	Definition
CNM	Carbon nanomaterials	NCC	Nanocrystalline cellulose
CNT	Carbon nanotubes	n^--CQD	Negative carbon quantum dot
COD	Carbon quantum dots	NMR	Nuclear magnetic resonance
DP	Degree of polymerization	NoVs	Noroviruses
DSC	Differential scanning calorimetry	NuPAGE LDS	Gel electrophoresis contains lithium dodecyl sulfate
DTA	Differential thermal analysis	Ox-MWCNTs	Oxidized multiwalled carbon nanotubes
ECL	Electroluminescence	Ox-SWCNTs	Oxidized single walled–carbon nanotubes (Ox-SWCNTs)
ECM	Extracellular matrix (ECM)	PBS	Phosphate-buffered saline
EDA	2,2′-ethylenedioxy-bis-ethylamine	PBST	Phosphate buffered saline with tween 20
ELISA	Enzyme-linked immunosorbent assay	P-CQD	Passivated-carbon quantum dot
EPA	3-ethoxypropylamine	p^+-CQDs	Positive carbon quantum dot
EPS	Extracellular polymeric substances (EPS)	PL	Photoluminescence
FGY	Fluorescence quantum yield	PS	Particle size
FTIR	Fourier transform infrared spectroscopy	QCE	Quantum confinement effect
GI.1, GII.4	Types of VLPs	ROS	Reactive oxygen species (ROS
GelCode	Blue stain for protein	SCS	Surficial charge status
HBGA	Histo-blood group antigens	SDS-PAGE	Gel electrophoresis containing sodium dodecyl sulphate polyacrylamide
HRP	Horseradish peroxidase	SEM	Scanning electron microscope
KDa	Kilodaltons, a molecular weight unit	TEM	Transmission electron microscope
LCD	Liquid crystal displays	TG	Terminal group of the surface molecule
LED	Light-emitting diodes	TGA	Thermogravimetric analysis
MCC	Microcrystalline cellulose	TMB	3,3′,5,5′-tetramethylbenzidine
MDR	Multidrug resistant	UV	Ultraviolet
MeOH	Methanol	VCN	Viable cell number
MIC	Minimum inhibitory concentration	VLP	Virus-like particles
MOPS SDS	Running buffer: 3-(N-morpholino) propanesulfonic acid	WB	Western blotting
MW	Molecular weight of the surface molecule	XRD	X-ray diffrection

References

1. Hindi, S.S.; Abohassan, R.A. Cellulosic microfibril and its embedding matrix within plant cell wall. *Int. J. Innov. Res. Sci. Eng. Technol.* **2016**, *5*, 2727–2734.
2. Hindi, S.S. Microcrystalline cellulose: The inexhaustible treasure for pharmaceutical industry. *Nanosci. Nanotechnol. Res.* **2017**, *4*, 22–31.
3. Hindi, S.S. Suitability of date palm leaflets for sulphated cellulose nanocrystals synthesis. *Nanosci. Nanotechnol. Res.* **2017**, *4*, 7–16. [CrossRef]
4. Hindi, S.S. Differentiation and synonyms standardization of amorphous and crystalline cellulosic products. *Nanosci. Nanotechnol. Res.* **2017**, *4*, 73–85.
5. Hindi, S.S. Nanocrystalline cellulose: Synthesis from pruning waste of *Zizyphus spina christi* and characterization. *Nanosci. Nanotechnol. Res.* **2017**, *4*, 106–114.
6. Hindi, S.S.; Raefat, A.; Abouhassan, R.A. Method for Making Nanoocrystalline Cellulose. U.S. Patent No. 10,144,786B2, 4 December 2018.
7. Hindi, S.S. Nanocrystalline Cellulose. U.S. Patent No. 11,161,918, 2 November 2021.
8. Hindi, S.S. Urchin-Shaped Nanocrystalline Material. U.S. Patent No. 11,242,410, 8 February 2022.
9. Hindi, S.S. Sulfate-Grafted Nanocrystalline Cellulose. U.S. Patent No. 11,242,411, 8 February 2022.

10. Hindi, S.S. A Method for Converting Micro- to Nanocrystalline Cellulose. U.S. Patent No. 10,808,045, 20 October 2020.
11. Marpongahtun, M.; Gea, S.; Muis, Y.; Andriayani, A.; Novita, T.; Piliang, A.F. Synthesis of carbon nanodots from cellulose nanocrystals oil palm empty fruit by pyrolysis. *J. Phys. Conf. Ser.* **2018**, *1120*, 012071. [CrossRef]
12. Zoppe, J.O.; Johansson, L.-S.; Seppälä, J. Manipulation of cellulose nanocrystal surface sulfate groups toward biomimetic nanostructures in aqueous media. *Carbohydr. Polym.* **2015**, *126*, 23–31. [CrossRef]
13. Dong, X.; Overton, C.M.; Tang, Y.; Darby, J.P.; Sun, Y.-P.; Yang, L. Visible light-activated carbon dots for inhibiting biofilm formation and inactivating biofilm-associated bacterial cells. *Front. Bioeng. Biotechnol.* **2021**, *9*, 786077. [CrossRef]
14. Dong, S.; Dong, S.; Yuan, Z.; Zhang, L.; Lin, Y.; Lu, C. Rapid screening of oxygen-states in carbon quantum dots by chemiluminescence probe. *Anal. Chem.* **2017**, *89*, 12520–12526. [CrossRef]
15. Dong, X.; Moyer, M.M.; Yang, F.; Sun, Y.P.; Yang, L. Carbon dots' antiviral functions against noroviruses. *Sci. Rep.* **2017**, *7*, 519. [CrossRef]
16. Mogharbel, A.T.; Abu-Melha, S.; Hameed, A.; Attar, R.M.S.; Alrefaei, A.F.; Almahri, A.; El-Metwaly, N. Anticancer and microbicide action of carbon quantum dots derived from microcrystalline cellulose: Hydrothermal versus infrared assisted techniques. *Arab. J. Chem.* **2023**, *15*, 104419. [CrossRef]
17. Zhang, B.; Xing, Y.; Li, Z.; Zhou, H.; Mu, Q.; Yan, B. Functionalized carbon nanotubes specifically bind to alpha-chymotrypsin's catalytic site and regulate its enzymatic function. *Nano Lett.* **2009**, *9*, 2280–2284. [CrossRef]
18. Nepal, D.; Geckeler, K.E. pH-sensitive dispersion and debundling of single-walled carbon nanotubes: Lysozyme as a tool. *Small* **2006**, *2*, 406–412. [CrossRef]
19. Sk, M.A.; Ananthanarayanan, A.; Huang, L.; Lim, K.H.; Chen, P. Revealing the tunable photoluminescence properties of graphene quantum dots. *J. Mater. Chem. C* **2014**, *2*, 6954–6960. [CrossRef]
20. Calvaresi, M.; Bottoni, A.; Zerbetto, F. Thermodynamics of binding between proteins and carbon nanoparticles: The case of C-60@Lysozyme. *J. Phys. Chem. C* **2015**, *119*, 28077–28082. [CrossRef]
21. Friedman, S.H.; DeCamp, D.L.; Sijbesma, R.P.; Srdanov, G.; Wudl, F.; Kenyon, G.L. Inhibition of the Hiv-1 protease by fullerene Derivatives—Model-building studies and experimental-verification. *J. Am. Chem. Soc.* **1993**, *115*, 6506–6509. [CrossRef]
22. Wang, Y.; Hu, A. Carbon quantum dots: Synthesis, properties and applications. *J. Mater. Chem.* **2014**, *2*, 6921–6939. [CrossRef]
23. Fernando, K.A.S.; Sahu, S.; Liu, Y.; Lewis, W.K.; Guliants, E.A.; Jafariyan, A.; Wang, P.; Bunker, C.E.; Sun, Y.-P. Carbon quantum dots and applications in photocatalytic energy conversion. *ACS Appl. Mater.* **2015**, *7*, 8363–8376. [CrossRef]
24. Gao, X.; Cui, Y.; Levenson, R.M.; Chung, L.W.K.; Nie, S. In vivo cancer targeting and imaging with semiconductor quantum dots. *Nat. Biotechnol.* **2004**, *22*, 969–976. [CrossRef]
25. Lim, S.Y.; Shen, W.; Gao, Z.Q. Carbon quantum dots and their applications. *Chem. Soc. Rev.* **2015**, *44*, 362–381. [CrossRef]
26. Luo, P.G.; Yang, F.; Yang, S.-T.; Sonkar, S.K.; Yang, L.; Broglie, J.J.; Liu, Y.; Sun, Y.-P. Carbon-based quantum dots for fluorescence imaging of cells and tissues. *Rsc. Adv.* **2014**, *4*, 10791–10807. [CrossRef]
27. Meziani, M.J.; Dong, X.; Zhu, L.; Jones, L.P.; LeCroy, G.E.; Yang, F.; Wang, S.; Wang, P.; Zhao, Y.; Yang, L.; et al. Visible-light-activated bactericidal functions of carbon "quantum" dots. *ACS Appl. Mater. Interfaces* **2016**, *8*, 10761–10766. [CrossRef]
28. LeCroy, G.E.; Yang, S.-T.; Yang, F.; Liu, Y.; Fernando, K.A.S.; Bunker, C.B.; Hu, Y.; Luo, P.G.; Sun, Y.-P. Functionalized carbon nanoparticles: Syntheses and applications in optical bioimaging and energy conversion. *Coordin. Chem. Rev.* **2016**, *320*, 66–81. [CrossRef]
29. LeCroy, G.E.; Sonkar, S.K.; Yang, F.; Veca, L.M.; Wang, P.; Tackett, K.N.; Yu, J.J.; Vasile, E.; Qian, H.; Liu, Y. Toward structurally defined carbon dots as ultracompact fluorescent probes. *ACS Nano.* **2014**, *8*, 4522–4529. [CrossRef]
30. Dong, X.; Liang, W.; Meziani, M.J.; Sun, Y.-P.; Yang, L. Carbon dots as potent antimicrobial Agents. *Theranostics* **2020**, *10*, 671–680. [CrossRef]
31. Jhonsi, M.A. *Carbon Quantum Dots for Bioimaging: State of the Art in Nano-Bioimaging, State of the Art in Nano-Bioimaging*; Ghamsari, M.S., Ed.; IntechOpen: London, UK, 2018; pp. 35–53.
32. Yang, F.; LeCroy, G.E.; Wang, P.; Liang, W.; Chen, J.; Fernando, K.S. Functionalization of carbon nanoparticles and defunctionalization toward structural and mechanistic elucidation of carbon quantum dots. *J. Phys. Chem. C* **2016**, *120*, 25604–25611. [CrossRef]
33. Liu, Y.; Wang, P.; Shiral Fernando, K.A.; LeCroy, G.E.; Maimaiti, H.; Harruff-Miller, B.A.; Lewis, W.K.; Bunker, C.E.; Hou, Z.L.; Sun, Y.P. Enhanced fluorescence properties of carbon dots in polymer films. *J. Mater. Chem. C* **2016**, *4*, 6967–6974. [CrossRef]
34. Lin, F.; Wang, Z.; Wu, F.-G. Carbon dots for killing microorganisms: An update since 2019. *J. Pharm.* **2022**, *15*, 1236. [CrossRef]
35. Wang, X.; Feng, Y.; Dong, P.; Huang, J. A mini review on carbon quantum dots: Preparation, properties, and electrocatalytic application. *Front. Chem.* **2019**, *7*, 671. [CrossRef]
36. Mahat, N.A.; Shamsudin, S.A.; Jullok, N.; Ma'Radzi, A.H. Carbon quantum dots embedded polysulfone membranes for antibacterial performance in the process of forward osmosis. *Desalination* **2020**, *493*, 114618. [CrossRef]
37. Koulivand, H.; Shahbazi, A.; Vatanpour, V.; Rahmandoost, M. Novel antifouling and antibacterial polyethersulfone membrane prepared by embedding nitrogen-doped carbon dots for efficient salt and dye rejection. *Mater. Sci. Eng. C* **2020**, *111*, 110787. [CrossRef]
38. Kousheh, S.A.; Moradi, M.; Tajik, H.; Molaei, R. Preparation of antimicrobial/ultraviolet protective bacterial nanocellulose film with carbon dots synthesized from lactic acid bacteria. *Int. J. Biol. Macromol.* **2020**, *155*, 216–225. [CrossRef]

39. Riahi, Z.; Rhim, J.W.; Bagheri, R.; Pircheraghi, G.; Lotfali, E. Carboxymethyl cellulose-based functional film integrated with chitosan-based carbon quantum dots for active food packaging applications. *Prog. Org. Coat.* **2022**, *166*, 106794. [CrossRef]
40. Tang, W.; Li, P.; Zhang, G.; Yang, X.; Yu, M.; Lu, H.; Xing, X. Antibacterial carbon dots derived from polyethylene glycol/polyethyleneimine with potent anti-friction performance as water-based lubrication additives. *J. Appl. Polym. Sci.* **2021**, *138*, e50620. [CrossRef]
41. Geng, B.; Li, P.; Fang, F.; Shi, W.; Glowacki, J.; Pan, D.; Shen, L. Antibacterial and osteogenic carbon quantum dots for regeneration of bone defects infected with multidrug-resistant bacteria. *Carbon* **2021**, *184*, 375–385. [CrossRef]
42. Mazumdar, A.; Haddad, Y.; Milosavljevic, V.; Michalkova, H.; Guran, R.; Bhowmick, S.; Moulick, A. Peptide-carbon quantum dots conjugate, derived from human retinoic acid receptor responder protein 2, against antibiotic-resistant Gram positive and Gram negative pathogenic bacteria. *Nanomaterials* **2020**, *10*, 325. [CrossRef]
43. Liu, S.; Quan, T.; Yang, L.; Deng, L.; Kang, X.; Gao, M.; Xia, Z.; Li, X.; Gao, D. N,Cl-codoped carbon dots from *Impatiens balsamina* L. stems and a deep eutectic solvent and their applications for Gram-positive bacteria identification, antibacterial activity, cell imaging, and ClO^- sensing. *ACS Omega* **2021**, *6*, 29022–29036. [CrossRef]
44. Shahshahanipour, M.; Rezaei, B.; Ensafi, A.A.; Etemadifar, Z. An ancient plant for the synthesis of a novel carbon dot and its applications as an antibacterial agent and probe for sensing of an anti-cancer drug. *Mater. Sci. Eng. C* **2019**, *98*, 826–833. [CrossRef]
45. Boobalan, T.; Sethupathi, M.; Sengottuvelan, N.; Kumar, P.; Balaji, P.; Gulyás, B.; Padmanabhan, P.; Selvan, S.T.; Arun, A. Mushroom-derived carbon dots for toxic metal ion detection and as antibacterial and anticancer agents. *ACS Appl. Nano Mater.* **2020**, *3*, 5910–5919. [CrossRef]
46. Ma, Y.; Zhang, M.; Wang, H.; Wang, B.; Huang, H.; Liu, Y.; Kang, Z. N-doped carbon dots derived from leaves with low toxicity via damaging cytomembrane for broad-spectrum antibacterial activity. *Mater. Today Commun.* **2020**, *24*, 101222. [CrossRef]
47. Surendran, P.; Lakshmanan, A.; Priya, S.S.; Geetha, P.; Rameshkumar, P.; Kannan, K.; Hegde, T.A.; Vinitha, G. Fluorescent carbon quantum dots from *Ananas comosus* waste peels: A promising material for NLO behaviour, antibacterial, and antioxidant activities. *Inorg. Chem. Commun.* **2021**, *124*, 108397. [CrossRef]
48. Genc, M.T.; Yanalak, G.; Aksoy, I.; Aslan, E.; Patır, I.H. Green carbon dots (GCDs) for photocatalytic hydrogen evolution and antibacterial applications. *ChemistrySelect* **2021**, *6*, 7317–7322. [CrossRef]
49. Saravanan, A.; Maruthapandi, M.; Das, P.; Luong, J.H.T.; Gedanken, A. Green synthesis of multifunctional carbon dots with antibacterial activities. *Nanomaterials* **2021**, *11*, 369. [CrossRef]
50. Eskalen, H.; Çeşme, M.; Kerli, S.; Özğan, Ş. Green synthesis of water-soluble fluorescent carbon dots from rosemary leaves: Applications in food storage capacity, fingerprint detection, and antibacterial activity. *J. Chem. Res.* **2021**, *45*, 428–435. [CrossRef]
51. Pandiyan, S.; Arumugam, L.; Srirengan, S.P.; Pitchan, R.; Sevugan, P.; Kannan, K.; Pitchan, G.; Hegde, T.A.; Gandhirajan, V. Biocompatible carbon quantum dots derived from sugarcane industrial wastes for effective nonlinear optical behaviour and antimicrobial activity applications. *ACS Omega* **2020**, *5*, 30363–30372. [CrossRef]
52. Qing, W.; Chen, K.; Yang, Y.; Wang, Y.; Liu, X. Cu^{2+}-doped carbon dots as fluorescence probe for specific recognition of Cr (VI) and its antimicrobial activity. *Microchem. J.* **2020**, *152*, 104262. [CrossRef]
53. Das, P.; Maruthapandi, M.; Saravanan, A.; Natan, M.; Jacobi, G.; Banin, E.; Gedanken, A. Carbon dots for heavy-metal sensing, pH-sensitive cargo delivery, and antibacterial applications. *ACS Appl. Nano Mater.* **2020**, *3*, 11777–11790. [CrossRef]
54. Zhao, X.; Wang, L.; Ren, S.; Hu, Z.; Wang, Y. One-pot synthesis of Forsythia@carbon quantum dots with natural anti-wood rot fungus activity. *Mater. Des.* **2021**, *206*, 109800. [CrossRef]
55. Wang, H.; Zhang, M.; Ma, Y.; Wang, B.; Shao, M.; Huang, H.; Liu, Y.; Kang, Z. Selective inactivation of Gram-negative bacteria by carbon dots derived from natural biomass: Artemisia argyi leaves. *J. Mater. Chem. B* **2020**, *8*, 2666–2672. [CrossRef]
56. Li, C.; Lin, F.; Sun, W.; Wu, F.G.; Yang, H.; Lv, R.; Zhu, Y.X.; Jia, H.R.; Wang, C.; Gao, G.; et al. Self-assembled rose bengalexo polysaccharide nanoparticles for improved photodynamic inactivation of bacteria by enhancing singlet oxygen generation directly in the solution. *ACS Appl. Mater. Interfaces* **2018**, *10*, 16715–16722. [CrossRef]
57. Lin, F.; Bao, Y.W.; Wu, F.G. Improving the phototherapeutic efficiencies of molecular and nanoscale materials by targeting mitochondria. *Molecules* **2018**, *23*, 3016. [CrossRef]
58. Moradlou, O.; Rabiei, Z.; Delavari, N. Antibacterial effects of carbon quantum dots@hematite nanostructures deposited on titanium against Gram-positive and Gram-negative bacteria. *J. Potoch. Photobio. A* **2019**, *379*, 144–149. [CrossRef]
59. Singh, P.; Pandit, S.; Beshay, M.; Mokkapati, V.R.S.S.; Garnaes, J.; Olsson, M.E.; Sultan, A.; Mackevica, A.; Mateiu, R.V.; Lütken, H.; et al. Anti-biofilm effects of gold and silver nanoparticles synthesized by the *Rhodiola rosea* rhizome extracts. *Artif. Cell Nanomed. Biotechnol.* **2018**, *46*, S886–S899. [CrossRef]
60. Ikuma, K.; Decho, A.W.; Lau, B.L.T. The Extracellular Bastions of Bacteria—A Biofilm Way of Life. *Nat. Educ. Knowl.* **2013**, *44*, 22.
61. Donlan, R.M. Biofilm Formation: A clinically relevant microbiological process. *Clin. Infect. Dis.* **2001**, *33*, 1387–1392. [CrossRef]
62. Wu, X.; Abbas, K.; Yang, Y.; Li, Z.; Tedesco, A.C.; Bi, H. Photodynamic anti-bacteria by carbon dots and their nano-composites. *J. Pharm.* **2022**, *15*, 487. [CrossRef]
63. Joo, K.I.; Lei, Y.; Lee, C.L.; Lo, J.; Xie, J.; Hamm-Alvarez, S.F.; Wang, P. Site-specific labelling of enveloped viruses with quantum dots for single virus tracking. *ACS Nano* **2008**, *2*, 1553–1562. [CrossRef]
64. Zheng, D.P.; Ando, T.; Fankhauser, R.L.; Beard, R.S.; Glass, R.I.; Monroe, S.S. Norovirus classification and proposed strain nomenclature. *Virol. J.* **2006**, *346*, 312–323. [CrossRef]

65. Bok, K.; Parra, G.I.; Mitra, T.; Abente, E.; Shaver, C.K.; Boon, D.; Engle, R.; Yu, C.; Kapikian, A.Z.; Sosnovtsev, S.V.; et al. Chimpanzees as an animal model for human norovirus infection and vaccine development. *Proc. Natl. Acad. Sci. USA* **2011**, *108*, 325–330. [CrossRef]
66. Patel, M.M.; Hall, A.J.; Vinje, J.; Parashar, U.D. Noroviruses: A comprehensive review. *J. Clin. Virol.* **2019**, *44*, 1–8. [CrossRef]
67. Bozkurt, H.; D'Souza, D.H.; Davidson, P.M. Thermal inactivation of human norovirus surrogates in spinach and measurement of its uncertainty. *J. Food Prot.* **2014**, *77*, 276–283. [CrossRef]
68. Vimont, A.; Fliss, I.; Jean, J. Efficacy and mechanisms of murine norovirus inhibition by pulsed-light technology. *Appl. Environ. Microbiol.* **2015**, *81*, 2950–2957. [CrossRef]
69. Todd, K.V.; Tripp, R.A. Human norovirus: Experimental models of infection. *Viruses* **2019**, *11*, 151. [CrossRef]
70. Liu, P.; Yuen, Y.; Hsiao, H.M.; Jaykus, L.A.; Moe, C. Effectiveness of liquid soap and hand sanitizer against Norwalk virus on contaminated hands. *Appl. Environ. Microbiol.* **2010**, *76*, 394–399. [CrossRef]
71. Park, S.; Park, H.H.; Kim, S.Y.; Kim, S.J.; Woo, K.; Ko, G. Antiviral properties of silver nanoparticles on a magnetic hybrid colloid. *Appl. Environ. Microbiol.* **2014**, *80*, 2343–2350. [CrossRef]
72. Broglie, J.J.; Alston, B.; Yang, C.; Ma, L.; Adcock, A.F.; Chen, W.; Yang, L. Antiviral activity of gold/copper sulphide core/shell nanoparticles against Human norovirus virus-like particles. *PLoS ONE* **2015**, *10*, e0141050. [CrossRef]
73. Gerrity, D.; Ryu, H.; Crittenden, J.; Abbaszadegan, M. Photocatalytic inactivation of viruses using titanium dioxide nanoparticles and low-pressure UV light. *J. Environ. Sci. Health A* **2008**, *43*, 1261–1270. [CrossRef]
74. Sun, Y.P.; Zhou, B.; Lin, Y.; Wang, W.; Fernando, K.A.; Pathak, P.; Meziani, M.J.; Harruff, B.A.; Wang, X.; Wang, H.; et al. Quantum-sized carbon dots for bright and colourful photoluminescence. *J. Am. Chem. Soc.* **2006**, *128*, 7756–7757. [CrossRef]
75. Vega, E.; Barclay, L.; Gregoricus, N.; Shirley, S.H.; Lee, D.; Vinjé, J. Genotypic and epidemiologic trends of norovirus outbreaks in the United States. 2009 to 2013. *J. Clin. Microbiol.* **2014**, *52*, 147–155. [CrossRef]
76. Reeck, A.; Kavanagh, O.; Estes, M.K.; Opekun, A.R.; Gilger, M.A.; Graham, D.Y.; Atmar, R.L. Serological correlate of protection against norovirus-induced gastroenteritis. *J. Infect. Dis.* **2010**, *202*, 1212–1218. [CrossRef]
77. Huang, P.; Farkas, T.; Marionneau, S.; Zhong, W.; Ruvoën-Clouet, N.; Morrow, A.L.; Altaye, M.; Pickering, L.K.; Newburg, D.S.; LePendu, J.; et al. Noroviruses bind to human ABO, Lewis, and secretor histo-blood group antigens: Identification of 4 distinct strain specific patterns. *Int. J. Infect. Dis.* **2003**, *188*, 19–31. [CrossRef]
78. Huang, P.; Farkas, T.; Zhong, W.; Tan, M.; Thornton, S.; Morrow, A.L.; Jiang, X. Norovirus and histo-blood group antigens: Demonstration of a wide spectrum of strain specificities and classification of two major binding groups among multiple binding patterns. *Virol. J.* **2005**, *79*, 6714–6722. [CrossRef]
79. Zhang, X.F.; Tan, M.; Chhabra, M.; Dai, Y.C.; Meller, J.; Jiang, X. Inhibition of histo-blood group antigen binding as a novel strategy to block norovirus infections. *PLoS ONE* **2013**, *8*, e69379. [CrossRef]
80. Singh, B.K.; Leuthold, M.M.; Hansman, G.S. Human noroviruses' fondness for histo-blood group antigens. *Virol. J.* **2015**, *89*, 2024–2040. [CrossRef]
81. Peng, H.; Travas Sejdic, J. Simple aqueous solution route to luminescent carbogenic dots from carbohydrates. *Chem. Mater.* **2009**, *21*, 5563–5565. [CrossRef]
82. Xu, X.; Ray, R.; Gu, Y.; Ploehn, H.J.; Gearheart, L.; Raker, K.; Scrivens, W.A. Electrophoretic analysis and purification of fluorescent single-walled carbon nanotube fragments. *J. Am. Chem. Soc.* **2004**, *126*, 12736–12737. [CrossRef]
83. Wang, L.; Chen, X.; Lu, Y.; Liu, C.; Yang, W. Carbon quantum dots displaying dual wavelength photoluminescence and electrochemiluminescence prepared by high-energy ball milling. *Carbon* **2015**, *94*, 472–478. [CrossRef]
84. Li, H.; He, X.; Kang, Z.; Huang, H.; Liu, Y.; Liu, J.; Lian, S.; Tsang, C.H.; Yang, X.; Lee, S.T. Water-soluble fluorescent carbon quantum dots and photocatalyst design. *Angew. Chem. Int. Ed.* **2010**, *49*, 4430–4434. [CrossRef]
85. Wu, M.; Yue, W.; Wu, W.; Hu, C.; Wang, X.; Zheng, J.-T.; Li, Z.; Jiang, B.; Qiu., J. Preparation of functionalized water-soluble photoluminescent carbon quantum dots from petroleum coke. *Carbon* **2014**, *78*, 480–489. [CrossRef]
86. Yan, Z.-Y.; Xiao, A.; Lu, H.; Liu, Z.; Chen, J.-Q. Determination of metronidazole by a flow injection chemiluminescence method using ZnO-doped carbon quantum dots. *New Carbon Mater.* **2014**, *29*, 216–224. [CrossRef]
87. Li, M.; Hu, C.; Yu, C.; Wang, S.; Zhang, P.; Qiu, J. Organic amine-grafted carbon quantum dots with tailored surface and enhanced photoluminescence properties. *Carbon* **2015**, *91*, 291–297. [CrossRef]
88. Zhang, W.; Shi, L.; Liu, Y.; Meng, X.; Xu, H.; Xu, Y.Q.; Liu, B.; Fang, X.; Li, H.; Ding, T. Supramolecular interactions via hydrogen bonding contributing to citric-acid derived carbon dots with high quantum yield and sensitive photoluminescence. *RSC Adv.* **2017**, *7*, 20345–20353. [CrossRef]
89. Liu, H.; Ye, T.; Mao, C. Fluorescent carbon nanoparticles derived from candle soot. *Angew. Chem. Int. Ed.* **2007**, *46*, 6473–6475. [CrossRef]
90. Tang, L.; Ji, R.; Cao, X.; Lin, J.; Jiang, H.; Li, X.; Teng, K.S.; Luk, C.M.; Zeng, S.; Hao, J.; et al. Deep ultraviolet photoluminescence of water-soluble self-passivated graphene quantum dots. *ACS Nano* **2012**, *6*, 5102–5110. [CrossRef]
91. Wang, Q.; Zheng, H.; Long, Y.; Zhang, L.; Gao, M.; Bai, W. Microwave–hydrothermal synthesis of fluorescent carbon dots from graphite oxide. *Carbon* **2011**, *49*, 3134–3140. [CrossRef]
92. Wang, J.; Xin, X.; Lin, Z. Cu_2ZnSnS_4 nanocrystals and graphene quantum dots for photovoltaics. *Nanoscale* **2011**, *3*, 3040–3048. [CrossRef]

93. Li, H.; He, X.; Liu, Y.; Yu, H.; Kang, Z.; Lee, S.T. Synthesis of fluorescent carbon nanoparticles directly from active carbon via a one-step ultrasonic treatment. *Mater. Res. Bull.* **2011**, *46*, 147–151. [CrossRef]
94. Jiang, H.; Chen, F.; Lagally, M.G.; Denes, F.S. New strategy for synthesis and functionalization of carbon nanoparticles. *Langmuir* **2009**, *26*, 1991–1995. [CrossRef]
95. Bourlinos, A.B.; Stassinopoulos, A.; Anglos, D.; Zboril, R.; Georgakilas, V.; Giannelis, E.P. Photoluminescent carbogenic dots. *Chem. Mater.* **2008**, *20*, 4539–4541. [CrossRef]
96. Liu, R.; Wu, D.; Liu, S.; Koynov, K.; Knoll, W.; Li, Q. An aqueous route to multicolor photoluminescent carbon dots using silica spheres as carriers. *Angew. Chem.* **2009**, *121*, 4668–4671. [CrossRef]
97. Zong, J.; Zhu, Y.; Yang, X.; Shen, J.; Li, C. Synthesis of photoluminescent carbogenic dots using mesoporous silica spheres as nanoreactors. *Chem. Commun.* **2011**, *47*, 764–766. [CrossRef]
98. Hamilton, I.P.; Li, B.; Yan, X.; Li, L.S. Alignment of colloidal graphene quantum dots on polar surfaces. *Nano Lett.* **2011**, *11*, 1524–1529. [CrossRef]
99. Mueller, M.L.; Yan, X.; Dragnea, B.; Li, L.S. Slow hot-carrier relaxation in colloidal graphene quantum dots. *Nano Lett.* **2011**, *11*, 56–60. [CrossRef]
100. Liu, R.; Wu, D.; Feng, X.; Müllen, K. Bottom-up fabrication of photoluminescent graphene quantum dots with uniform morphology. *J. Am. Chem. Soc.* **2011**, *133*, 15221–15223. [CrossRef]
101. Lu, J.; Yeo, P.S.; Gan, C.K.; Wu, P.; Loh, K.P. Transforming C60 molecules into graphene quantum dots. *Nat. Nanotechnol.* **2011**, *6*, 247–252. [CrossRef]
102. Vinci, J.C.; Ferrer, I.M.; Seedhouse, S.J.; Bourdon, A.K.; Reynard, J.M.; Foster, B.A.; Bright, F.V.; Colón, L.A. Hidden properties of carbon dots revealed after HPLC fractionation. *J. Phys. Chem. Lett.* **2012**, *4*, 239–243. [CrossRef]
103. Tao, H.; Yang, K.; Ma, Z.; Wan, J.; Zhang, Y.; Kang, Z.; Liu, Z. In vivo NIR fluorescence imaging, biodistribution, and toxicology of photoluminescent carbon dots produced from carbon nanotubes and graphite. *Small* **2012**, *8*, 281–290. [CrossRef]
104. Zheng, X.T.; Than, A.; Ananthanaraya, A.; Kim, D.H.; Chen, P. Graphene quantum dots as universal fluorophores and their use in revealing regulated trafficking of insulin receptors in adipocytes. *ACS Nano* **2013**, *7*, 6278–6286. [CrossRef]
105. Shan, F.; Fu, L.; Chen, X.; Xie, X.; Liao, C.; Zhu, Y.; Xia, H.; Zhang, J.; Yan, L.; Wang, Z.; et al. Waste-to-wealth: Functional biomass carbon dots based on bee pollen waste and application. *Chin. Chem. Lett.* **2022**, *33*, 2942–2948. [CrossRef]
106. Huang, G.; Chen, X.; Wang, C.; Zheng, H.; Huang, Z.; Chen, D.; Xie, H. Photoluminescent carbon dots derived from sugarcane molasses: Synthesis, properties, and applications. *RSC Adv.* **2017**, *7*, 47840–47847. [CrossRef]
107. Vandarkuzhali, S.A.A.; Jeyalakshmi, V.; Sivaraman, G.; Singaravadivel, S.; Krishnamurthy, K.R.; Viswanathan, B. Highly fluorescent carbon dots from pseudo-stem of banana plant: Applications as nanosensor and bio-imaging agents. *Sens. Actuators B Chem.* **2017**, *252*, 894–900. [CrossRef]
108. Sachdev, A.; Gopinath, P. Green synthesis of multifunctional carbon dots from coriander leaves and their potential application as antioxidants, sensors and bioimaging agents. *Analyst* **2015**, *140*, 4260–4269. [CrossRef] [PubMed]
109. D'souza, S.L.; Chettiar, S.S.; Koduru, J.R.; Kailasaa, S.K. Synthesis of fluorescent carbon dots using *Daucus carota* subsp. *Sativus* roots for mitomycin drug delivery. *Optik* **2018**, *158*, 893–900.
110. Amin, N.; Afkhami, A.; Hosseinzadeh, L.; Madrakian, T. Green and cost-effective synthesis of carbon dots from date kernel and their application as a novel switchable fluorescence probe for sensitive assay of zoledronic acid drug in human serum and cellular imaging. *Anal. Chim. Acta* **2018**, *1030*, 183–193. [CrossRef] [PubMed]
111. Kasibabu, B.S.B.; D'souza, S.L.; Jha, S.; Kailasa, S.K. Imaging of bacterial and fungal cells using fluorescent carbon dots prepared from *carica papaya* juice. *J. Fluoresc.* **2015**, *25*, 803–810. [CrossRef]
112. Mehta, V.N.; Jha, S.; Kailasa, S.K. One-pot green synthesis of carbon dots by using *Saccharum officinarum* juice for fluorescent imaging of bacteria (*Escherichia coli*) and yeast (*Saccharomyces cerevisiae*) cells. *Mater. Sci. Eng. C* **2014**, *38*, 20–27. [CrossRef]
113. Shen, J.; Shang, S.; Chen, X.; Wang, D.; Cai, Y. Facile synthesis of fluorescence carbon dots from sweet potato for Fe^{3+} sensing and cell imaging. *Mater. Sci. Eng. C* **2017**, *76*, 856–864. [CrossRef]
114. Cheng, C.; Shi, Y.; Li, M.; Xing, M.; Wu, Q. Carbon quantum dots from carbonized walnut shells: Structural evolution, fluorescence characteristics, and intracellular bioimaging. *Mater. Sci. Eng. C* **2017**, *79*, 473–480. [CrossRef]
115. Hu, Y.; Zhang, L.; Li, X.; Liu, R.; Lin, L.; Zhao, S. Green preparation of S and N Co-doped carbon dots from water chestnut and onion as well as their use as an off-on fluorescent probe for the quantification and imaging of coenzyme A. *ACS. Sustain. Chem. Eng.* **2017**, *5*, 4992–5000. [CrossRef]
116. Mehta, V.N.; Jha, S.; Basu, H.; Singhal, R.K.; Kailasa, S.K. One-step hydrothermal approach to fabricate carbon dots from apple juice for imaging of mycobacterium and fungal cells. *Sens. Actuators. B Chem.* **2015**, *213*, 434–443. [CrossRef]
117. Atchudan, R.; Edison, T.N.J.I.; Chakradhar, D.; Perumal, S.; Shim, J.J.; Lee, Y.R. Facile green synthesis of nitrogen-doped carbon dots using *Chionanthus retusus* fruit extract and investigation of their suitability for metal ion sensing and biological applications. *Sens. Actuators B Chem.* **2017**, *246*, 497–509. [CrossRef]
118. Yang, M.; Meng, X.; Li, B.; Ge, S.; Lu, Y. N, S co-doped carbon dots with high quantum yield: Tunable fluorescence in liquid/solid and extensible applications. *J. Nanopart. Res.* **2017**, *19*, 217. [CrossRef]
119. Feng, X.; Jiang, Y.; Zhao, J.; Miao, M.; Cao, S.; Fang, J.; Shi, L. Easy synthesis of photoluminescent N-doped carbon dots from winter melon for bio-imaging. *RSC Adv.* **2015**, *5*, 31250–31254. [CrossRef]

120. Huang, H.; Lv, J.J.; Zhou, D.L.; Bao, N.; Xu, Y.; Wang, A.J.; Feng, J.J. One-pot green synthesis of nitrogen-doped carbon nanoparticles as fluorescent probes for mercury ions. *RSC. Adv.* **2013**, *3*, 21691–21696. [CrossRef]
121. Mewada, A.; Pandey, S.; Shinde, S.; Mishra, N.; Oza, G.; Thakur, M.; Sharon, M.; Sharon, M. Green synthesis of biocompatible carbon dots using aqueous extract of *Trapa bispinosa* peel. *Mater. Sci. Eng. C* **2013**, *33*, 2914–2917. [CrossRef]
122. Atchudan, R.; Edison, T.N.J.I.; Lee, Y.R. Nitrogen-doped carbon dots originating from unripe peach for fluorescent bioimaging and electrocatalytic oxygen reduction reaction. *J. Colloid Interface Sci.* **2016**, *482*, 8–18. [CrossRef]
123. Atchudan, R.; Edison, T.N.J.I.; Sethuraman, M.G.; Lee, Y.R. Efficient synthesis of highly fluorescent nitrogen-doped carbon dots for cell imaging using unripe fruit extract of *Prunus mume*. *Appl. Surf. Sci.* **2016**, *384*, 432–441. [CrossRef]
124. Zhao, S.; Lan, M.; Zhu, X.; Xue, H.; Ng, T.-W.; Meng, X.; Lee, C.-S.; Wang, P.; Zhang, W. Green synthesis of bifunctional fluorescent carbon dots from garlic for cellular imaging and free radical scavenging. *ACS Appl. Mater. Interfaces* **2015**, *7*, 17054–17060. [CrossRef]
125. Bandi, R.; Gangapuram, R.B.R.; Dadigala, R.; Eslavath, R.; Singh, S.S.; Guttena, V. Facile and green synthesis of fluorescent carbon dots from onion waste and their potential applications as sensor and multicolour imaging agents. *RSC Adv.* **2016**, *6*, 28633–28639. [CrossRef]
126. Tripathi, K.M.; Tran, T.S.; Tung, T.T.; Losic, D.; Kim, T. Water soluble fluorescent carbon nanodots from biosource for cells imaging. *J. Nanomater.* **2017**, *2017*, 7029731. [CrossRef]
127. Hoan, B.T.; Tam, P.D.; Pham, V.H. Green synthesis of highly luminescent carbon quantum dots from lemon juice. *J. Nanotechnol.* **2019**, *2019*, 2852816. [CrossRef]
128. Paul, S.; Banerjee, S.L.; Khamrai, M.; Samanta, S.; Singh, S.; Kundu, P.P.; Anup, K.; Ghosh, A.K. Hydrothermal synthesis of gelatine quantum dots for high-performance biological imaging applications. *Photochem. Photobiol. B* **2020**, *212*, 112014. [CrossRef]
129. Yu, J.; Song, N.; Zhang, Y.-K.; Zhong, S.-X.; Wang, A.-J.; Chen, J. Green preparation of carbon dots by Jinhua bergamot for sensitive and selective fluorescent detection of Hg^{2+} and Fe^{3+}. *Sens. Actuators B Chem.* **2015**, *214*, 29–35. [CrossRef]
130. Ramanan, V.; Thiyagarajan, S.K.; Raji, K.; Suresh, R.; Sekar, R.; Ramamurthy, P. Outright green synthesis of fluorescent carbon dots from eutrophic algal blooms for in vitro imaging. *ACS Sustain. Chem. Eng.* **2016**, *4*, 4724–4731. [CrossRef]
131. Kumawat, M.K.; Thakur, M.; Gurung, R.B.; Srivastava, R. Graphene quantum dots for cell proliferation, nucleus imaging, and photoluminescent sensing applications. *Sci. Rep.* **2017**, *7*, 15858. [CrossRef]
132. Balajia, M.; Jegatheeswarana, S.; Nithyaa, P.; Boomib, P.; Selvamc, S.; Sundrarajana, M. Photoluminescent reduced graphene oxide quantum dots from latex of *Calotropis gigantea* for metal sensing, radical scavenging, cytotoxicity, and bioimaging in *Artemia salina*: A greener route. *J. Photochem. Photobiol.* **2018**, *178*, 371–379.
133. Gu, D.; Shang, S.; Yu, Q.; Shen, J. Green synthesis of nitrogen-doped carbon dots from lotus root for Hg (II) ions detection and cell imaging. *Appl. Surf. Sci.* **2016**, *390*, 38–42. [CrossRef]
134. Kumawat, M.K.; Thakur, M.; Gurung, R.B.; Srivastava, R. Graphene quantum dots from *Mangifera indica*: Application in near-infrared bioimaging and intracellular nanothermometry. *ACS Sustain. Chem. Eng.* **2017**, *5*, 1382–1391. [CrossRef]
135. Sivasankaran, U.; Jesny, S.; Jose, A.R.; Kumar, K.G. Fluorescence determination of glutathione using tissue paper-derived carbon dots as fluorophores. *Anal. Sci.* **2017**, *33*, 281–285. [CrossRef]
136. Park, S.Y.; Lee, H.U.; Park, E.S.; Lee, S.C.; Lee, J.W.; Jeong, S.W.; Kim, C.H.; Lee, Y.C.; Huh, Y.S.; Lee, J. Photoluminescent green carbon nanodots from food-waste-derived sources: Large-scale synthesis, properties, and biomedical applications. *ACS. Appl. Mater. Interfaces* **2014**, *6*, 3365–3370. [CrossRef]
137. Ding, Z.; Li, F.; Wen, J.; Wang, X.; Sun, R. Gram-scale synthesis of single-crystalline graphene quantum dots derived from lignin biomass. *Green Chem.* **2018**, *20*, 1383–1390. [CrossRef]
138. Yang, X.; Zhuo, Y.; Zhu, S.; Luo, Y.; Feng, Y.; Dou, Y. Novel and green synthesis of high-fluorescent carbon dots originated from honey for sensing and imaging. *Biosen. Bioelectron.* **2014**, *60*, 292–298. [CrossRef]
139. Singh, A.; Eftekhari, E.; Scott, J.; Kaur, J.; Yambem, S.; Leusch, F.; Wellings, R.; Gould, T.; Ostrikov, K.; Sonar, P.; et al. Carbon dots derived from human hair for ppb level chloroform sensing in water. *Sustain. Mater. Technol.* **2020**, *25*, e00159. [CrossRef]
140. Janus, L.; Piątkowski, M.; Radwan-Praglowska, J.; Bogdal, D.; Matysek, D. Chitosan-based carbon quantum dots for biomedical applications: Synthesis and characterization. *Nanomaterials* **2019**, *9*, 274. [CrossRef]
141. Koshizawa, T. Degradation of wood cellulose and cotton linters in phosphoric acid. *Japan TAPPI J.* **1960**, *14*, 455–458. [CrossRef]
142. Sadeghifar, H.; Filpponen, I.; Clarke, S.P.; Brougham, D.F.; Argyropoulos, D.S. Production of cellulose nanocrystals using hydrobromic acid and click reactions on their surface. *J. Mater. Sci.* **2011**, *46*, 7344–7355. [CrossRef]
143. Sucaldito, M.R.; Camacho, D.H. Characteristics of unique HBr-hydrolyzed cellulose nanocrystals from freshwater green algae (*Cladophora rupestris*) and its reinforcement in starch-based film. *Carbohydr. Polym.* **2017**, *169*, 315–323. [CrossRef]
144. Trache, D.; Donnot, A.; Khimeche, K.; Benelmir, R.; Brosse, N. Physico-chemical properties and thermal stability of microcrystalline cellulose isolated from Alfa fibres. *Carbohydr. Polym.* **2014**, *104*, 223–230. [CrossRef]
145. Zhang, Y.; Xu, Y.; Yue, X.; Dai, L.; Ni, Y. Isolation and characterization of microcrystalline cellulose from bamboo pulp through extremely low acid hydrolysis. *J. Wood Chem. Technol.* **2019**, *39*, 242–254. [CrossRef]
146. El-Sakhawy, M.; Hassan, M.L. Physical and mechanical properties of microcrystalline cellulose prepared from agricultural residues. *Carbohydr. Polym.* **2007**, *67*, 1–10. [CrossRef]
147. Kupiainen, L.; Ahola, J.; Tanskanen, J. Kinetics of formic acid-catalyzed cellulose hydrolysis. *BioResources* **2014**, *9*, 2645–2658. [CrossRef]

148. Zambrano, L.-F.; Villasana, Y.; Bejarano, M.L.; Luciani, C.; Niebieskikwiat, D.; Cueva, D.F.; Aguilera, D.; Orejuela, L.M. Optimization of Microcrystalline Cellulose Isolation from Cocoa Pod Husk via Mild Oxalic Acid Hydrolysis: A Response Surface Methodology Approach. Available online: https://ssrn.com/abstract=4307097 (accessed on 17 March 2023).
149. Nurhadi, B.; Angeline, A.; Sukri, N.; Masruchin, N.; Arifin, H.R.; Saputra, R.A. Characteristics of microcrystalline cellulose from nata de coco: Hydrochloric acid versus maleic acid hydrolysis. *J. Appl. Polym. Sci.* **2021**, *139*, 51576. [CrossRef]
150. Trusovs, S. Microcrystalline Cellulose. U.S. Patent 6392034 B1, 21 May 2002.
151. Nguyen, X.T. Process for Preparing Microcrystalline Cellulose. U.S. Patent 7005514 B2, 28 February 2006.
152. Kassaye, S.; Pant, K.K.; Sapna, J. Hydrolysis of cellulosic bamboo biomass into reducing sugars via a combined alkaline solution and ionic liquid pretreament steps. *Renew. Energy* **2017**, *104*, 177–184. [CrossRef]
153. DeLong, E.A. Method of Producing Level off DP Microcrystalline Cellulose and Glucose from Lignocellulosic Material. U.S. Patent 4,645,541, 19 July 1989.
154. Prosvirnikov, D.B.; Safin, R.G.; Zakirov, S.R. Microcrystalline cellulose based on cellulose containing raw material modified by steam explosion treatment. In *Solid State Phenomena*; Trans Tech Publications Ltd.: Wollerau, Switzerland, 2018; Volume 284.
155. Prosvirnikov, D.B.; Timerbaev, N.F.; Safin., R.G. Microcrystalline cellulose from lignocellulosic material activated by steam explosion treatment and mathematical modeling of the processes accompanying its preparation. *Mater. Sci. Forum* **2019**, *945*, 911–918.
156. Ha, E.Y.; Landi, C.D. Method for Producing Microcrystalline Cellulose. U.S. Patent 5,769,934, 23 June 1998.
157. Hanna, M.; Biby, G.; Miladinov, V. Production of Microcrystalline Cellulose by Reactive Extrusion. U.S. Patent 6,228,213, 8 May 2001.
158. Merci, A.; Urbano, M.V.E.; Grossmann, C.A.; Tischer, S.; Mali, S. Properties of microcrystalline cellulose extracted from soybean hulls by reactive extrusion. *Food Res. Int.* **2015**, *73*, 38–43. [CrossRef]
159. Stupińska, H.; Iller, E.; Zimek, Z.; Wawro, D.; Ciechańska, D.; Kopania, E.; Palenik, J.; Milczarek, S.; Steplewski, W.; Krzyzanowska, G. An environment-friendly method to prepare microcrystalline cellulose. *Fibres Text. East. Eur.* **2007**, *5–6*, 167–172.
160. Pranger, L.; Rina, T. Biobased nanocomposites prepared by in situ polymerization of furfuryl alcohol with cellulose whiskers or montmorillonite clay. *Macromolecules* **2008**, *41*, 8682–8687. [CrossRef]
161. Abdul Khalil, H.P.S.; Davoudpour, Y.; Islam, M.N.; Mustapha, A.; Sudesh, K.; Dungani, R.; Jawaid, M. Production and modification of nanofibrillated cellulose using various mechanical processes: A review. *Carbohydr. Polym.* **2014**, *99*, 649–665. [CrossRef]
162. Usuda, M. Hydrolysis of cellulose in concentrated phosphoric acid: Effect of functional groups on the rate of hydrolysis. *Kogyo Kagaku Zasshi.* **1967**, *70*, 349–352. [CrossRef]
163. Yu, H.; Qin, Z.; Liang, B.; Liu, N.; Zhou, Z.; Chen, L. Facile extraction of thermally stable cellulose nanocrystals with a high yield of 93% through hydrochloric acid hydrolysis under hydrothermal conditions. *J. Mater. Chem. A* **2013**, *1*, 3938–3944. [CrossRef]
164. Li, B.; Xu, W.; Kronlund, D.; Määttänen, A.; Liu, J.; Smått, J.H.; Peltonen, J.; Willför, S.; Mu, X.; Xu, C. Cellulose nanocrystals prepared via formic acid hydrolysis followed by TEMPO-mediated oxidation. *Carbohydr. Polym.* **2015**, *133*, 605–612. [CrossRef]
165. Li, D.; Henschen, J.M.E. Esterification and hydrolysis of cellulose using oxalic acid dihydrate in a solvent-free reaction suitable for preparation of surface-functionalised cellulose nanocrystals with high yield. *Green Chem.* **2017**, *19*, 5564–5567. [CrossRef]
166. Chen, L.; Zhu, J.Y.; Carlos, B.; Kitin, P.B.; Elder, T.J. Highly thermal-stable and functional cellulose nanocrystals and nanofibrils produced using fully recyclable organic acids. *Green Chem.* **2016**, *18*, 3835–3843. [CrossRef]
167. Filson, P.B.; Dawson-Andoh, B.E. Sono-chemical preparation of cellulose nanocrystals from lignocellulose derived materials. *Biores. Technol.* **2009**, *100*, 2259–2264. [CrossRef]
168. Filson, P.B.; Dawson-Andoh, B.E.; Schwegler-Berry, D. Enzymatic-mediated production of cellulose nanocrystals from recycled pulp. *Green Chem.* **2009**, *11*, 1808–1814. [CrossRef]
169. Sacui, I.A.; Nieuwendaal, R.C.; Burnett, D.J.; Stranick, S.J.; Jorfi, M.; Weder, C.; Foster, E.J.; Olsson, R.T.; Gilman, J.W. Comparison of the properties of cellulose nanocrystals and cellulose nanofibrils isolated from bacteria, tunicate, and wood processed using acid, enzymatic, mechanical, and oxidative methods. *ACS Appl. Mater. Interfaces* **2014**, *6*, 6127–6138. [CrossRef]
170. Montanari, S.; Roumani, M.; Heux, L.; Vignon, M.R. Topochemistry of carboxylated cellulose nanocrystals resulting from TEMPO-mediated oxidation. *Macromolecules* **2005**, *38*, 1665–1671. [CrossRef]
171. Habibi, Y.; Goffin, A.-L.; Schiltz, N.; Duquesne, E.; Dubois, P.; Dufresne, A. Bionanocomposites based on poly (ε-caprolactone)-grafted cellulose nanocrystals by ring-opening polymerization. *J. Mater. Chemist.* **2008**, *18*, 5002–5010. [CrossRef]
172. Kaushik, A.; Mandeep, S.; Gaurav, V. Green nanocomposites based on thermoplastic starch and steam exploded cellulose nanofibrils from wheat straw. *Carbohydr. Polym.* **2010**, *82*, 337–345. [CrossRef]
173. Zhang, J.; Yuan, X.; Jiang, L.; Wu, Z.; Chen, X.; Wang, H.; Wang, H.; Zeng., G. Highly efficient photocatalysis toward tetracycline of nitrogen doped carbon quantum dots sensitized bismuth tungstate based on interfacial charge transfer. *J. Colloid Interface Sci.* **2018**, *511*, 296–306. [CrossRef]
174. Chen, D.; Zhuang, X.; Zhai, J.; Zheng, Y.; Lu, H.; Chen, L. Preparation of highly sensitive Pt nanoparticles-carbon quantum dots/ionic liquid functionalized graphene oxide nanocomposites and application for H_2O_2 detection. *Sens. Actuators B* **2018**, *255*, 1500–1506. [CrossRef]
175. Zhang, R.; Chen, W. Nitrogen-doped carbon quantum dots: Facile synthesis and application as a "turn-off" fluorescent probe for detection of Hg^{2+} ions. *Biosens. Bioelectron.* **2014**, *55*, 83–90. [CrossRef] [PubMed]

176. Dong, Y.; Pang, H.; Yang, H.B.; Guo, C.; Shao, J.; Chi, Y.; Li, C.M.; Yu, T. Carbon-based dots co-doped with nitrogen and sulfur for high quantum yield and excitation-independent emission. *Angew. Chem.* **2013**, *125*, 7954–7958. [CrossRef]
177. Liu, J.; Liu, X.; Luo, H.; Gao, Y. One-step preparation of nitrogen-doped and surface-passivated carbon quantum dots with high quantum yield and excellent optical properties. *RSC Adv.* **2014**, *4*, 7648. [CrossRef]
178. Yu, C.; Xuan, T.; Chen, Y.; Zhao, Z.; Liu, X.; Lian, G.; Li, H. Gadolinium-doped carbon dots with high quantum yield as an effective fluorescence and magnetic resonance bimodal imaging probe. *J. Alloy. Compd.* **2016**, *688*, 611–619. [CrossRef]
179. Niu, W.-J.; Li, Y.; Zhu, R.-H.; Shan, D.; Fan, Y.-R.; Zhang, X.-J. Ethylenediamine-assisted hydrothermal synthesis of nitrogen-doped carbon quantum dots as fluorescent probes for sensitive biosensing and bioimaging. *Sens. Actuators B* **2015**, *218*, 229–236. [CrossRef]
180. Zhu, S.; Meng, Q.; Wang, L.; Zhang, J.; Song, Y.; Jin, H.; Zhang, K.; Sun, H.; Wang, H.; Yang, B. Highly photoluminescent carbon dots for multicolour patterning, sensors, and bioimaging. *Angew. Chem. Int. Ed.* **2013**, *52*, 3953–3957. [CrossRef]
181. Gao, Z.; Lin, Z.Z.; Chen, X.; Lai, Z.; Huang, Z. Carbon dots-based fluorescent probe for trace Hg^{2+} detection in water sample. *Sens. Actuators B* **2016**, *222*, 965–971. [CrossRef]
182. Su, H. Facile synthesis of N–rich carbon quantum dots from porphyrins as efficient probes for bioimaging and biosensing in living cells. *Int. J. Nanomed.* **2017**, *12*, 7375–7391.
183. Qian, Z.; Ma, J.; Shan, X.; Feng, H.; Shao, L.; Chen., J. Highly luminescent N-doped carbon quantum dots as an effective multifunctional fluorescence sensing platform. *Chem. Eur. J.* **2014**, *20*, 2254–2263. [CrossRef]
184. Li, M.; Yu, C.; Hu, C.; Yang, W.; Zhao, C.; Wang, S.; Zhang, M.; Zhao, J.; Wang, X.; Qiu, J. Solvothermal conversion of coal into nitrogen-doped carbon dots with singlet oxygen generation and high quantum yield. *Chem. Eng. J.* **2017**, *320*, 570–575. [CrossRef]
185. Pierrat, P.; Wang, R.; Kereselidze, D.; Lux, M.; Didier, P.; Kichler, A.; Pons, F.; Lebeau, L. Efficient in vitro and in vivo pulmonary delivery of nucleic acid by carbon dot-based nanocarriers. *Biomaterials* **2015**, *51*, 290–302. [CrossRef]
186. Lei, Z.; Xu, S.; Wan, J.; Wu, P. Facile synthesis of N-rich carbon quantum dots by spontaneous polymerization and incision of solvents as efficient bioimaging probes and advanced electrocatalysts for oxygen reduction reaction. *Nanoscale* **2016**, *8*, 2219–2226. [CrossRef]
187. Dong, Y.; Wang, R.; Li, H.; Shao, J.-W.; Chi, Y.; Lin, X.; Chen, G. Polyamine-functionalized carbon quantum dots for chemical sensing. *Carbon* **2012**, *50*, 2810–2815. [CrossRef]
188. Niu, J.; Gao, H. Synthesis and drug detection performance of nitrogen-doped carbon dots. *J. Lumin.* **2014**, *149*, 159–162. [CrossRef]
189. Khan, W.U.; Wang, D.; Zhang, W.; Tang, Z.; Ma, X.; Ding, X.; Du, S.; Wang, Y. High quantum yield green-emitting carbon dots for Fe (III) detection, biocompatible fluorescent ink and cellular imaging. *Sci. Rep.* **2017**, *7*, 14866. [CrossRef]
190. Dong, Y.; Cai, J.; Fang, Q.; You, X.; Chi, Y. Extraction of electrochemiluminescent oxidized carbon quantum dots from activated carbon. *Chem. Mater.* **2010**, *22*, 5895–5899. [CrossRef]
191. Iravani, S.; Varma, R.S. Green synthesis, biomedical and biotechnological applications of carbon and graphene quantum dots: A review. *Environ. Chem. Lett.* **2020**, *18*, 703–727. [CrossRef]
192. Xu, H.; Yang, X.; Li, G.; Zhao, C.; Liao, X. Green synthesis of fluorescent carbon dots for selective detection of tartrazine in food samples. *J. Agric. Food Chem.* **2015**, *63*, 6707–6714. [CrossRef]
193. Singh, I.; Arora, R.; Dhiman, H.; Pahwa, R. Carbon quantum dots: Synthesis, characterization and biomedical applications. *Turk. J. Pharm. Sci.* **2018**, *15*, 219–230. [CrossRef]
194. Li, X.; Chen, W.; Zhan, Q.; Dai, L.; Sowards, L.; Pender, M.; Naik, R.R. Direct measurements of interactions between polypeptides and carbon nanotubes. *J. Phys. Chem. B* **2006**, *110*, 12621–12625. [CrossRef] [PubMed]
195. Acharya, A.P.; Nafisi, P.M.; Gardner, A.; Mackay, J.L.; Kundu, K.; Kumar, S.; Murthy, N.A. A fluorescent peroxidase probe increases the sensitivity of commercial ELISAs by two orders of magnitude. *Chem. Commun.* **2013**, *49*, 10379–10381. [CrossRef] [PubMed]
196. Abu Rabe, D.I.; Al-Awak, M.M.; Yang, F.; Okonjo, P.A.; Dong, X.; Teisl, L.R.; Wang, P.; Tang, Y.; Pan, N.; Sun, Y.P.; et al. The dominant role of surface functionalization in carbon dots' photo-activated antibacterial activity. *Int. J. Nanomed.* **2019**, *23*, 2655–2665. [CrossRef] [PubMed]
197. Dong, X.; Ge, L.; Abu Rabe, D.I.; Mohammed, O.O.; Wang, P.; Tang, Y.; Kathariou, S.; Yang, L.; Sun, Y.-P. Photoexcited state properties and antibacterial activities of carbon dots relevant to mechanistic features and implications. *Carbon* **2020**, *170*, 137–145. [CrossRef]
198. Hutson, A.M.; Atmar, R.L.; Marcus, D.M.; Estes, M.K. Norwalk virus–like particle hemagglutination by binding to h histo–blood group antigens. *Virol. J.* **2003**, *77*, 405–415. [CrossRef]
199. Lindesmith, L.; Moe, C.; Marionneau, S.; Ruvoen, N.; Jiang, X.; Lindblad, L.; Stewart, P.; LePendu, J.; Baric, R. Human susceptibility and resistance to Norwalk virus infection. *Nat. Med.* **2003**, *9*, 548–553. [CrossRef]
200. Tan, M.; Jiang, X. Norovirus and its histo-blood group antigen receptors: An answer to a historical puzzle. *Trends Microbi.* **2005**, *13*, 285–293. [CrossRef]
201. Tan, M.; Jiang, X. Norovirus-host interaction: Implications for disease control and prevention. *Expert. Rev. Mol. Med.* **2007**, *9*, 1–22. [CrossRef]
202. Harrington, P.R.; Lindesmith, L.; Yount, B.; Moe, C.L.; Baric, R.S. Binding of Norwalk virus-like particles to ABH histo-blood group antigens is blocked by antisera from infected human volunteers or experimentally vaccinated mice. *Virol. J.* **2002**, *76*, 12335–12343. [CrossRef]

203. LoBue, A.D.; Lindesmith, L.; Yount, B.; Harrington, P.R.; Thompson, J.M.; Johnston, R.E.; Moe, C.L.; Baric, R.S. Multivalent norovirus vaccines induce strong mucosal and systemic blocking antibodies against multiple strains. *Vaccine* **2006**, *24*, 5220–5234. [CrossRef]
204. Hale, A.; Mattick, K.; Lewis, D.; Estes, M.; Jiang, X.; Green, J.; Eglin, R.; Brown, D. Distinct epidemiological patterns of Norwalk-like virus infection. *J. Med. Virol.* **2000**, *62*, 99–103. [CrossRef]
205. Hardy, M.E.; Tanaka, T.N.; Kitamoto, N.; White, L.J.; Ball, J.M.; Jiang, X.; Estes, M.K. Antigenic mapping of the recombinant Norwalk virus capsid protein using monoclonal antibodies. *Virol. J.* **1996**, *217*, 252–261. [CrossRef]
206. D'Souza, D.H.; Su, X.W.; Roach, A.; Harte, F. High-pressure homogenization for the inactivation of human enteric virus surrogates. *J. Food Prot.* **2009**, *72*, 2418–2422. [CrossRef]
207. Prasad, B.V.; Hardy, M.E.; Dokland, T.; Bella, J.; Rossmann, M.G.; Estes, M.K. X-ray crystallographic structure of the Norwalk virus capsid. *Science* **1999**, *286*, 287–290. [CrossRef]
208. Chen, Y.; Tan, M.; Xia, M.; Hao, N.; Zhang, X.C.; Huang, P.; Jiang, X.; Li, X.; Rao, Z. Crystallography of a Lewis-binding Norovirus, elucidation of strain-specificity to the polymorphic human histoblood group antigens. *PLoS Pathog.* **2011**, *7*, e1002152. [CrossRef]
209. Tan, M.; Xia, M.; Chen, Y.; Bu, W.; Hegde, R.S.; Meller, J.; Li, X.; Jiang, X. Conservation of carbohydrate binding interfaces—Evidence of human HBGA selection in Norovirus evolution. *PLoS ONE* **2009**, *4*, e5058. [CrossRef]
210. Hou, J.; Wang, W.; Zhou, T.; Wang, B.; Li, H.; Ding, L. Synthesis and formation mechanistic investigation of nitrogen-doped carbon dots with high quantum yields and yellowish-green fluorescence. *Nanoscale* **2016**, *8*, 11185–11193. [CrossRef]
211. Molaei, J.M. Carbon quantum dots and their biomedical and therapeutic applications: A review. *RSC Adv.* **2019**, *9*, 6460. [CrossRef]

polymers

MDPI

Article

Cellulose Nanofibers/Pectin/Pomegranate Extract Nanocomposite as Antibacterial and Antioxidant Films and Coating for Paper

Enas Hassan [1], Shaimaa Fadel [1], Wafaa Abou-Elseoud [1,2], Marwa Mahmoud [3] and Mohammad Hassan [1,2,*]

1 Cellulose and Paper Department, National Research Centre, 33 El-Buhouth Street, Dokki, Giza 12622, Egypt
2 Advanced Materials and Nanotechnology Group, Centre of Excellence for Advanced Sciences, National Research Centre, 33 El-Buhouth Street, Dokki, Giza 12622, Egypt
3 Food Technology Department, National Research Centre, 33 El-Buhouth Street, Dokki, Giza 12622, Egypt
* Correspondence: ml.hassan@nrc.sci.eg

Abstract: Bio-based polymer composites find increasing research and industrial interest in different areas of our life. In this study, cellulose nanofibers (CNFs) isolated from sugar beet pulp and nanoemulsion prepared from sugar beet pectin and pomegranate extract (PGE) were used for making films and used as coating with antioxidant and antimicrobial activities for paper. For Pectin/PGE nanoemulsion preparation, different ratios of PGE were mixed with pectin using ultrasonic treatment; the antibacterial properties were evaluated to choose the formula with the adequate antibacterial activity. The antioxidant activity of the nanoemulsion with the highest antimicrobial activity was also evaluated. The nanoemulsion with the optimum antibacterial activity was mixed with different ratios of CNFs. Mechanical, greaseproof, antioxidant activity, and antibacterial properties of the CNFs/Pectin/PGE films were evaluated. Finally, the CNFs/Pectin/PGE formulation with the highest antibacterial activity was tested as a coating material for paper. Mechanical, greaseproof, and air porosity properties, as well as water vapor permeability and migration of the coated layer from paper sheets in different media were evaluated. The results showed promising applicability of the CNFs/Pectin/PGE as films and coating material with antibacterial and antioxidant activities, as well as good stability for packaging aqueous, fatty, and acidic food products.

Keywords: sugar beet pulp; cellulose nanofibers; pectin; pomegranate; nanoemulsion; antibacterial; antioxidant; papermaking; coating; migration

Citation: Hassan, E.; Fadel, S.; Abou-Elseoud, W.; Mahmoud, M.; Hassan, M. Cellulose Nanofibers/Pectin/Pomegranate Extract Nanocomposite as Antibacterial and Antioxidant Films and Coating for Paper. *Polymers* **2022**, *14*, 4605. https://doi.org/10.3390/polym14214605

Academic Editor: Shuangquan Yao

Received: 14 October 2022
Accepted: 27 October 2022
Published: 30 October 2022

1. Introduction

Specialty paper products are important category of paper with wide commercial production and uses. Among these specialty paper products are those with antibacterial activity which find increasing use in packaging and hygienic products. Research in that area is progressing in increasing rates in order to reach economic, eco-friendly, and safe additives or chemical modification routes. Additives with antibacterial activity extracted from bio-based materials or non-food non-feed residues are of special interest since they have the benefits of being widely available, cheap, eco-friendly, and generally safe, e.g., having no health hazard. Among these residues is pomegranate peels, which are rich in antibacterial and antioxidants extractives, Hundreds of tons of pomegranate peels are available every year at food companies all over the world. Pomegranate peel extractives (PGE) are widely studied and used in food, pharmaceutical, and medicinal applications due to their powerful antibacterial, antifungal, virucidal, and antiviral activity [1–4]. PGE is rich in many antioxidant, antibacterial, and antiproliferative compounds [5]. To improve the solubility of PGE and its application in aqueous solutions, their use in forms of emulsions is considered a good practice. Ethanol-extracted PGE can be formulated into nano-size emulsions using proper emulsifying agents [6–10]. The nano-size emulsions have the privilege of a very high surface area, and thus could be used effectively in smaller amounts than the micro-emulsions. Emulsifying agents can be of synthetic origin such as polyglycerol polyricinoleate and Tween 80, which are used for the preparation of nanoemulsion from pomegranate extract [8]. Nevertheless, the use of bio-based surfactants are preferred when it comes to food, food packaging, or cosmetic products. For this reason, maltodextrin and whey protein isolate [6,10], sodium carboxymethyl cellulose [7,11], and chitosan [9] have been used for preparation of PGE emulsions. In this aspect, sugar beet pectin in particular has high emulsifying activity when compared to pectin isolated from other food residues, such as citrus and apple peels, due to its specific chemical composition, the presence of a high degree of acetylation, and low molecular weight [11]. On the other hand, pectin from other resources, such as citrus and apple peels, have poor emulsifying properties but have other features such as thickening, gelling, and film forming properties thanks to their much higher molecular weight and much lower degree of esterification than that of SBP [12,13].

It is worth mentioning here that sugar beet pulp (SBP) is a common agricultural residue in different areas of the world, as 20% of the world production of sugar comes from sugar beet [14]. After sugar extraction, SBP residue is rich cellulose, hemicellulloses, and pectin, in addition to other minor components.

Although sugar beet pectin has been widely studied for making micro-emulsions [15–17], a few studies have reported on the preparation of nano-emulsion [18,19]. To the best of our knowledge, there have been no previous studies regarding use of SBP pectin for making emulsions from PGE.

In addition to the specific properties of pectin isolated from SBP, CNFs with elementary fibrils (width ~5 nm) could be easily isolated from SBP after pectin extraction due to the unique cell wall structure of sugar beet where most of the tissue is parenchymal, which is characterized by only a very thin primary wall and loosely organized cellulose nano-size microfibrils embedded in a matrix of hemicelluloses and pectin [20]. Several publications have studied the isolation of CNFs from SBP using different methods and pretreatments [11].

The well-known unique properties of CNFs such as the high specific surface area, the nanometer wide and micrometer length dimensions, the combination of unique intrinsic mechanical strength with good flexibility properties, and the ability of strong hydrogen bonding along the nanofibers or with other different matrices motivate their application in different areas including papermaking. CNFs can improve mechanical, barrier (air, moisture, oil, and thermal), and printability properties, and reduce paper weight products [21–28]. Nevertheless, due to the negative effect of CNFs on the drainage of water during paper sheet formation as a result of their very small size, very high water holding capacity, and ability to fill empty spaces between the pulp fibers and clog pores of the wet web [27,29,30], applying CNFs as a surface coating after paper sheet formation instead of applying them as a paper additive has resulted in increasing interest for improving paper properties and the production of novel paper products [31–38]. Usually, paper coating materials are used in very small quantities and applied as a very thin layer, yet they bring about significant effects on paper properties. Specifically, CNFs isolated from SBP were successfully used as a coating material for paper sheets to improve mechanical and oil-proof properties, as well as air permeation resistance [31].

While there is no use of pectin from any resources in papermaking so far, its use with CNFs has been reported in alginate scaffolds for biomedical use [39], to improve water resistance of soybean protein [40], as a co-carrier to improve water redispersibility of spray-dried CNFs in water [41], to prepare aerogel with improved mechanical properties [42], and for the preparation of printing inks [43,44].

However, both pectin and CNFs have no antibacterial activity due to their polysaccharide nature. Therefore, their use together in products with antibacterial properties requires the addition of an antibacterial agent. For pectin, adding PGE to it has been used for the preparation of antibacterial edible coatings of fruits [45,46] or in food products such as jam and juice [46,47]; pectin used in the aforementioned studies was isolated from pineapple, orange, pomegranate, and banana. However, there is no previous work published so far on using PGE with SBP pectin for making nanoemulsion or use of their mixture in papermaking.

Based on the above-mentioned limitations of pectin and CNFs for use in antibacterial paper products, the current manuscript studies the preparation of nanocomposite from pectin/PGE nanoemulsion and CNFs for use as a bio-based and environmentally friendly films and coating for different applications such as food packaging materials.

In fact, one of the most appealing solutions and modern approaches for preventing microbial or virus contamination and their spreading is through designing appropriate surface coating, i.e., fabrication of antimicrobial or antivirus surfaces [48]. This can be achieved through adding a specific antibacterial or antivirus agent, such as pomegranate extract, in a coating matrix; this is known as an active antimicrobial or antivirus coating approach. Pomegranate extract contains different small molecules with antimicrobial and antivirus activities which work with different mechanisms to de-activate viruses and bacteria [2–4,49–55].

In the current work, CNFs were used as the film forming material to impart strength, pectin was the emulsifying agent to form nanoemulsions from pomegranate extract to make it compatible with CNFs, and pomegranate extract was the antibacterial agent.

2. Materials and Methods

2.1. Raw Material and Reagents

Wet-pressed SBP (~20 wt.% solid content) was kindly supplied by the Alnubariah Company for Sugar, Alexandria, Egypt. It was directly dried in an oven with hot air circulation at 50 °C for 12 h. The chemical composition of the SBP was 38.02% α-cellulose, 18.2% pentosans, 3.85% lignin, 2.77% ash, 19.40% galacturonic acid content, and 10.13% protein content, as determined according to standard methods of chemical analysis [56–58].

Sodium chlorite (technical grade 80%), glacial acetic acid, sulfuric acid, sodium thiosulfate, potassium bromide, potassium bromate, hydrochloric acid, acetic acid, citric acid, and sodium hydroxide were of analytical grade. They were purchased from Fisher Scientific U.K. Ldt (Loughborough, UK) and used as received. Polyamideamine-epichlorohydrin (PAE) crosslinking agent was commercial grade (solid content ~33 wt.%, Solines, Wilmington, DE, USA). PAE solution was diluted to 1 wt.% with distilled water before use. *E. coli* EMCCN 3060 and *S. aureus* EMMCN 3057 bacteria were kindly supplied from the Egyptian Microbial Culture Collection Network (EMCCN) at the National Research Centre, Giza, Egypt.

2.2. Extraction of Pectin

Extraction of pectin by acid hydrolysis was carried out as previously published by Abou-Elseoud et al., as follows [59]: SBP was suspended in water at a liquor ratio of 1:15 and acidified to pH 1 with sulfuric acid. It was then heated for 2 h at 85 °C under mechanical stirring. The residue was then separated from the soluble compounds by vacuum filtration, washed with distilled water, and kept in the fridge at 4 °C until use for isolation of the nanofibers.

To isolate pectin from the filtrate, it was centrifuged in 50-mL centrifuge tubes at 10,000 rpm for 10 min to remove fines, and pectin was precipitated by the addition of ethanol at volume ratio of 3:1 ethanol to filtrate; the mixture was left for 2 h and the precipitated pectin was centrifuged at 10,000 rpm for 20 min, washed with 70% ethanol/water mixture, centrifuged again, and dried at 40 °C for 48 h. The chemical composition of the extracted pectin was 72.6% galacturonic acid, 10.64% neutral sugars, 10.5% protein, and 0.5% ferulic acid [59].

2.3. Isolation of Cellulose Nanofibers (CNFs)

Isolation of CNFs from the de-pectinated SBP was carried out according to the previously published protocol [31]. In brief, after bleaching with acetic acid/chlorite mixture, the bleached de-pectinated SBP was suspended in water at 2 wt.% consistency and subjected to high shear mixing using ESR-500x laboratory high shear homogenizer (ELE, Shanghai, China) at 10,000 rpm for 15 min. The suspension was then passed twice through a two-chamber Homolab 2.2 high-pressure homogenizer (FBF, Parma, Italy). The CNF suspension was then kept in fridge at 4 °C until use. The chemical composition of isolated CNFs was 87.8% α-cellulose, 6.96% pentosans, 0.45% lignin, 1.10% ash, 2.44% galacturonic acid content, and 0.31% protein content, as determined according to standard methods of chemical analyses [56–58].

A JEM-2100 high-resolution transmission electron microscope (HRTEM) (JEOL, Tokyo, Japan) was used for characterizing the microstructure of the isolated CNFs after being stained with phosphotungstic acid solution.

2.4. Extraction and Characterization of Pomegranate Extract (PGE)

Pomegranate peels were washed with water, oven-dried in an oven with hot air circulation at 40 °C for 24 h, and then ground to pass through a 20-μm sieve. The extraction was carried out using 70/30 (v/v) ethanol/water mixture in a Soxhlet for 8 h at 85 °C. The solvent containing the extract was then evaporated at 65 °C using a Rotavapor R-210 rotary evaporator (BÜCHI Labortechnik AG, Flawil, Switzerland) under vacuum. To complete the drying, the highly viscous extract was poured in a glass Petri dish and dried in an oven under vacuum for 18 h at 65 °C. Serial dilution was made from the extract and kept for further analysis.

2.5. Determination of Individual and Total Phenolics

The Folin–Ciocalteu assay, adapted from Ramful et al. [60], was used for determination of total phenolics. In addition, the individual phenolic compounds were determined by high-performance liquid chromatography (HPLC) using an Agilent 1260 series (Agilent, Santa Clara, CA, USA). The separation was carried out using a C18 column (4.6 mm × 250 mm i.d., 5 μm). The mobile phase consisted of water (A) and 0.02% trifluoro-acetic acid in acetonitrile (B) at a flow rate 1 mL/min. The mobile phase was programmed consecutively in a linear gradient as follow: 0–5 min (80% A); 0–5 min (80% A); 5–8 min (40% A); 8–12 min (50% A); 12–14 min (80% A); and 14–16 min (80% A). The multi-wavelength detector was monitored at 280 nm. The injection volume was 10 μL for each of the sample solutions. The column temperature was maintained at 35 °C.

2.6. DPPH Radical Scavenging Activity

The effect of extracts on 1,1-diphenyl-2-picrylhydrazyl (DPPH) free radical was estimated according to the procedure described by Aboelsoued et al. [61]. The absorbance was measured at 517 nm using Jenway 7305 UV-visible spectrophotometer (Jenway, Staffordshire, England). The control was conducted with ethanol instead of the sample. DPPH scavenging capacity was calculated by using the following equation:

$$Scavenging\ activity\ (\%) = \frac{Ac - As}{Ac} \times 100$$

where Ac and As are the absorbance at 517 nm of the control and sample, respectively.

L-ascorbic acid solutions as standards were also analyzed by DPPH and ABTS methods. The total antioxidant values of citrus samples were expressed as mg g^{-1} dry weight L-ascorbic acid equivalent antioxidant capacity (VCEAC).

2.7. Determination of Ferric Reducing Power (FRAP) Assay

The FRAP assay is based on the ability of phenolics to reduce Fe^{3+} to Fe^{2+} [62]. To prepare the FRAP reagent, 0.1 M acetate buffer (pH 3.6), 10 mM TPTZ, and 20 mM ferric chloride (10:01:01, v/v/v) were mixed. Then, 20 µL of previously diluted extract were added to 150 µL of reagent. The absorbance was measured at 593 nm using Jenway 7305 UV-visible spectrophotometer (Jenway, Staffordshire, England).. The analysis was performed in triplicate, using an aqueous Trolox solution as standard, and the results were expressed as lmoles Trolox equivalents/100 g of fresh weight sample.

2.8. Preparation and Characterization of Pectin/Pomegranate Extract (Pectin/PGE) Emulsion

Pectin/PGE emulsions containing 2.5, 5, 10, 15, and 20 wt.% of PGE (based on the oven dry weight of pectin) were prepared by ultrasonic treatment under cooling at 4 °C in an ice bath for 2 min using a UP 400 Hielscher ultrasonic processor (Hielscher Ultrasonics GmbH, Teltow, Germany); A 1-cm diameter probe was used at an amplitude of 75%. The concentration of Pectin/PGE in water was 4.5%.

TEM of the emulsions was carried out using JEM-2100 HRTEM (JEOL, Tokyo, Japan) for characterizing the microstructure of the emulsion after being stained with phosphotungstic acid solution.

The antimicrobial activity of the prepared Pectin/PGE emulsions was evaluated by minimal inhibitory concentration test (MIC) as previously described [63]. Gram-positive bacteria (*Staphylococcus aureus*) and Gram-negative bacteria (*Escherichia coli*) were used as test organisms. A pre-culture of bacteria was grown in Tryptic Soy Broth medium overnight at 37 °C and a serial dilution were made from each strain until obtained dilutions of 1×10^5 and 1×10^8 CFU/mL for *Staphylococcus aureus* and *Escherichia coli*, respectively. Next, 100 µL of both tested bacteria were added to the test tubes containing 9.9 mL of sterile Tryptic Soy Broth medium and exposed to 0.1 g of the different Pectin/PGE emulsions, which were oven-dried at 65 °C under vacuum; a neat pectin sample was tested as a blank. All samples were then incubated at 37 °C with shaking at 140 rpm for 24 h. After a 24 h incubation, a series of dilutions were prepared by the addition of 1 mL of each culture to 9 mL of sterile 0.3 mM phosphate buffer (pH 6.8), followed by seeding 100 µL of each culture onto an agar plate. The plates were incubated at 37 °C for 24 h and the surviving cells counted. The antimicrobial activity was expressed as a reduction of the bacterial colonies after contact with the test specimen and compared to the number of bacterial colonies from the blank sample (neat pectin). The percentage reduction (inhibition) was calculated using the following equation:

$$\% \text{ Reduction} = ((B - A)/B) \times 100$$

where A is the surviving cells (CFU—colony forming units) for the plates containing the treated substrate and B is the surviving cells from the control.

2.9. Preparation and Characterization of CNFs/Pectin/PGE Films

CNFs/Pectin/PGE films containing different ratios of Pectin/20%PGE emulsion were prepared by casting suspension mixtures in a 9-cm diameter Teflon petri dish. Glycerol was added at a fixed ratio of 25% of all samples to get good film formation without damage due to shrinkage. The ratios of Pectin/20%PGE were 2.5, 5, 7.5, 10, 15, and 20 wt.% of the oven dry weight of CNFs plus glycerol. Then, 2% of PAE wet strength agent (based on total weight of the films) was added to all samples. The suspensions were dried at 40 °C for 18 h in an oven with hot-air circulating. The produced films were conditioned at 50% relative humidity for 48 h at 25 °C before testing.

Tensile strength properties were measured using LR10 K Lloyd instrument (Lloyd Instruments, Fareham, UK) with a 1 kN load cell at 25 °C using a crosshead speed of 2 mm/min. Strips with 10 × 90 mm width by length, respectively, were used and the distance between the grips was 20 mm. Five specimens from each sample were measured and the results averaged.

Anti-bacterial activity of the films was tested using the disk diffusion method [63]; Gram-positive *Staphylococcus aureus* and Gram-negative *Escherichia coli* bacteria were used as test organisms. A loopful from each stock strain was transferred into 10 mL of Tryptic Soy Broth medium with 0.6% yeast extract and incubated at 37 °C overnight. Then, 100 μL from each strain was seeded on the surface of Tryptic Soy agar plates, and ~1-cm^2 of the films was placed onto the inoculated surfaces and then incubated at 37 °C for 24 h to detect the bacterial inhibition zones. The experiment was performed in triplicate.

2.10. Coating of Paper Sheet

CNFs/20%Pectin/PGE suspension with 3% solid content containing 2% of PAE (based on total dry weight of the suspension) was used for coating commercial wrapping paper sheets, which have a basis weight of ~30 g/m^2. The paper sheets were fixed over a glass plate and coated with the suspension using ZUA 2000 coater (Zehntner GmbHTesting Instruments, Sissach, Switzerland), coating was carried out using a gap of 500 μm.

Coated paper sheets were dried in air circulation oven at 80 °C for 15 min for crosslinking of PAE. The coated paper samples were conditioned at 50% relative humidity for 48 h at 25 °C before testing. The amount of CNFs/Pectin/PGE coating was determined gravimetrically as g/m^2 from the difference in basis weight of coated and uncoated paper sheets, as follows:

$$\text{Amount of coating } (\text{g/m}^2) = \\ \text{Basis weight of coated paper sheet in g/m}^2 - \text{Basis weight of uncoated paper sheet in g/m}^2$$

2.11. Characterization of Paper Sheets

The surface and cross-section of the coated paper was examined by scanning electron microscopy (SEM) using an FEI Quanta 200 scanning electron microscope (FEI Company BV, Eindhoven, The Netherlands) with an acceleration voltage of 20 kV. Paper sheet samples were coated with gold using a sputter coater system (Edwards Sputter Coater, Sussex, UK) before testing.

Tensile strength testing of paper sheets was carried out using LR10K Lloyd universal testing machine (Lloyd Instruments, Fareham, UK) with a 1 KN load cell at a constant crosshead speed of 2 mm/min according to TAPPI T494 (TAPPI 2006). Porosity was measured using a Gurley air permeability tester 4110 (W. & L.E. Gurley, Troy, NY, USA) according to ASTM D726-58. A greaseproof test was carried out using turpentine oil (TAPPI standard T454). A water vapor permeability test was carried out according to the ASTM standard (ASTM E96); all the tests were performed in triplicate at an atmospheric pressure (1 atm) and the results were averaged. WVP was calculated according to the following equation:

$$\text{WVP } (\text{g}\cdot\text{m}^{-1}\text{s}^{-1}\text{Pa}^{-1}) = (m\cdot e)/(A\cdot t\cdot p)$$

where m is the mass increase (in g) of the $CaCl_2$, A is the area of the film, and t is the exposure time in the chamber. The thickness of the film is e and p is the partial water vapor pressure difference across both of the film specimens corresponding to 0–60% RH, i.e., 1875 Pa.

The overall migration test was carried out according to EU Regulation Nr. 10/2011. Three stimulants were used in the test which represents water, fatty, and acidic conditions: 10% v/v ethanol in water, 50% v/v ethanol in water, and 3% acetic acid solution. All samples were compared to a blank water sample (Millipore water with resistivity 18.5 MΩ) as a reference. As the single sided cell method was used, results were calculated considering the area of only one surface of the test specimen. After 10 days at 40 °C, the samples were picked up from contact, and the extraction aqueous solutions were collected in flasks, heated in a rotary evaporator under vacuum to dryness, and weighed until constant weight (EN 1186-5-single side contact in cell test) [64].

2.12. Statistical Analysis

The results of tested samples were presented as the mean ± SD using Microsoft Excel. For the total phenolics and antioxidant activity testing, the data were statistically analyzed by one-way ANOVA using SPSS software version 20. Significance was considered at a level of 0.05.

3. Results and Discussion

3.1. Composition of PGE

The components of the ethanol/water PGE were characterized by HPLC. Table 1 shows the components separated by the HPLC column and their percentages; the HPLC chromatogram is attached in the Supplementary Information Figure S1. The major components in the PGE are in accordance with previous studies on PGE extracted using ethanol solutions [65].

Table 1. Composition of pomegranate extract (PGE)obtained by high-performance liquid chromatography (HPLC).

Constituent	Concentration (μg/g)
Chlorogenic acid	21.95
Gallic acid	11.18
Ellagic acid	9.51
Catechin	5.55
Coffeic acid	3.06
Methyl gallate	0.50
Naringenin	0.36
Pyro catechol	0.081
Rutin	0.066
Cinnamic acid	0.042

3.2. Pectin/PGE Emulsions

Using PGE extracted with ethanol with hydrophilic polymers, such as cellulose, requires emulsification with suitable emulsifier, which acts as a compatibilizer between the relatively hydrophobic groups of PGE and the hydrophilic hydroxyl groups at the cellulose surface.

3.2.1. Particle Size of Pectin/PGE Emulsions

SBP pectin is characterized by a high ability to form micro- and nano-emulsion thanks to the presence of high content of acetyl groups at its polysaccharide chains [11]. In the current work, pectin could form nanoemulsions from the PGE, as shown in the TEM images in Figure 1. The size of the separate emulsion particles was in the range from ~25 to 50 nm for the emulsions with different PGE loading, i.e., the size emulsion particles was not dependent on the concentration of PGE used. It was also noticed that the nanoparticles form several aggregates, but their size was still in the submicron range and less than ~200 nm. Results of particle size analysis by a laser in Figure 2 and Table 2 show a similar trend of the results, e.g., emulsion particles size not dependent on their concentration in pectin, with diameters in the range from 193 to 208 nm. This reflects the high ability of pectin to form emulsions in the nano-diameter range, even at a high content of PGE (20%). The higher mean diameter results in the case of particle size analysis than that in the case of using TEM could be due to the different principals of measuring instruments used, and the formation of aggregates from the nanoemulsion particles. Cumulative particle size analysis in Table 2 also shows that 90% of the particles had diameters less than ~296, 330, 284, 292, and 323 nm for pectin emulsions containing 2.5, 5, 10, 15, and 20% of PGE, respectively, again indicating the ability of sugar beet pectin to emulsify high ratios of PGE.

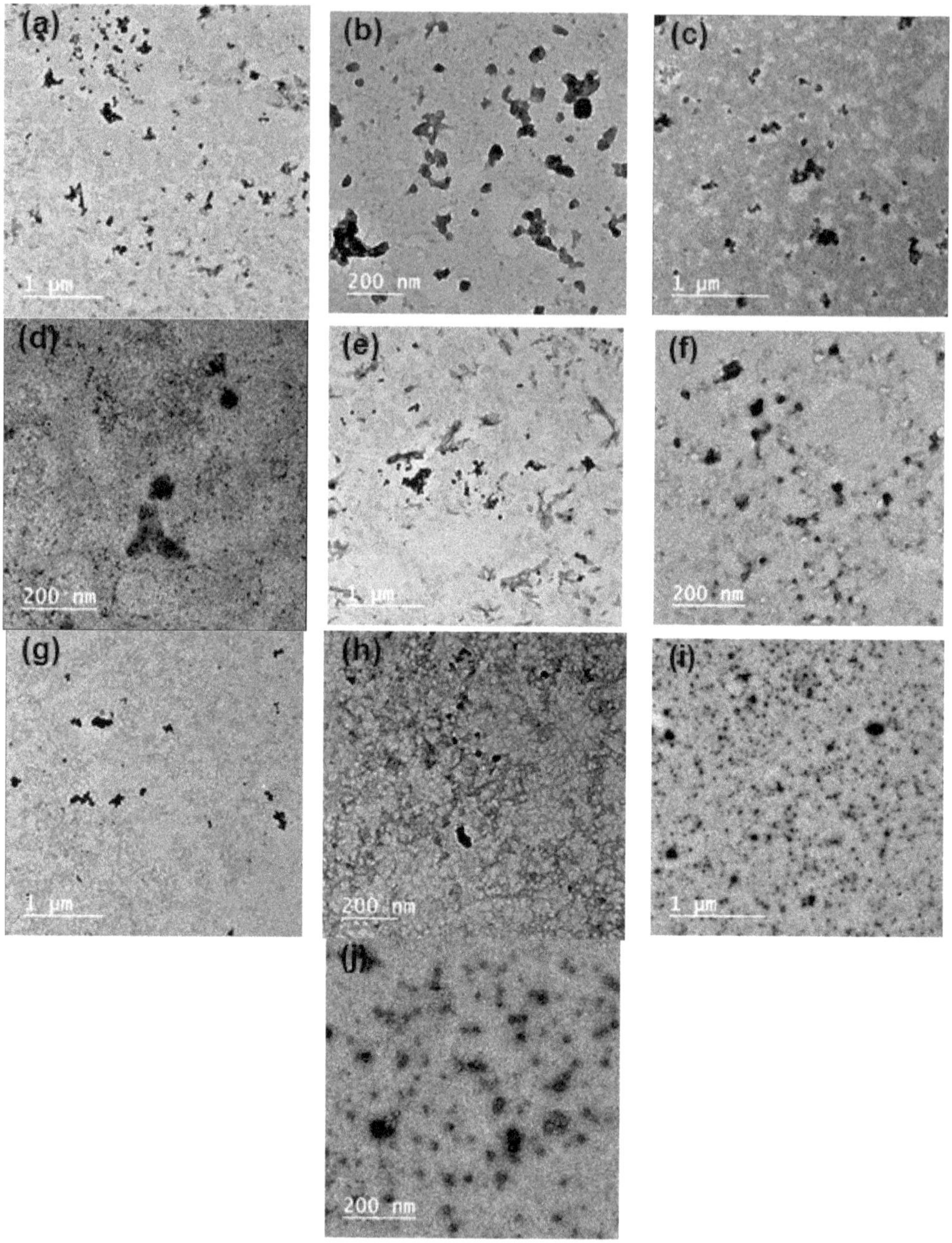

Figure 1. Transmission electron microscopy TEM images of the prepared emulsions from Pectin and pomegranate extract (PGE) at different magnifications: (**a**,**b**) Pectin/2.5% PGE, (**c**,**d**) Pectin/5% PGE, (**d**,**e**) Pectin 10% PGE, (**f**,**g**) Pectin/15% PGE, and (**h**,**i**) Pectin/20% PGE.

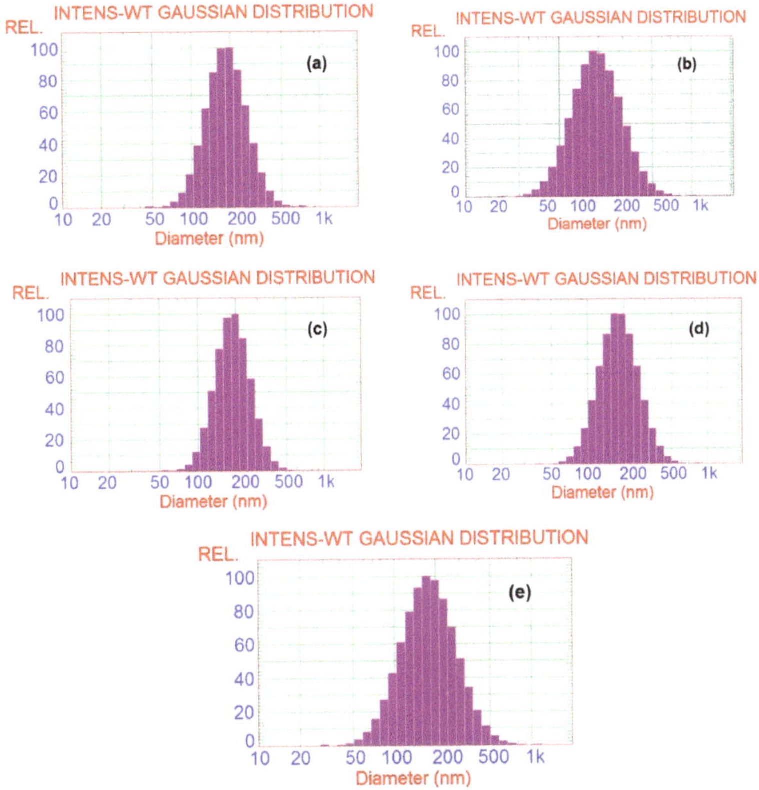

Figure 2. Particle size analysis histograms of Pectin/PGE emulsions with different PGE loadings: (**a**) 2.5% PGE, (**b**) 5% PGE, (**c**) 10% PGE, (**d**) 15% PGE, and (**e**) 20% PGE.

Table 2. Particle size analysis of Pectin/PGE emulsions with different PGE loadings.

	Mean Diameter (nm)	Variance (P.I)	Cumulative Analysis of Particle Size
Pectin/2.5% PGE	198.0 ± 72.3	0.13	25% of distribution < 144.7 nm 50% of distribution < 185.1 nm 75% of distribution < 236.8 nm 80% of distribution < 251.7 nm 90% of distribution < 295.6 nm 99% of distribution < 432.8 nm
Pectin/5% PGE	208.7 ± 89.9	0.186	25% of distribution < 142.1 nm 50% of distribution < 190.0 nm 75% of distribution < 254.2 nm 80% of distribution < 273.1 nm 90% of distribution < 330.2 nm 99% of distribution < 518.0 nm
Pectin/7.5% PGE	198.1 ± 64.0	0.104	25% of distribution < 151.2 nm 50% of distribution < 187.9 nm 75% of distribution < 233.7 nm 80% of distribution < 246.7 nm 90% of distribution < 284.3 nm 99% of distribution < 398.4 nm

Table 2. *Cont.*

	Mean Diameter (nm)	Variance (P.I)	Cumulative Analysis of Particle Size
Pectin/10% PGE	193.2 ± 73.2	0.144	25% of distribution < 139.2 nm 50% of distribution < 179.7 nm 75% of distribution < 232.0 nm 80% of distribution < 247.2 nm 90% of distribution < 292.0 nm 99% of distribution < 433.9 nm
Pectin/20% PGE	197.9 ± 92.4	0.218	25% of distribution < 129.4 nm 50% of distribution < 177.4 nm 75% of distribution < 243.0 nm 80% of distribution < 262.8 nm 90% of distribution < 322.7 nm 99% of distribution < 525.6 nm

3.2.2. Antibacterial Activity of Pectin/PGE Emulsion

Since pectin is a polysaccharide without antibacterial activity, one of the aims of the current work was preparing pectin emulsion with antibacterial activity for different applications. PGE extract is known to have strong antibacterial properties against Gram-positive and Gram-negative bacteria thanks to the presence of several polyphenolic constituents in the PGE, including chlorogenic acid, catechins, and gallic and ellagic acid [66,67]. Several mechanisms for the antibacterial activity of PGE were proposed, including the binding of the polyphenols to the bacterial protein [49–51], destroying the synthesis of bacterial RNA and DNA [52,53], and destabilization of the outer bacterial membrane [54,55]. In the current study, the antibacterial activity of the PGE/pectin emulsions was studied to ensure the potential use of the prepared nanoemulsion as an antimicrobial mixture and that the emulsification did not affect the antimicrobial activity of PGE.

Table 3 shows the % inhibition of Pectin/PGE nanoemulsion with different PGE loadings. As seen in the table, even at the lowest PGE concentration used, ~83 and 80% inhibition against *E. coli* and *S. aureus* could be achieved, respectively. There was no increase in the % inhibition at PGE loading >15%, where % inhibition was ~92% and 99.5% against *E. coli* and *S. aureus*, respectively. The observed higher activity of PGE extract against the *S. aureus* than that of *E. coli* is in agreement with the results of previous studies [5,58–68].

Table 3. Inhibition of *E. coli* and *S. aureus* bacteria by Pectin/PGE emulsion with different PGE loadings.

	% Inhibition	
Sample	***S. aureus***	***E. coli***
Pectin + 2.5% PGE	80.0 ± 4.38	83.3 ± 7.07
Pectin + 5% PGE	84.6 ± 4.35	83.3 ± 7.07
Pectin + 10% PGE	95.4 ± 4.35	86.7 ± 4.71
Pectin + 15% PGE	99.2 ± 0.22	91.7 ± 2.36
Pectin + 20% PGE	99.7 ± 0.22	91.7 ± 4.71

3.2.3. Antioxidant Activity of Pectin/PGE Emulsion

As mentioned above, PGE is rich in polyphenolic compounds which give it a strong antioxidant and antibacterial activity. The total phenolics (such as gallic acid), antioxidant assay by the DPPH (2,2-diphenyl-1-picryl-hydrazyl hydrate) free radical method, and Trolox equivalent's antioxidant capacity (TPTZ) of PGE, Pectin, and Pectin/PGE nanoemulsion were determined, and the results are shown in Table 4.

Table 4. Total phenolics and antioxidant activity profile of PGE, Pectin, and Pectin/PEG emulsion.

Samples	DPPH Antioxidant Activity (mg Vitamin C/g Sample)	TPTZ (μg Trolox eq/g Sample)	Total Phenolic Compounds (mg Gallic Acid eq/g Sample)
PGE	785.23 ± 1.94 [c]	3387.6 ± 35.55 [c]	88.65 ± 4.91 [f]
Pectin	3.49 ± 0.03 [b]	93.19 ± 0.01 [b]	10.07 ± 0.12 [c]
Pectin/PGE emulsion	4.34 ± 0.01 [b]	100.9 ± 0.28 [b]	9.79 ± 2.13 [e]

Means in each column with different letters are significantly different ($p < 0.05$) ± SD. Analysis performed using ANOVA and the Duncan test.

As shown in the table, PGE had a high content of total phenolics, which is in accordance with the previous finding of Aboelsoued et al. [61]. On the other hand, pectin extracted from sugar beet had a significant amount of phenolics, reaching 10.07 mg gallic acid/g of pectin ($p < 0.05$). The Pectin/PGE nanoemulsion also showed antioxidant activity but the value of total phenolics was low since PGE represents 20% of the dry weight of Pectin/PGE, and also the emulsion had a concentration of 4.5% in water. The antioxidant activity was measured using both DPPH as mg ascorbic acid per gram sample and TPTZ as μg Trolox eq/g sample methods, in accordance to screening the activities of samples. PGE had the highest DPPH radical scavenging activity (785.23 ±1.94 mg ascorbic acid per gram sample, $p < 0.05$). On the other hand, pectin and Pectin/PGE nanoemulsion had DPPH antioxidant activities of 4.34 ± 0.01 and 3.49 ± 0.03 mg ascorbic acid per gram, respectively. The reducing power activity method TPTZ (FRAP) showed the same trend as the DPPH results. Xiong et al. [69] suggested that the antioxidant activity of pectin could be due to the higher content of electrophilic groups, which could accelerate the release of hydrogen from OH groups of the different sugars in pectin and act as an electron donor.

HPLC analysis results of Pectin/PGE in Table 5 were in accordance with antioxidants and total phenolics results mentioned above. Supplementary Information Figure S2 shows the chromatogram of Pectin/PGE nanoemulsion. Some of the phenolics in PGE could not be detected in the HPLC chromatogram of the Pectin/PGE nanoemulsion due to their originally low concentration in PGE. In addition, some phenolics such as syringic acid, ferulic acid, apigenin, and hesperetin phenolics appeared in the Pectin/PGE nanoemulsion chromatogram, which could have originated from pectin.

Table 5. Phenolic compounds of PGE and Pectin/PGE emulsion determined by high-performance liquid chromatography (HPLC).

Phenolic Compounds	PGE mg/g	Pectin/PGE Emulsion mg/mL
Gallic acid	11.18	0.06
Chlorogenic acid	21.95	0.012
Catechin	5.55	0
Methyl gallate	0.50	0.001
Coffeic acid	3.06	0.008
Syringic acid	0	0.003
Pyro catechol	0.081	0
Rutin	0.066	0.002
Ellagic acid	9.51	0
Ferulic acid	0.000	0.003
Naringenin	0.36	0
Querectin	0.000	0.002
Cinnamic acid	0.042	0
Apigenin	0	0.001
Hesperetin	0	0.002

3.3. CNFs/Pectin/PGE films

CNFs isolated from the de-pectinated SBP were used for making films with Pectin/PGE emulsion; CNFs had ~5 nm thickness and were several microns in length, as shown in the TEM image in Figure 3 [31]. The Pectin/PGE emulsion chosen for mixing with CNFs was that containing 20% of PGE since it gave the highest antibacterial activity toward the studied Gram-positive and Gram-negative bacteria. CNF films containing different ratios of the chosen Pectin/PGE emulsion were prepared. The ratios of the Pectin/PGE emulsion ranged from 2.5% to 20% based on the dry weight of the CNFs and the fixed ratio of PAE (2 wt.% based on the weight of the total film) was added. Films with good formation and homogeneity were obtained. It should be pointed out that glycerol was added to the films as a plasticizer in order to study the effect of Pectin/PGE emulsion, because without the addition of glycerol, the films suffered from significant shrinkage and cracking.

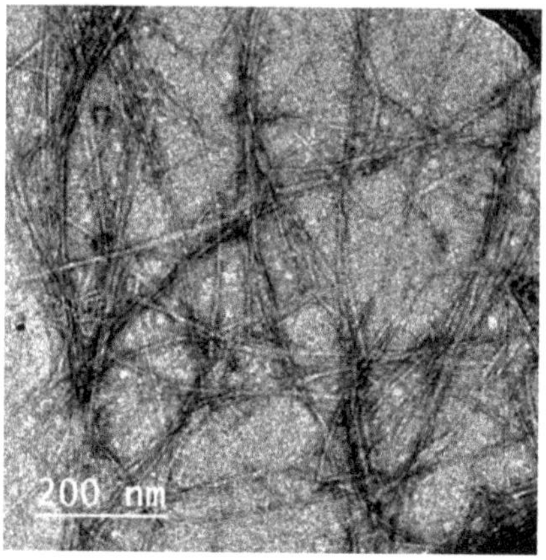

Figure 3. TEM image of cellulose nanofibers (CNFs) isolated from de-pectinated sugar beet pulp (SBP).

3.3.1. Tensile strength properties of CNFs/Pectin/PGE films

The effect of Pectin/PGE on the tensile strength properties of CNFs is shown in Figure 4. Tensile strength of cellulose fiber films is controlled by fiber-to-fiber shear strength, fiber tensile strength, and fiber pull-out work from the cellulose sheet [70]. Fiber-to-fiber bonding has a more positive effect than the length of fibers [71]. As Figure 4 shows, the presence of Pectin/PGE in the CNFs matrix did not affect its tensile strength up to the addition of 10% of the former. This indicates that the presence of Pectin/PGE particles acted as links between CNFs through hydrogen bonding and compensated the expected loss in tensile strength due to decreasing the CNF content. Pectin contains carboxylic groups which are highly polar and their presence in the films could induce stronger hydrogen bonding that improves the shear strength of fiber–fiber bonds [72]. However, at Pectin/PGE loadings of 15 and 20%, the tensile strength of the films decreased by about 19 to 32% compared to blank CNF film. This means that, at high Pectin/PGE loadings (15–20%), the fiber-to-fiber shear strength significantly decreased and exceeded the effect of hydrogen bonding between pectin and CNFs, and finally lead to the decreasing tensile strength of the films.

Regarding Young's modulus of the films, it was not affected by Pectin/PGE addition until addition of 10% was reached, i.e., the films became stiffer with an addition $\geq$10% of Pectin/PGE. The strain at the maximum load also showed a significant decrease at Pectin/PGE loadings $\geq$10%.

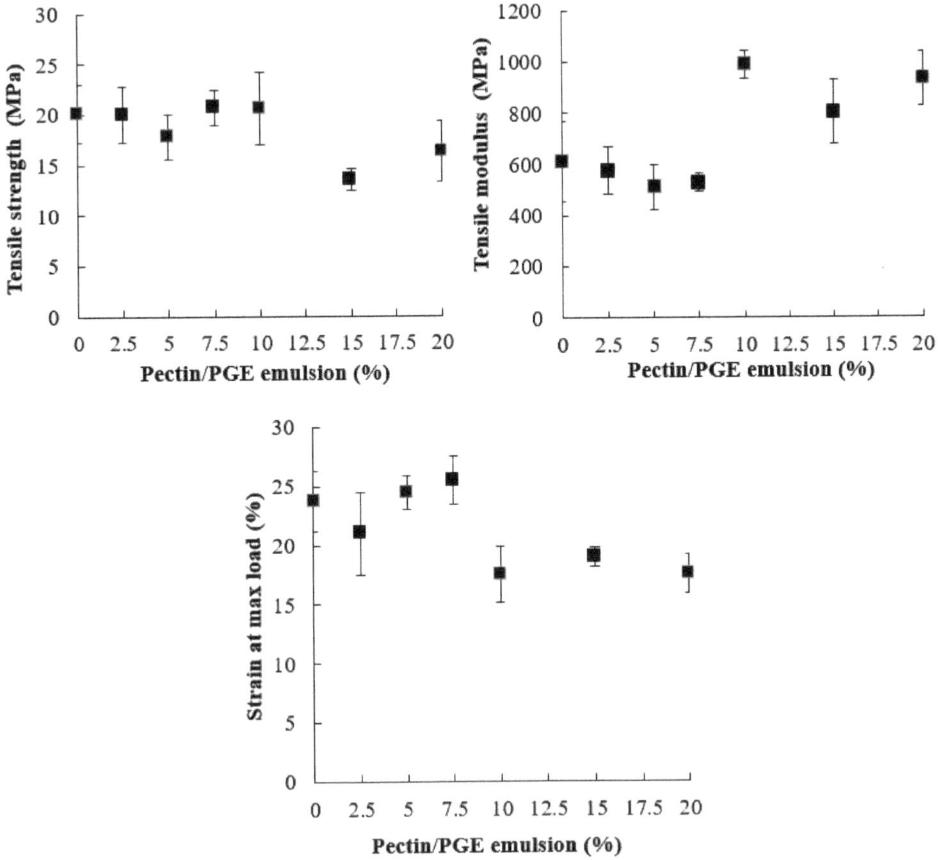

Figure 4. Tensile strength properties of CNFs/Pectin/PGE films with different loadings of PGE.

3.3.2. Greaseproof properties of CNFs/Pectin/PGE Films

An important requirement for films used in the packaging of oily or fatty products is the greaseproof property. Polymers rich in hydrophilic groups, such as cellulose and pectin, can form an extensive network of hydrogen bonding and have very low interaction with oil and grease [36]. In addition, CNF films have a very dense structure which, along with extensive hydrogen bonding, makes the penetration of oils and fats rather difficult [36]. In the current work, the grease resistance of CNF films containing different ratios of Pectin/PGE nanoemulsion was studied to investigate the effect of the presence of the nanoemulsion particles on that property. As shown in Table 6, all films showed good greaseproof property, especially with the addition of $\geq 5\%$ of Pectin/PGE nanoemulsion. According to the standard method used, time more than 30 min for the permeation of oil across the films qualifies them as high greaseproof material. The remarkable increase in greaseproof property as a result of increasing the Pectin/PGE content could be attributed to the presence of more pectin, which contains carboxylic groups originating from the galacturonic building units in its structure (~72 wt.% of pectin is galacturonic acid). The carboxylic groups have higher water affinity than hydroxyl groups due to higher polarity of the formers. When the nanoemulsion of Pectin/PGE formed, the polar groups of pectin are directed outward, and thus the outer surface of the nanoemulsion particles is highly polar.

Table 6. Greaseproof results of CNFs/Pectin/PGE films.

Sample	Film Thickness (mm)	Time for Oil Penetration through Paper Cross Section (min)
CNFs	0.166 ± 0.004	14 ± 2.0
CNFs/2.5% Pectin/PGE	0.186 ± 0.009	16 ± 1.8
CNFs/5% Pectin/PGE	0.164 ± 0.008	>45
CNFs/7.5% Pectin/PGE	0.189 ± 0.009	>45
CNFs/10% Pectin/PGE	0.186 ± 0.008	>45
CNFs/15% Pectin/PGE	0.174 ± 0.007	>45
CNFs/20% Pectin/PGE	0.188 ± 0.007	>45

3.3.3. Antibacterial Activity of CNFs/Pectin/PGE Films

The antibacterial activity of CNFs/Pectin/PGE containing different ratios of Pectin/PGE emulsions was studied in order to obtain films with appropriate antibacterial properties. The disc diffusion technique, which is a widely used quantitative method for studying antimicrobial properties of films, was used against *S. aureus and E. coli* bacteria, which are the most popular bacteria contaminants in foodstuff. The images of the test are shown in Figures 5 and 6. The method measures the diameter of the non-infected zone (so called the inhibition zone) in mm around the sample due to the release of the antibacterial agent from the sample to the agar media used in the test. As the images in the figures show, blank film made from CNFs and crosslinking agent did not show any antibacterial activity, and spreading of the *E. coli* and *S. aureus* bacteria took place on the film. In the case of testing CNFs/Pectin/PGE films against *S. aureus*, films with Pectin/PGE content from 2.5 to 5% showed the spread of the bacterium in the whole Petri dish and on the surface of the films, while that containing ≥7.5% Pectin/PGE exhibited a clear surface and the bacterium could not grow on the film. In the case of testing the films against *E. coli*, a bacterium-clear surface of the films could be seen for the sample containing ≥10% of Pectin/PGE.

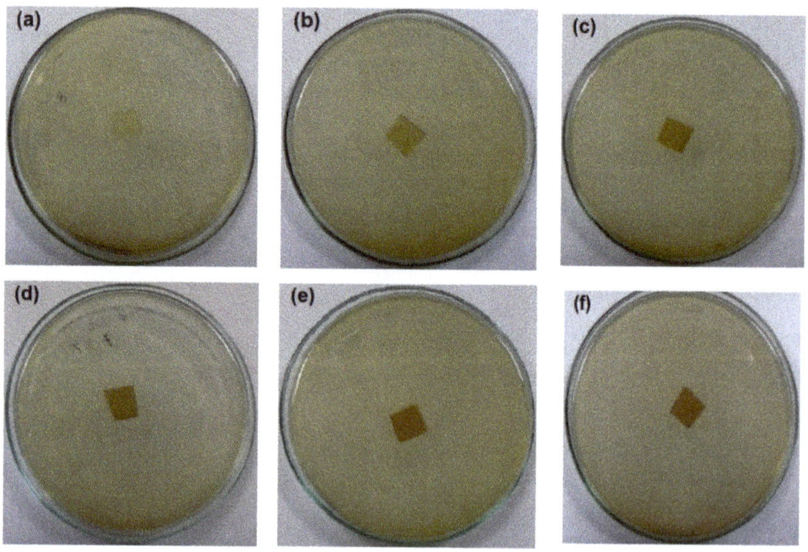

Figure 5. *Cont.*

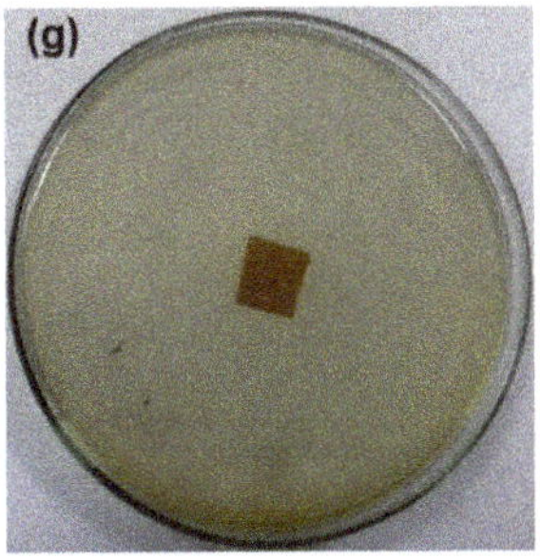

Figure 5. Images of the antibacterial test of CNFs/Pectin/PGE films against *S. aureus*: (**a**) blank CNFs, (**b**) CNFs + 2.5% Pectin/PGE, (**c**) CNFs + 5% Pectin/PGE, (**d**) CNFs + 7.5% Pectin/PGE, (**e**) CNFs + 10% Pectin/PGE, (**f**) CNFs + 10% Pectin PGE, and (**g**) CNFs + 20% Pectin/PGE.

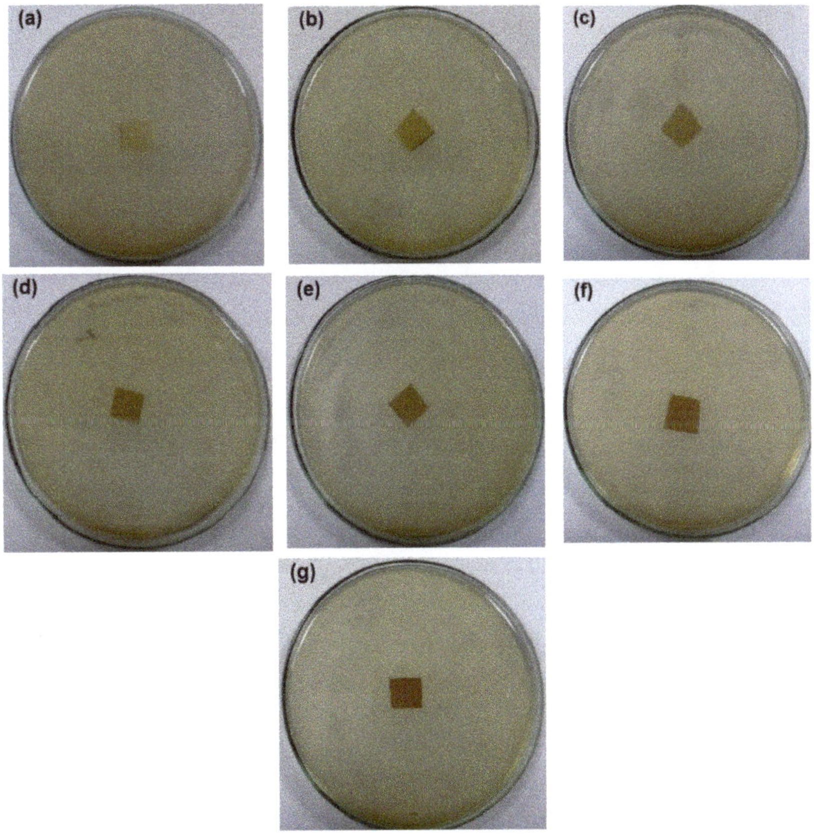

Figure 6. Images of the antibacterial test of CNFs/Pectin/PGE films against *E. coli*: (**a**) blank CNFs, (**b**) CNFs + 2.5% Pectin/PGE, (**c**) CNFs + 5% Pectin/PGE, (**d**) CNFs + 7.5% Pectin/PGE, (**e**) CNFs + 10% Pectin/PGE, (**f**) CNFs + 15% Pectin PGE, and (**g**) CNFs + 20% Pectin/PGE.

It should be pointed out here that the films did not show a clear inhibition zone around them in the test. In our study, since the antibacterial agent (PGE) was emulsified with pectin and both are embedded in the CNF matrix, PGE components could not release in highly sufficient amounts from the film to the surroundings during the test, and therefore no clear inhibition zone formed. However, there was no growth of the studied bacteria on the surface of the films, i.e., an antibacterial surface could be obtained.

3.3.4. Total Phenolics and Antioxidant Activity of CNFs/Pectin/PGE Film

Table 7 shows total phenolics and antioxidant activity of CNFs/20% pectin/PGE film; data of those of pectin and Pectin/PGE nanoemulsion are also added to the table. It is interesting to see that CNF film had antioxidant activity and a small amount of total phenolics. This could be due to the attachment of residual pectin to the isolated CNFs [33]. CNFs/20% Pectin/PGE film showed antioxidant activity but, as expected, it was lower than that of Pectin/PEG nanoemulsion since the film contains 20% of Pectin/PEG.

Table 7. Total phenolics and antioxidant activity profile of CNFs film, Pectin/PEG emulsion, and CNFs/Pectin/PGE film.

Samples	DPPH Antioxidant Activity (mg Vitamin C/g Sample)	TPTZ (µg Trolox eq/g Sample)	Total Phenolic Compounds (mg Gallic Acid eq/g Sample)
CNFs/Pectin/PGE film	1.60 ± 0.02 a	10.93 ± 0.2 a	4.75 ± 0.53 b
CNFs film	1.97 ± 0.04 a	8.81 ± 0.08 a	2.55 ± 0.12 a
Pectin/PGE emulsion	4.34 ± 0.01 b	100.9 ± 0.28 b	9.79 ± 2.13 e

Means in each column with different letters are significantly different ($p < 0.05$) ± SD. Analysis performed using ANOVA and the Duncan test.

3.4. *Paper Sheets Coated with CNFs/Pectin/PGE Emulsion*

Coating of paper is an important industrial application since it can impart new properties on paper products using only small amounts of coating materials, which are usually more expensive than blank paper sheets. Using CNFs for paper coating is very interesting from both a scientific and industrial point of view due to the unique properties of the thin layer of CNFs formed at the paper surface. In fact, coating using CNFs is more preferred than the addition of CNFs to pulp before the making of paper sheets due to the negative effect of CNFs on paper sheet formation and drying because of the very high water affinity of CNFs, possible loss through the sieve of the machine, and also the lack of homogenous distribution in the final paper sheets [36].

In the current work, commercial paper sheets used for wrapping were coated with CNFs/20% Pectin/PGE mixture since this formulation gave good antibacterial properties, as seen in the previous section. The paper sheets used had a thickness of ~0.04 mm and a basis weight of ~30 g/m^2. The coating layer of the applied CNFs/20% Pectin/PGE had a basis weight of ~5.8 g/m^2; it was selected after a preliminary trial to obtain good and homogenous coverage of paper sheets. SEM images of paper sheets before and after coating are shown in Figure 7; the images showed good coverage of the paper sheets with the dense plastic-like layer of the CNFs/Pectin/PGE; thickness of the coated layer was about 4 µm, as shown from the cross section image (Figure 7c).

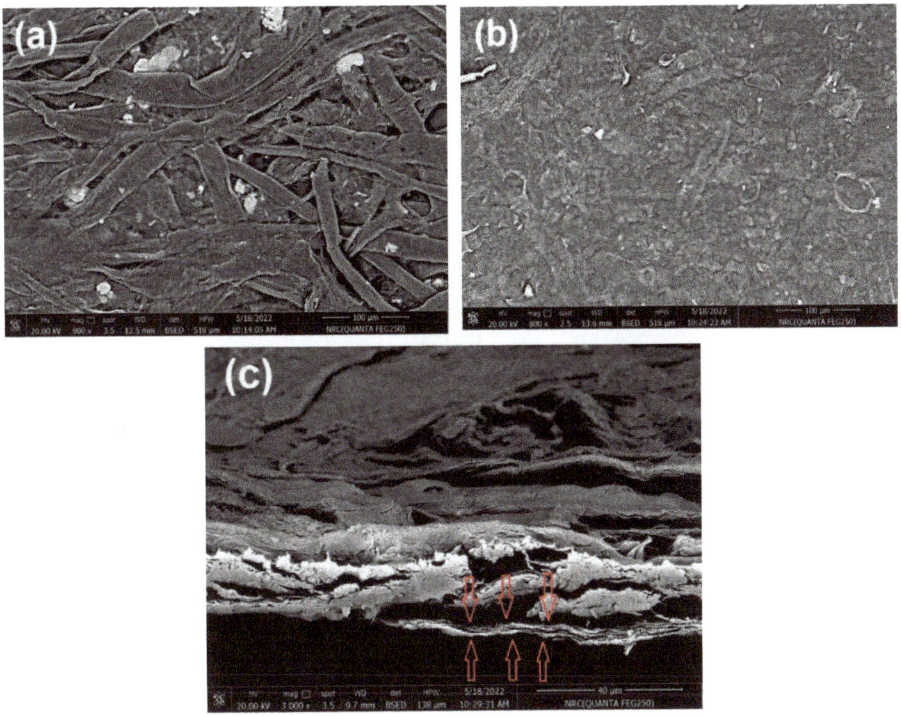

Figure 7. Scanning electron microscopy (SEM) of (**a**) the surface of a paper sheet before coating, (**b**) the surface of a paper sheet after coating with CNFs/20% Pectin/PGE, and (**c**) the cross-section of a paper sheet after coating with CNFs/20% Pectin/PGE.

3.4.1. Mechanical Properties of Coated Paper Sheets

The effect of coating paper sheets with CNFs/20% Pectin/PGE on tensile strength properties was studied, and the results are shown in Table 8. As the results show, coating of paper sheets with the small thickness layer of CNFs/20% Pectin/PGE resulted in a slight decrease (~7.5%) in tensile strength in the machine direction (MD) of paper sheets, but no significant change occurred in the cross direction (CD). The Young's modulus of coated paper sheets decreased in both CD and MD; the decrease was about 28% and 33% in the MD and CD of paper sheets, respectively, which means that the paper sheets became more stretchable. This was clear from the increase in strain at the maximum load of paper sheets in both the MD and CD directions. The decrease in tensile strength properties in spite of applying the CNFs/20% Pectin/PGE layer could be attributed to the effect of wetting and drying paper sheets during coating [73].

Table 8. Tensile strength properties of blank and paper sheets coated with CNFs/20% Pectin/PGE mixture.

Sample	Tensile Strength (MPa)		Young's Modulus (GPa)		Strain at Max. Load (%)	
	MD *	CD *	MD *	CD *	MD *	CD *
Blank paper sheets	28.14 ± 1.58	15.99 ± 1.98	5.91 ± 0.61	3.99 ± 0.24	1.53 ± 0.25	1.30 ± 0.17
CNFs/Pectin/PGE coated paper sheets	26.02 ± 1.22	15.35 ± 1.48	4.21 ± 0.19	2.69 ± 0.26	1.65 ± 0.23	1.70 ± 0.19

* CD and MD are the cross and machine direction of paper sheets, respectively.

In addition to testing tensile strength properties of the coated paper sheet in the dry conditions, the wet tensile strength of paper sheets with and without coating was studied using the ASTM method where folded strips were subjected to tensile force when immersed in water. The results showed that the wet tensile strength of a blank paper sheet was 3.26 ± 0.14 and 1.70 ± 0.11 N in the MD and CD directions of paper sheets, respectively while that of coated paper sheets was 3.53 ± 0.53 and 2.13 ± 0.24 N in the MD and CD directions of paper sheets, respectively. These results mean an increase in the wet tensile strength of paper sheet of about 8.3 and 25% in the MD and CD directions, respectively, by the very thin layer of CNFs/20% Pectin/PGE.

The improved wet tensile strength of coated paper sheets could be attributed to the formation of a tight network of crosslinked CNFs at the surface of paper sheets which can reduce the penetration of water. Previous results showed the significantly lower water absorption of sheets made from CNFs than that from the pulp fibers [25,74]. In addition to the aforementioned reason, the addition of a crosslinking agent to the CNFs/Pectin/PGE mixture could also result in the crosslinking of the fibers of the paper sheet, which thus became more resistant to water.

3.4.2. Physical Properties of Coated Paper Sheets

In spite of the small thickness of applied coating with CNFs, all previous work showed its effectiveness in changing properties of paper sheets and also imparting them with new properties [31]. In the current work, the effect of the CNFs/20% Pectin/PGE coating on air porosity, water vapor permeability, and grease resistance of paper sheets was studied; the results are shown in Table 9. Regarding air porosity, due to the very tight structure of the CNFs/20% Pectin/PGE film formed on the surface of paper sheets, the porosity was significantly decreased, and the time required to pass 100 mL of air increased by about 7 times. Such a decrease in air porosity is also responsible for the high barrier properties of CNFs films for gases including oxygen [75–77]. It is interesting to see that air porosity of the paper sheets coated with CNFs/20% Pectin/PGE is higher than that obtained in a previous work using only a neat CNF coating with about the same thickness [31]. In that study, the time required for passing 100 mL of air across a paper sheet with a 4-µm CNF coating (~4.5 g/m^2) was about 160 s. This, in fact, means that the presence of pectin/PGE emulsion between the CNFs network exerted more of a barrier for the passing of air.

Table 9. Effect of CNFs/20% Pectin/PGE coating on the physical properties of paper sheets.

Sample	Porosity (s/100 mL)	Water Vapor Permeability ($gm^{-1}s^{-1}Pa^{-1}$) × 10^{-11}	Grease Resistance (Time for Oil Sorption through Paper Cross Section in Minutes)
Blank paper sheets	36 ± 2	2.15 ± 0.091	Immediate sorption across paper thickness
CNFs/Pectin/PGE coated paper sheets	280 ± 8	2.10 ± 0.19	2.8 ± 0.31

Regarding the moisture barrier property (water vapor permeability) of paper coated with CNFs/20% Pectin/PGE, the situation is different because of the high hydrophilic nature of CNFs which work in an opposite direction to the effect of a very tight structure of CNF coating. In humid conditions, CNF swelling increases the tendency for the permeability of moisture [78]. The presence of Pectin/PGE emulsion, where the hydrophilic functional groups of the emulsified nanoparticles are to the outside and encapsulating the hydrophobic PGE, could also increase the tendency for water vapor permeability. As a result of the aforementioned two opposite effects, the net result found in the current work is that there was no significant effect of the CNFs/20% Pectin/PGE coating on water vapor permeability of paper sheets. A similar effect was found before in the case of coating paper sheets with neat CNFs [31].

Regarding the grease resistance of CNFs/20% Pectin/PGE coated paper, it was found that the coating significantly improved the grease resistance of paper sheets but not to the level that meets the required standards. In fact, the greaseproof property of CNFs originates from the extensive hydrogen bonding network, which result in high cohesive energy density and low interaction between CNFs and grease [36,79].

3.4.3. Migration Testing

Coated paper used as packaging materials should have stability of the coated layer, especially if they are in contact with aqueous, fat, or acidic products. In the case of adding functional additives to the coating layer, their release upon contact with different stimulants could be beneficial. In the current work, testing the stability of the CNFs/20% Pectin/PGE layer coated on paper sheets was carried out in 10% alcohol/water, 50% alcohol/water, and 3% acetic acid/water solutions, which represent aqueous, fat, and acidic stimulants; the test was carried out at 40 °C for 10 days, which represent the conditions for long-term storage at or below room temperature, including 15 min of heating up to 100 °C or 70 °C for up to 2 h [64]. Paper sheets coated with CNFs only were also tested as controls. As the results in Table 10 show, blank paper sheets showed constant weight loss upon their contact with the different stimulants (~1.8–2 mg/dm^2). According to the standard method used, this loss is lower than the maximum limit allowed for the migration of coated material (<10 mg/dm^2). On the other hand, paper sheets coated with CNFs/20% Pectin/PGE showed more weight loss as a result of immersion in the different stimulants. The highest weight loss was in aqueous stimulant (3.46 mg/dm^2), while the lowest loss was in the fat stimulant (2.3 mg/dm^2); acidic stimulant resulted in a loss of 2.95 mg/dm^2. Loss of Pectin/PGE with their antimicrobial and antioxidant activities as a result of contact with the different stimulants could be beneficial in packaging applications. The small values of the weight loss indicate the slow release and these very little amounts of Pectin/PGE could migrate into the food in contact with the coated paper. These very small amounts will not affect the color, taste, and properties of the food. In addition, the antibacterial properties, along with the antioxidant ones, could lead to increasing the shelf life of the food product. Testing of the prepared coated paper in packaging of different food products is planned to be studied in future work.

Table 10. Overall migration test results of CNFs/20% Pectin/PGE coated paper sheets.

Sample	Weight Loss (mg/dm^2) upon Contact with Stimulant		
	10% Alcohol	50% Alcohol	3% Acetic acid
Blank paper sheets	1.960 ± 0.02	1.87 ± 0.05	1.83 ± 0.07
CNFs/Pectin/PGE coated paper sheets	3.46 ± 0.03	2.31 ± 0.05	2..95 ± 0.06

4. Conclusions

With the aim to prepare films and coating materials with antioxidant and antibacterial activities from renewable bio-based resources, CNFs/Pectin/PGE nanocomposites were successfully prepared. Pectin and CNFs were isolated from sugar beet pulp, and pomegranate extract (PGE) was extracted from pomegranate peels. The isolated sugar beet pectin could be used as an emulsifier to PGE to prepare nanoemulsions with antimicrobial and antioxidant activities, as well as good compatibility with CNFs. The size of the Pectin/PGE nanoemulsion particles was not dependent on the concentration of PGE, while the antimicrobial activity at the different PGE loadings (from 2.5 to 20%) ranged from 83 to 92% and from 80 to 99.7% against *E. coli* and *S. aureus*, respectively. Homogenous films with antimicrobial and antioxidant activity, as well as flexibility, good mechanical properties, and high greaseproof properties could be prepared. Presence of Pectin/PGE nanoemulsion particles in CNF films improved their greaseproof property thanks to the hydrophilic outer surface of the nanoemulsion. The surface of the CNFs/Pectin/PGE films acquired antibacterial properties against Gram-positive and Gram-negative bacteria.

However, because PGE was emulsified in pectin and the Pectin/PGE was embedded within the crosslinked CNFs network, the release of the pectin/PGE was in very small amounts. This could be beneficial for the packaging of food products since the very slow release means an insignificant effect on the packaged food regarding its quality properties such as color or taste, and at the same time, could extend the shelf life by the antimicrobial effect.

In addition to the films prepared, the CNFs/Pectin/PGE mixture could be successfully used as a coating mixture for the commercial wrapping of paper sheets. The micrometer-scale coating (~4 μm) applied to paper sheets could impart them with a high barrier to air and improved greaseproof property, without affecting the water vapor permeability of paper sheets. The applied CNFs/Pectin/PGE layer can increase the wet tensile strength of paper sheets, i.e., higher stability in wet conditions, but did not improve these properties in dry conditions. The coated CNFs/Pectin/PGE film showed good adhesion to the paper sheet surface in different solutions representing aqueous, alcoholic, and acidic media at 40 °C for 10 days. Based on the obtained results, it could be concluded that antimicrobial and antioxidant films and coating for wrapping paper could be prepared and used safely in contact with different food products, extending the shelf life as well as keeping good quality attributes.

Supplementary Materials: The following supporting information can be downloaded at: https://www.mdpi.com/article/10.3390/polym14214605/s1, Figure S1: High-performance liquid chromatography (HPLC) chromatogram of pomegranate extract (PGE); Figure S2. High-performance liquid chromatography (HPLC) chromatogram of Pectin/PGE emulsion.

Author Contributions: Conceptualization, M.H. and E.H.; methodology, W.A.-E., S.F., M.M., E.H. and M.H.; investigation, W.A.-E., S.F., M.M., E.H. and M.H.; data curation, W.A.-E., S.F., M.M., E.H. and M.H.; writing—original draft preparation, M.H.; writing—review and editing, W.A.-E., S.F., M.M., E.H. and M.H.; supervision, M.H. All authors have read and agreed to the published version of the manuscript.

Funding: This research was partially funded by STDF, Egypt, project number 25848 entitled: Utilization of Sugar Beet Pulp Residues in Production of Pectin and Microcrystalline Cellulose.

Institutional Review Board Statement: Not applicable.

Data Availability Statement: All data are presented in the manuscript.

Acknowledgments: The authors would like to thank the National Research Centre for kindly providing all the facilities for carrying out the work mentioned in the manuscript.

Conflicts of Interest: The authors declare no conflict of interest.

References

1. Xiang, Q.; Li, M.; Wen, J.; Ren, F.; Yang, Z.; Jiang, X.; Chen, Y. The bioactivity and applications of pomegranate peel extract: A review. *J. Food Biochem.* **2022**, *46*, e14105. [CrossRef]
2. Salles, T.S.; Meneses, M.D.F.; Caldas, L.A.; Sá-Guimarães, T.E.; de Oliveira, D.M.; Ventura, J.A.; Azevedo, R.C.; Kuster, R.M.; Soares, M.R.; Ferreira, D.F. Virucidal and antiviral activities of pomegranate (*Punica granatum*) extract against the mosquito-borne Mayaro virus. *Parasites Vectors* **2021**, *14*, 443. [CrossRef]
3. Houston, D.M.J.; Bugert, J.J.; Denyer, S.P.; Heard, C.M. Potentiated virucidal activity of pomegranate rind extract (PRE) and punicalagin against *Herpes simplex* virus (HSV) when co-administered with zinc (II) ions, and antiviral activity of PRE against HSV and aciclovir-resistant HSV. *PLoS ONE* **2017**, *12*, e0179291.
4. Annalisa, T.; Antonio, C.; Luciano, P.; Emilia, P.; Daniela, I.; Giuliana, G.; Ivan, C.; Paola, V.; Fabio, A. Pomegranate Peel Extract as an Inhibitor of SARS-CoV-2 Spike Binding to Human ACE2 Receptor (in vitro): A Promising Source of Novel Antiviral Drugs. *Front. Chem.* **2021**, *9*, 638187.
5. Peršurić, Ž.; Martinović, L.S.; Malenica, M.; Gobin, I.; Pedisić, S.; Dragović-Uzelac, V.; Pavelić, S.K. Assessment of the Biological Activity and Phenolic Composition of Ethanol Extracts of Pomegranate (*Punica granatum* L.) Peels. *Molecules* **2020**, *25*, 5916. [CrossRef]
6. Rashid, R.; Masoodi, F.A.; Wani, S.M.; Manzoor, S.; Gull, A. Ultrasound assisted extraction of bioactive compounds from pomegranate peel, their nanoencapsulation and application for improvement in shelf life extension of edible oils. *Food Chem.* **2022**, *385*, 132608. [CrossRef]

7. Hady, E.; Youssef, M.; Aljahani, A.H.; Aljumayi, H.; Ismail, K.A.; El-Damaty, E.-S.; Sami, R.; El-Sharnouby, G. Enhancement of the Stability of Encapsulated Pomegranate (*Punica granatum* L.) Peel Extract by Double Emulsion with Carboxymethyl Cellulose. *Crystals* **2022**, *12*, 622. [CrossRef]
8. Sanhueza, L.; García, P.; Giménez, B.; Benito, J.M.; Matos, M.; Gutiérrez, G. Encapsulation of Pomegranate Peel Extract (*Punica granatum* L.) by Double Emulsions: Effect of the Encapsulation Method and Oil Phase. *Foods* **2022**, *11*, 310. [CrossRef]
9. Gull, A.; Bhat, N.; Wani, S.M.; Masoodi, F.A.; Amin, T.; Ganai, S.A. Shelf life extension of apricot fruit by application of nanochitosan emulsion coatings containing pomegranate peel extract. *Food Chem.* **2021**, *349*, 129149. [CrossRef]
10. Kori, A.H.; Mahesar, S.A.; Sherazi, S.T.H.; Khatri, U.A.; Laghari, Z.H.; Panhwar, T. Effect of process parameters on emulsion stability and droplet size of pomegranate oil-in-water. *Grasas Aceites* **2021**, *72*, e410. [CrossRef]
11. Hassan, M.L.; Berglund, L.; Abou Elseoud, W.S.; Hassan, E.A.; Oksman, K. Effect of pectin extraction method on properties of cellulose nanofibers isolated from sugar beet pulp. *Cellulose* **2021**, *28*, 10905–10920. [CrossRef]
12. Freitas, C.M.P.; Coimbra, J.S.R.; Souza, V.G.L.; Sousa, R.C.S. Structure and Applications of Pectin in Food, Biomedical, and Pharmaceutical Industry: A Review. *Coatings* **2021**, *11*, 922. [CrossRef]
13. Kumar, A.; Chaudhary, R.K.; Singh, R.; Singh, S.P.; Wang, S.Y.; Hoe, Z.Y.; Pan, C.T.; Shiue, Y.L.; Wei, D.Q.; Kaushik, A.C. Nanotheranostic applications for detection and targeting neurodegenerative diseases. *Front. Neurosci.* **2020**, *14*, 305. [CrossRef]
14. Stevanato, P.; Chiodi, C.; Broccanello, C.; Concheri, G.; Biancardi, E.; Pavli, O.; Skaracis, G. Sustainability of the sugar beet crop. *Sugar Tech* **2019**, *21*, 703–716. [CrossRef]
15. Niu, H.; Chen, X.; Luo, T.; Chen, H.; Fu, X. The interfacial behavior and long-term stability of emulsions stabilized by gum arabic and sugar beet pectin. *Carbohydr. Polym.* **2022**, *291*, 119623. [CrossRef]
16. Yang, Y.; Chen, D.; Yu, Y.; Huang, X. Effect of ultrasonic treatment on rheological and emulsifying properties of sugar beet pectin. *Food Sci. Nutr.* **2020**, *8*, 4266–4275. [CrossRef]
17. Liu, Z.; Pi, F.; Guo, X.; Guo, X.; Yu, S. Characterization of the structural and emulsifying properties of sugar beet pectins obtained by sequential extraction. *Food Hydrocoll.* **2019**, *88*, 31–42. [CrossRef]
18. Artiga-Artigas, M.; Reichert, C.; Salvia-Trujillo, L.; Zeeb, B.; Martín-Belloso, O.; Weiss, J. Protein/Polysaccharide Complexes to Stabilize Decane-in-Water Nanoemulsions. *Food Biophys.* **2020**, *15*, 335–345. [CrossRef]
19. Saberi, A.H.; Zeeb, B.; Weiss, J.; McClements, D.J. Tuneable stability of nanoemulsions fabricated using spontaneous emulsification by biopolymer electrostatic deposition. *J. Colloid Interface Sci.* **2015**, *455*, 172–178. [CrossRef]
20. Dinand, E.; Chanzy, H.; Vignon, R. Suspensions of cellulose microfibrils from sugar beet pulp. *Food Hydrocoll.* **1999**, *13*, 275–283. [CrossRef]
21. Jele, T.B.; Lekha, P.; Sithole, B. Role of cellulose nanofibrils in improving the strength properties of paper: A review. *Cellulose* **2022**, *29*, 55–81. [CrossRef]
22. Ewnetu Sahlie, M.; Zeleke, T.S.; Aklog Yihun, F. Water Hyacinth: A sustainable cellulose source for cellulose nanofiber production and application as recycled paper reinforcement. *J. Polym. Res.* **2022**, *29*, 230. [CrossRef]
23. Balea, A.; Fuente, E.; Concepcion Monte, M.; Merayo, N.; Campano, C.; Negro, C.; Blanco, A. Industrial application of nanocelluloses in papermaking: A review of challenges, technical solutions, and market perspectives. *Molecules* **2020**, *25*, 526. [CrossRef]
24. Hassan, M.L.; Bras, J.; Mauret, E.; Fadel, S.M.; Hassan, E.A.; El-Wakil, N.A. Palm rachis microfibrillated cellulose and oxidized-microfibrillated cellulose for improving paper sheets properties of unbeaten softwood and bagasse pulps. *Ind. Crops Prod.* **2015**, *64*, 9–15. [CrossRef]
25. Hassan, E.A.; Hassan, M.L.; Oksman, K. Improving bagasse pulp paper sheet properties with microfibrillated cellulose isolated from xylanase-treated bagasse. *Wood Fiber Sci.* **2011**, *43*, 76–82.
26. Hutton-Prager, B.; Ureña-Benavides, E.; Parajuli, S.; Adenekan, K. Investigation of cellulose nanocrystals (CNC) and cellulose nanofibers (CNF) as thermal barrier and strengthening agents in pigment-based paper coatings. *J. Coat. Technol. Res.* **2022**, *19*, 337–346. [CrossRef]
27. Tajik, M.; Jalali Torshizi, H.; Resalati, H.; Hamzeh, Y. Effects of cellulose nanofibrils and starch compared with polyacrylamide on fundamental properties of pulp and paper. *Int. J. Biol. Macromol.* **2021**, *192*, 618–626. [CrossRef]
28. Wang, J.; Wu, Y.; Chen, W.; Wang, H.; Dong, T.; Bai, F.; Li, X. Cellulose nanofibrils with a three-dimensional interpenetrating network structure for recycled paper enhancement. *Cellulose* **2022**, *29*, 3773–3785. [CrossRef]
29. Merayo, N.; Balea, A.; de la Fuente, E.; Blanco, Á.; Negro, C. Synergies between cellulose nanofibers and retention additives to improve recycled paper properties and the drainage process. *Cellulose* **2017**, *24*, 2987–3000. [CrossRef]
30. Balea, A.; Blanco, Á.; Monte, M.C.; Merayo, N.; Negro, C. Effect of Bleached Eucalyptus and Pine Cellulose Nanofibers on the Physico-Mechanical Properties of Cartonboard. *BioResources* **2016**, *11*, 8123–8138. [CrossRef]
31. Fadel, S.M.; Abou-Elseoud, W.S.; Hassan, E.A.; Ibrahim, S.; Hassan, M.L. Use of sugar beet cellulose nanofibers for paper coating. *Ind. Crops. Prod.* **2022**, *180*, 114787. [CrossRef]
32. Al-Gharrawi, M.Z.; Wang, J.; Bousfield, D.W. Improving water vapor barrier of cellulose based food packaging using double layer coatings and cellulose nanofibers. *Food Packag. Shelf Life* **2022**, *33*, 100895. [CrossRef]
33. De Oliveira, M.L.C.; Mirmehdi, S.; Scatolino, M.V.; Júnior, M.G.; Sanadi, A.R.; Damasio, R.A.P.; Tonoli, G.H.D. Effect of overlapping cellulose nanofibrils and nanoclay layers on mechanical and barrier properties of spray-coated papers. *Cellulose* **2022**, *29*, 1097–1113. [CrossRef]

34. Khlewee, M.; Al-Gharrawi, M.; Bousfield, D. Modeling the penetration of polymer into paper during extrusion coating. *J. Coat. Technol. Res.* **2022**, *19*, 25–34. [CrossRef]
35. Tarrés, Q.; Aguado, R.; Pèlach, M.À.; Mutjé, P.; Delgado-Aguilar, M. Electrospray Deposition of Cellulose Nanofibers on Paper: Overcoming the Limitations of Conventional Coating. *Nanomaterials* **2022**, *12*, 79. [CrossRef]
36. Hubbe, M.A.; Ferrer, A.; Tyagi, P.; Yin, Y.; Salas, C.; Pal, L.; Rojas, O.J. Nanocellulose in thin films, coatings, and plies for packaging applications: A Review. *BioResources* **2017**, *12*, 2143–2233. [CrossRef]
37. Mousavi, S.M.M.; Afra, E.; Tajvidi, M.; Bousfield, D.W.; Dehghani-Firouzabadi, M. Cellulose nanofiber/carboxymethyl cellulose blends as an efficient coating to improve the structure and barrier properties of paperboard. *Cellulose* **2017**, *24*, 3001–3014. [CrossRef]
38. Brodin, F.W.; Gregersen, O.W.; Syverud, K. Cellulose nanofibrils: Challenges and possibilities as a paper additive or coating material—A review. *Nord. Pulp Pap. Res. J.* **2014**, *29*, 156–166. [CrossRef]
39. Balahura, L.-R.; Dinescu, S.; Balaș, M.; Cernencu, A.; Lungu, A.; Vlăsceanu, G.M.; Iovu, H.; Costache, M. Cellulose nanofiber-based hydrogels embedding 5-FU promote pyroptosis activation in breast cancer cells and support human adipose-derived stem cell proliferation, opening new perspectives for breast tissue engineering. *Pharmaceutics* **2021**, *13*, 1189. [CrossRef]
40. Cai, L.; Li, Y.; Lin, X.; Chen, H.; Gao, Q.; Li, J. High-performance adhesives formulated from soy protein isolate and bio-based material hybrid for plywood production. *J. Clean. Prod.* **2022**, *353*, 131587. [CrossRef]
41. Sungsinchai, S.; Niamnuy, C.; Wattanapan, P.; Charoenchaitrakool, M.; Devahastin, S. Spray drying of non-chemically prepared nanofibrillated cellulose: Improving water redispersibility of the dried product. *Int. J. Biol. Macromol.* **2022**, *207*, 434–442. [CrossRef]
42. Wu, W.; Wu, Y.; Lin, Y.; Shao, P. Facile fabrication of multifunctional citrus pectin aerogel fortified with cellulose nanofiber as controlled packaging of edible fungi. *Food Chem.* **2022**, *374*, 131763. [CrossRef]
43. Pitton, M.; Fiorati, A.; Buscemi, S.; Melone, L.; Farè, S.; Contessi Negrini, N. 3D Bioprinting of Pectin-Cellulose Nanofibers Multicomponent Bioinks. *Front. Bioeng. Biotechnol.* **2021**, *9*, 732689. [CrossRef]
44. Cernencu, A.I.; Lungu, A.; Stancu, I.-C.; Serafim, A.; Heggset, E.; Syverud, K.; Iovu, H. Bioinspired 3D printable pectin-nanocellulose ink formulations. *Carbohydr. Polym.* **2019**, *220*, 12–21. [CrossRef]
45. Shiekh, K.A.; Ngiwngam, K.; Tongdeesoontorn, W. Polysaccharide-based active coatings incorporated with bioactive compounds for reducing postharvest losses of fresh fruits. *Coatings* **2022**, *12*, 8. [CrossRef]
46. Ghorbani, E.; Dabbagh Moghaddam, A.; Sharifan, A.; Kiani, H. Emergency Food Product Packaging by Pectin-Based Antimicrobial Coatings Functionalized by Pomegranate Peel Extracts. *J. Food Qual.* **2021**, *2021*, 6631021. [CrossRef]
47. Chacko, C.M.; Estherlydia, D. Sensory, physicochemical and antimicrobial evaluation of jams made from indigenous fruit peels. *Carpathian J. Food Sci. Technol.* **2013**, *5*, 69–75.
48. Lishchynskyi, O.; Shymborska, Y.; Stetsyshyn, Y.; Raczkowska, J.; Skirtach, A.G.; Peretiatko, T.; Budkowski, A. Passive antifouling and active self-disinfecting antiviral surfaces. *Chem. Eng. J.* **2022**, *446*, 137048. [CrossRef]
49. Yang, W.; Liu, F.; Xu, C.; Sun, C.; Yuan, F.; Gao, Y. Inhibition of the Aggregation of Lactoferrin and (−)-epigallocatechin Gallate in the Presence of Polyphenols, Oligosaccharides, and Collagen Peptide. *J. Agric. Food Chem.* **2015**, *63*, 5035–5045. [CrossRef]
50. Nakayama, M.; Shimatani, K.; Ozawa, T.; Shigemune, N.; Tomiyama, D.; Yui, K.; Katsuki, M.; Ikeda, K.; Nonaka, A.; Miyamoto, T. Mechanism for the antibacterial action of epigallocatechin gallate (EGCg) on Bacillus subtilis. *Biosci. Biotechnol. Biochem.* **2015**, *79*, 845–854. [CrossRef]
51. Mori, A.; Nishino, C.; Enoki, N.; Tawata, S. Antibacterial activity and mode of action of plant flavonoids against *Proteus vulgaris* and *Staphylococcus aureus*. *Phytochemistry* **1987**, *26*, 2231–2234. [CrossRef]
52. Lou, Z.; Wang, H.; Rao, S.; Sun, J.; Ma, C.; Li, J. p-Coumaric acid kills bacteria through dual damage mechanisms. *Food Control* **2012**, *25*, 550–554. [CrossRef]
53. Zhao, W.H.; Hu, Z.Q.; Okubo, S.; Hara, Y.; Shimamura, T. Mechanism of synergy between epigallocatechin gallate and betalactams against methicillin-resistant *Staphylococcus aureus*. *Antimicrob. Agents Chemother.* **2001**, *45*, 1737–1742. [CrossRef]
54. Ollila, F.; Halling, K.; Vuorela, P.; Vuorela, H.; Slotte, J.P. Characterization of flavonoid–biomembrane interactions. *Arch. Biochem. Biophys.* **2002**, *399*, 103–108. [CrossRef]
55. Pirzadeh, M.; Caporaso, N.; Rauf, A.; Shariati, M.A.; Yessimbekov, Z.; Khan, M.U.; Imran, M.; Mubarak, M.S. Pomegranate as a source of bioactive constituents: A review on their characterization, properties and applications. *Crit. Rev. Food Sci. Nutr.* **2021**, *61*, 982–999. [CrossRef]
56. Browning, B.L. *Methods of Wood Chemistry. Volume II*; Wiley: New York, NY, USA, 1967; p. 489.
57. Meseguer, I.; Aguilar, M.; González, M.J.; Martínez, C. Extraction and colorimetric quantification of uronic acids of the pectic fraction in fruit and vegetables. *J. Food Compos. Anal.* **1998**, *11*, 285–291. [CrossRef]
58. Sàez-Plaza, P.; Michałowski, T.; Navas, M.J.; Asuero, A.G.; Wybraniec, S. An Overview of the Kjeldahl Method of Nitrogen Determination. Part I. Early History, Chemistry of the procedure, and titrimetric finish. *Crit. Rev. Anal. Chem.* **2013**, *43*, 178–223. [CrossRef]
59. Abou-Elseoud, W.S.; Hassan, E.A.; Hassan, M.L. Extraction of pectin from sugar beet pulp by Enzymatic and ultrasound-assisted treatments. *Carbohydr. Polym. Technol. Appl.* **2021**, *2*, 100042. [CrossRef]
60. Ramful, D.; Tarnus, E.; Aruoma, O.I.; Bourdon, E.; Bahorun, T. Polyphenol composition, vitamin C content and antioxidant capacity of Mauritian citrus fruit pulps. *Food Res. Int.* **2011**, *44*, 2088–2099. [CrossRef]

61. Aboelsoued, D.; Abo-Aziza, F.A.M.; Mahmoud, M.H.; Abdel Megeed, K.N.; Abu El Ezz, N.M.T.; Abu-Salem, F.M. Anticryptosporidial effect of pomegranate peels water extract in experimentally infected mice with special reference to some biochemical parameters and antioxidant activity. *J. Parasit. Dis.* **2019**, *43*, 215–228. [CrossRef]
62. Barros, H.R.M.; Ferreira, T.A.P.C.; Genovese, M.I. Antioxidant capacity and mineral content of pulp and peel from commercial cultivars of citrus from Brazil. *Food Chem.* **2012**, *134*, 1892–1898. [CrossRef] [PubMed]
63. Balouiri, M.; Sadiki, M.; Ibnsouda, S.K. Methods for in vitro evaluating antimicrobial activity: A review. *J. Pharm. Anal.* **2016**, *6*, 71–79. [CrossRef] [PubMed]
64. Bhunia, K.; Sablani, S.S.; Tang, J.; Rasco, B. Migration of chemical compounds from packaging polymers during microwave, conventional heat treatment, and storage. *Compr. Rev. Food Sci. Food Saf.* **2013**, *12*, 523–545. [CrossRef] [PubMed]
65. Leesombun, A.; Sariya, L.; Taowan, J.; Nakthong, C.; Thongjuy, O.; Boonmasawai, S. Natural antioxidant, antibacterial, and antiproliferative activities of ethanolic extracts from *Punica granatum* L. Tree barks mediated by extracellular sSignal-regulated kinase. *Plants* **2022**, *11*, 2258. [CrossRef] [PubMed]
66. Pagliarulo, C.; De Vito, V.; Picariello, G.; Colicchio, R.; Pastore, G.; Salvatore, P.; Volpe, M.G. Inhibitory effect of pomegranate (*Punica granatum* L.) polyphenol extracts on the bacterial growth and survival of clinical isolates of pathogenic Staphylococcus aureus and Escherichia coli. *Food Chem.* **2016**, *190*, 824–831. [CrossRef] [PubMed]
67. Bandele, O.J.; Clawson, S.J.; Osheroff, N. Dietary polyphenols as topoisomerase II poisons: B ring and C ring substituents determine the mechanism of enzyme-mediated DNA cleavage enhancement. *Chem. Res. Toxicol.* **2008**, *21*, 1253–1260. [CrossRef]
68. Yassin, M.T.; Mostafa, A.A.; Askar, A.A.A. In vitro evaluation of biological activities and phytochemical analysis of different solvent extracts of *Punica granatum* L. (Pomegranate) peels. *Plants* **2021**, *10*, 2742. [CrossRef]
69. Xiong, B.; Zhang, W.; Wu, Z.; Liu, R.; Yang, C.; Hui, A.; Huang, X.; Xian, Z. Preparation, characterization, antioxidant and anti-inflammatory activities of acid-soluble pectin from okra (*Abelmoschus esculentus* L.). *Int. J. Biol. Macromol.* **2021**, *181*, 824–834. [CrossRef]
70. Larsson, P.T.; Lindström, T.; Carlsson, L.A.; Fellers, C. Fiber length and bonding effects on tensile strength and toughness of kraft paper. *J. Mater. Sci.* **2018**, *53*, 3006–3015. [CrossRef]
71. Carlsson, L.A.; Lindström, T. A shear-lag approach to the tensile strength of paper. *Compos. Sci. Technol.* **2005**, *65*, 183–189. [CrossRef]
72. Laine, J.; Lindström, T.; Glad-Nordmark, G.; Risinger, G. Studies on topochemical modification of cellulosic fibres. Part 1. Chemical conditions for the attachment of carboxymethyl cellulose onto fibres. *Nord. Pulp Pap. Res. J.* **2000**, *15*, 520–526. [CrossRef]
73. Çiçekler, M.; Şahin, H.T. The effects of wetting-drying on bleached kraft paper properties. *J. Bartin Fac. For.* **2020**, *22*, 436–446.
74. Abdul Khalil, H.P.S.; Davoudpour, Y.; Islam, M.N.; Mustapha, A.; Sudesh, K.; Dungani, R.; Jawaid, M. Production and modification of nanofibrillated cellulose using various mechanical processes: A review. *Carbohydr. Polym.* **2014**, *99*, 649–665. [CrossRef] [PubMed]
75. Ilyas, R.A.; Azmi, A.; Nurazzi, N.M.; Atiqah, A.; Atikah, M.S.N.; Ibrahim, R.; Norrrahim, M.N.F.; Asyraf, M.R.M.; Sharma, S.; Punia, S.; et al. Oxygen permeability properties of nanocellulose reinforced biopolymer nanocomposites. *Mater. Today Proc.* **2021**, *52*, 2414–2419. [CrossRef]
76. Nair, S.S.; Zhu, J.; Deng, Y.; Ragauskas, A.J. High performance green barriers based on nanocellulose. *Sustain. Chem. Process* **2014**, *2*, 23. [CrossRef]
77. Wang, J.; Gardner, D.J.; Stark, N.M.; Bousfield, D.W.; Tajvidi, M.; Cai, Z. Moisture and oxygen barrier properties of cellulose nanomaterial-based films. *ACS Sustain. Chem. Eng.* **2018**, *6*, 49–70. [CrossRef]
78. Ferrer, A.; Pal, L.; Hubbe, M. Nanocellulose in packaging: Advances in barrier layer technologies. *Ind. Crop. Prod.* **2017**, *95*, 574–582. [CrossRef]
79. Österberg, M.; Vartiainen, J.; Lucenius, J.; Hippi, U.; Seppälä, J.; Serimaa, R.; Laine, J. A fast method to produce strong NFC films as a platform for barrier and functional materials. *ACS Appl. Mater. Interfaces* **2013**, *5*, 4640–4647. [CrossRef]

polymers

MDPI

Article

Pressure-Steam Heat Treatment-Enhanced Anti-Mildew Property of Arc-Shaped Bamboo Sheets

Xingyu Liang [1,2,†], Yan Yao [1,†], Xiao Xiao [1,2,3], Xiaorong Liu [1,2], Xinzhou Wang [1,4,*] and Yanjun Li [1,4,*]

1 Jiangsu Co-Innovation Center of Efficient Processing and Utilization of Forest Resources, Nanjing Forestry University, Nanjing 210037, China
2 Hangzhou ZhuangYi Furniture Co., Ltd., Hangzhou 311251, China
3 Bamboo Engineering and Technology Research Center, State Forestry and Grassland, Nanjing 210037, China
4 Dareglobal Technologies Group Co., Ltd., Danyang 212310, China
* Correspondence: xzwang@njfu.edu.cn (X.W.); lalyj@njfu.edu.cn (Y.L.)
† These authors contributed equally to this manuscript.

Abstract: Bamboo is one of the most promising biomass materials in the world. However, the poor anti-mildew property and poor dimensional stability limits its outdoor applications. Current scholars focus on the modification of bamboo through heat treatment. Arc-shaped bamboo sheets are new bamboo products for special decoration in daily life. In this paper, we reported pressure-steam heat treatment and explored the effect of pressure-steam on the micro-structure, crystallinity index, anti-mildew, chemical composition, physical properties, and mechanical properties of bamboo via X-ray diffractometer (XRD), scanning electron microscopy (SEM), Fourier-transform infrared (FTIR), wet chemistry method and nanoindentation (NI). Herein, saturated-steam heat treatment was applied for modified moso bamboo for enhancing the anti-mildew properties and mechanical properties of moso bamboo. Results showed that with the introduction of saturated steam, the content of hemicellulose and cellulose decreased, while the lignin-relative content increased significantly. The anti-mildew property of moso bamboo was enhanced due to the decomposition of polysaccharide. Last, the modulus of elasticity and hardness of treated moso bamboo cell walls were enhanced after saturated-steam heat treatment. For example, the MOE of the treated moso bamboo cell wall increased from 12.7 GPa to 15.7 GPa. This heat treatment strategy can enhance the anti-mildew property of moso bamboo and can gain more attention from entrepreneurs and scholars.

Keywords: bamboo; pressure-steam heat treatment; bamboo cell wall; anti-mildew property

Citation: Liang, X.; Yao, Y.; Xiao, X.; Liu, X.; Wang, X.; Li, Y. Pressure-Steam Heat Treatment-Enhanced Anti-Mildew Property of Arc-Shaped Bamboo Sheets. *Polymers* **2022**, *14*, 3644. https://doi.org/10.3390/polym14173644

Academic Editor: Shuangquan Yao

Received: 18 July 2022
Accepted: 31 August 2022
Published: 2 September 2022

1. Introduction

Bamboo is an indispensable and ideal material for alleviating the shortage of wood resources in modern society [1]. Bamboos are widely used in the fields of construction, decoration, building and daily necessities [2]. Currently, bamboo dominates the wood-processing industry due to its excellent mechanical property, short growth cycle, and easy harvest [3]. For outdoor application, bamboo-based products such as bamboo scriber, flattened bamboo boards, bamboo flooring boards, and so on, are easily affected by fungi, UV, and water due to the abundance of polysaccharide and poor dimensional stability [4]. Thus, the poor anti-mildew property and poor dimensional stability limits their outdoor applications. Shortage of woody materials, enhancing utilization of bamboo materials, and the move towards an enhancement of dimensional stability and mechanical properties of moso bamboo have prompted the development of bamboo modification [5]. Thus, heat treatment has attracted more attention from scholars and scientists due to its eco-friendliness and cost-effectiveness in comparison to chemical modification and other modification methods. In addition, the research on the biodegradability of bamboo or woody materials is a research gap in current research [6].

As we known, bamboo culms consist of hemicellulose, cellulose, lignin, ash, and so on [7–9]. The main chemical composition mentioned above contains amorphous phases. When exposed to high-temperature and high-pressure conditions, these chemical components exhibit viscoelastic and plastic behaviors. Until now, chemical agent impregnation, chemical modification, and thermal modification can effectively enhance the mechanical properties and anti-ant and anti-mildew properties of bamboo samples. Unfortunately, the utilization of chemical agents is not environment-friendly and can be harmful to the human body when these bamboo-based products are applied in our daily life. Over the years, dry steam, inert gas (nitrogen), and water were usually used as a heat treatment medium with a treatment temperature between 150 °C and 230 °C and a treatment time of 1–6 h. The heat treatment reduces the hygroscopic property of the bamboo, consequently reducing the hygroscopic property of the bamboo, thus reducing its shrinkage and swelling properties and improving the dimensional stability when the treatment is above 150 °C. At 180 °C or higher, heat treatment can significantly improve the anti-fungi property of bamboo. However, high temperature also leads to a decrease in its mechanical properties and also to the decrement of moisture content in bamboo [10]. Fortunately, pressure-steam is an effective and environment-friendly heat treatment medium in bamboo processing factories. Pressure-steam is the steam that is in equilibrium with heated liquid water at the same pressure, which has not been heated more than the boiling point for that pressure. In addition, pressure-steam can provide pressure, high temperature, and high moisture content in sealed equipment that can modified bamboo quickly. Bamboo is an anisotropic biomass material, which consists of many cells that are oriented in the axial and radial directions. Therefore, knowledge about the specific molecular mechanical phenomena at the cellular and subcellular levels is of great importance for understanding the effects of thermal modification. Additionally, the visco-elasticity of bamboo restricts its application in large structures that require long-term loading. Thus, it is of great interest to investigate the hardness and modulus of the elasticity of bamboo cell walls as a function of the parameters of thermal treatment. Traditional work focused on the effects of heat treatment medium on the macro-mechanical properties such as shrinkage ratio, bending strength, density, equilibrium moisture content (emc), and so on. Although they performed excellent work on the bamboo thermal modification, they did not explore the micro-mechanical properties of bamboo cell walls. Nanoindentation (NI) is a useful technology that can analyze the nano-mechanics of bamboo from the cell-wall level. For analyzing the effects of pressure-steam heat treatment on the nano-mechanics of bamboo samples, NI was used. In addition, for outdoor application, the anti-mildew properties of treated bamboo samples are also important [11–13]. Thus, the anti-mildew properties of the untreated bamboo and treated bamboo samples were also investigated in this paper. Thus, the effect of pressure-steam on the micro-mechanical and anti-mildew properties of moso bamboo is still not clear.

In this paper, 6-year-old bamboo was thermally modified with pressure-steam under different temperatures for the same duration. Additionally, we reported pressure-steam heat treatment and explored the effect of pressure-steam on the micro-structure, crystallinity index, anti-mildew, chemical composition, physical properties, and mechanical properties of bamboo via X-ray diffractometer (XRD), scanning electron microscopy (SEM), Fourier-transform infrared (FTIR), wet chemistry method and nanoindentation (NI). In addition, we tested the anti-mildew property of the control and pressure-steam-treated bamboo samples in this manuscript.

2. Materials and Methods

2.1. Materials and Pressure-Steam Heat Treatment

Six-year-old moso bamboo (Phyllostachys heterocycla) was collected from Gaoan city, Jiangxi, China. The bamboo specimens were collected from the upper layer. The initial moisture content of the specimens was 90%.

Bamboo specimens with dimensions of 1050 × 850 × 10 mm^3 (length × width × thickness) were prepared for pressure-steam treatment. Then, the arc-shaped bamboo sheets were transferred into the pressure-steam equipment (12R3426-1, Hangzhou Rongda Boiler Container Co., Ltd., Hangzhou, China) for saturated-steam heat treatment at different temperatures for the same duration. Sample A presented the untreated bamboo sample in this manuscript. Sample B and sample C presented the treated bamboo at 160 °C and 180 °C and same duration (12 min), respectively. The heat treatment process of bamboo are shown in Figure 1A–D.

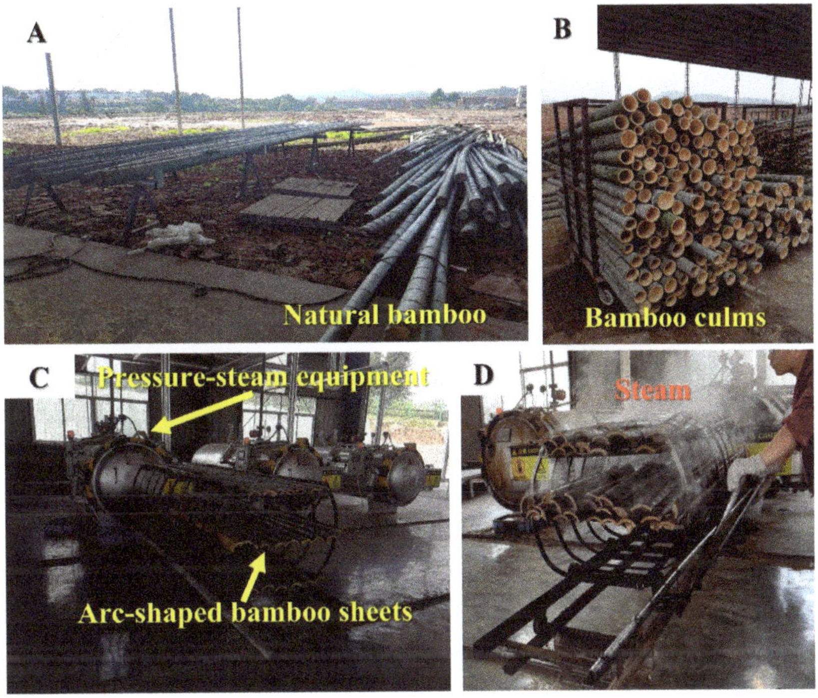

Figure 1. Heat treatment process of bamboo: (**A**) Natural Bamboo in factory; (**B**) Bamboo culms; (**C**) Pressure-steam equipment and arc-shaped bamboo sheets; (**D**) the bamboo sample after saturated-steam heat treatment.

2.2. Chemical Components

The change in chemical components (hemicellulose, cellulose, and lignin) were determined by the wet chemistry method. The bamboo specimens were ground into powder with an average dimension of 30–60 mesh. NREL'LAPS were applied for determining the change in chemical components [14].

2.3. Scanning Electron Microscopy

Natural bamboo and pressure-steam-treated bamboo specimens with average dimensions of 5 × 5 × 1 mm^3 (length × width × thickness) were polished for the SEM (FEI Quanta 200, Eindhoven, Holland) observation. The micro-structure of the control and pressure-steam-treated bamboo was placed in a vacuum environment and then observed by scanning electron microscopy.

2.4. Measurement of Cellulose Crystallinity Degree

The crystallinity index of the control and pressure-steam-treated bamboo samples was determined by X-ray diffractometer (Ultima IV, Tokyo, Japan) with a rate of 2°/min ranging from 5° to 40°. The crystallinity degree can be calculated as below [15–17]:

$$CrI = (I_{002} - I_{am})/I_{002} \times 100\% \tag{1}$$

where CrI represents the crystallinity index, I_{am} represents the minimum intensity of the amorphous, and I_{002} represents the maximum intensity of the diffraction.

2.5. FTIR

The natural bamboo and softened treated bamboo were powdered into 100–200 mesh size and dried under vacuum at 80 °C for 12 h. The powder from different samples was used for Fourier-transform infrared (FTIR) spectroscopy analysis by Nicolet iS50 FT-IR spectrometer (Thermo Fisher Scientific, Waltham, MA, USA). Data in the wave number range of 4000 cm^{-1}–500 cm^{-1} were collected in ATR mode with 64 scans and a resolution of 4 cm^{-1}.

2.6. EMC, Density, and Bending Strength of Bamboo Samples after Pressure-Steam Heat Treatment

The equilibrium moisture content, bending strength, and oven-density were tested according to GB/T 15780-1995 "Experimental method for physical and mechanical properties of bamboo". In detail, the bamboo strips with average dimensions of 160 × 10 × 8 mm^3 (longitudinal × tangential × radial) were prepared by splitting, cross-cutting, and outer-inner layer removing for mechanical testing. Twelve repeated samples of each experimental set were tested in the 3-point bending test. In the density test, the bamboo samples with average sizes of 10 (length) × 10 (width) × t mm (bamboo wall thickness) were prepared from bamboo culms for testing the density by oven-drying method.

2.7. Nano-Indentation Method

The instrument used for the experiments in this paper is the G200 Nano-indenter from Agilent, Santa Clara, CA, USA. The instrument combines the nanoindentation head and in-situ scanning imaging functions to achieve precise positioning of indentations at the nanoscale, improving the reliability and accuracy of experiments and thus enabling fine characterization of the mechanical properties of bamboo cell walls.

The experiments were carried out using the XP system quasi-static continuous stiffness testing (CSM) technique with a Berkovich trigonal diamond indenter (Micro Star Inc., League City, TX, USA). The experimental parameters were set to 100 N/m for contact stiffness, 0.5 nm/s for thermal drift, 0.05 s^{-1} for constant strain rate loading, and 45 Hz for simple harmonic force. The test location of bamboo cell walls and typical Nano-indentation load-depth curves are shown in Figure 2A,B.

During the nanoindentation test, the indenter is first pressed into the material and then held when the set maximum load is reached. During the holding period, the material at the bottom of the indenter begins to creep and deform, which is reflected in the fact that the depth of indentation continues to increase over time during the holding period.

$$H = \frac{P_{max}}{A} \tag{2}$$

where P_{max} is the peak load, and A is that the projected contact space of the indents at peak load. The hardness of different treated bamboo specimens can be calculated as following:

$$Er = \frac{\sqrt{\pi}}{2\beta}\frac{S}{\sqrt{A}} \tag{3}$$

where *Er* is the combined elastic modulus of both the sample and indenter; *S* is initial unloading stiffness; and β is a correction factor correlated to indenter geometry (β = 1.034).

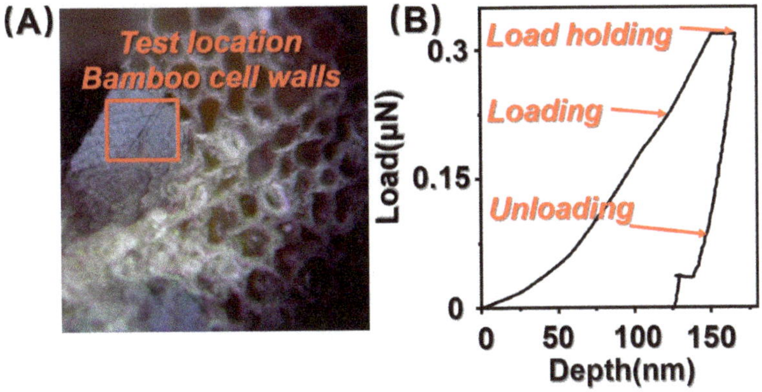

Figure 2. (**A**) Test location of bamboo cell walls; (**B**) typical Nano-indentation (NI) load-depth curves.

2.8. Analytical Procedures

The differences between the control and treated bamboo samples were determined by statistical software SPSS (25.0, IBM, Washington, USA) through Tukey's tests. Additionally, 8 replicates were used to calculate the means for chemical compositions, crystallinity index, modulus of rupture, modulus of elasticity, and hardness in bamboo specimens.

3. Results and Discussion

3.1. Micro-Morphology of the Control and Pressure-Steam Treated Bamboo Samples

The SEM images of all bamboo samples are shown in Figure 3. The surface morphology of the control was smooth. With the increase of pressure-steam treatment temperature, the lumen volumes of bamboo are gradually distorted and deformed during the treatment. It can be attributable to the high temperature and high pressure [18]. In other words, the high-pressure condition led to the decomposition of chemical composition in bamboo cell walls [19–21].

3.2. XRD, FTIR, Chemical Composition Analysis

As shown in Figure 4A. To verify the degradation effect of high-temperature saturated steam on the cellular material of the bamboo material, the chemical composition of the bamboo process fibers before and after the treatment was determined by the wet chemical method. As can be seen from Figure 4, the main chemical composition of the untreated bamboo material was 41.3% (cellulose), 21.6% (hemicellulose), and 21.2% (lignin). The cellulose and hemicellulose content decreased by 2.5% and 42.5%, respectively, while the lignin content increased by 4.9%. Hemicellulose is a polysaccharide substance with a low degree of polymerization, relatively poor thermal and chemical stability, and is prone to decomposition reactions such as deacetylation of polysaccharide substances under the action of high temperature and high humidity [15]. In the chemical composition analysis, it was found that at high-temperature saturated-steam vapor, the molecular chains in the non-crystalline region of cellulose are prone to breakage, exposing more. The chemical composition of the cellulose is analyzed and found to be. The hydroxyl groups on the short molecular chains undergo a "bridging reaction" to form ether bonds, which rearrange and crystallize. The relative crystallinity of the treated bamboo process fibers is increased [18]. The results from FTIR confirmed the decomposition of hemicellulose.

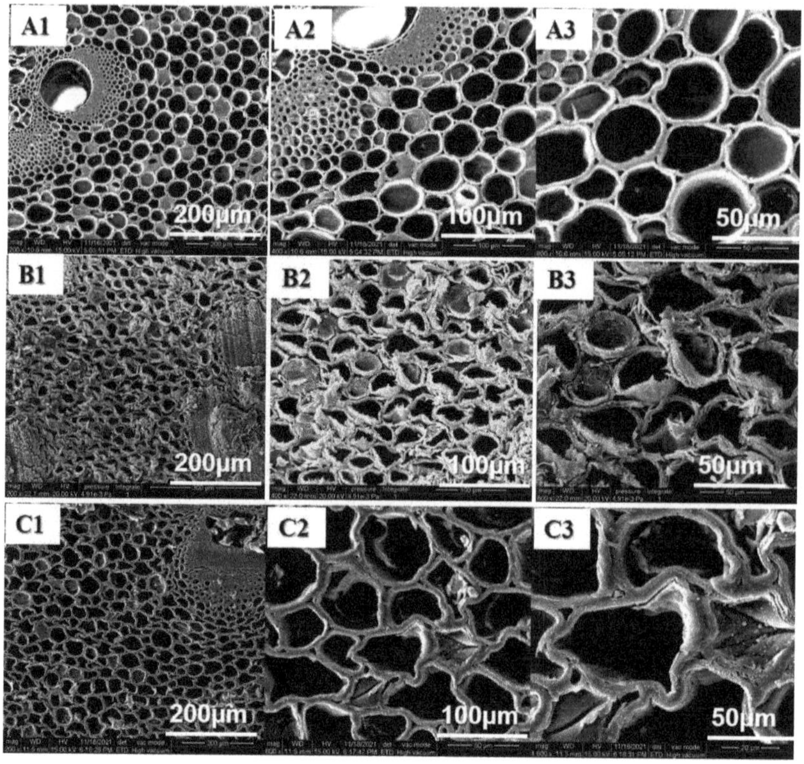

Figure 3. SEM images of different bamboo samples: (**A1–A3**) Untreated bamboo sample; (**B1–B3**) Softened treated bamboo samples (160 °C/12 min); (**C1–C3**) Softened treated bamboo board at 180 °C/12 min.

In order to further investigate the effect of hydrothermal modification on the functional groups of bamboo fiber, FTIR was applied for this purpose. As shown in Figure 4B, 1238 cm^{-1} is the C-O strength vibrations peak, while the 1731 cm^{-1} is attributed to the C-O and C=O stretching vibrations of hemicelluloses. The intensity of peaks at 1238 cm^{-1} and 1731 cm^{-1} decreased. They are the absorption peak of hemicellulose; the decrement of intensity of peaks at 1238 cm^{-1} and 1731 cm^{-1} represented the decomposition of hemicellulose. During the hydrothermal modification, the cleavage of acetyl groups to acetic acid occurs, which promotes the degradation of hemicellulose further during the thermal treatment. The peak at 1590 cm^{-1} and 1328 cm^{-1} assigned to the aromatic skeletal vibrations and C=O stretching of lignin and the C=O groups linked to the aromatic skeleton barely changed in this study and decreased in the spectra of thermal-treated bamboo with the increasing temperature, which may result from the possible condensation reactions of lignin during the thermal treatment. The intensity of the absorption peak at 898 cm^{-1} decreased significantly. This may be attributed to a small degradation of the "amorphous region" in cellulose due to the high temperature and high pressure [22].

As is visible in Figure 4C,D, the crystallinity was calculated by Segal's formula. Obviously, treatment parameters contribute positively to the crystallinity index. The higher the treatment temperature, the higher the cellulose crystallinity degree. The mean crystallinity index (CrI) values of the control were 42.7%. Hydrothermal treatment at 160 °C for 12 min has a higher CrI than the control. For example, the crystallinity index increased from 42.7% to 52.3%. The increment of CrI can be attributed to the cellulose becoming more crystalline [23–25]. Hemicelluloses have a character of amorphous nature, so the hemicellulose can be more easily hydrolyzed than lignin and cellulose. Additionally, the decomposition of the amorphous parts in cellulose maybe also contribute to this phenomenon. Statistical analysis demonstrated that there is a significant difference between the crystallinity indexes of bamboo samples treated over 160 °C and untreated bamboo samples [26–29].

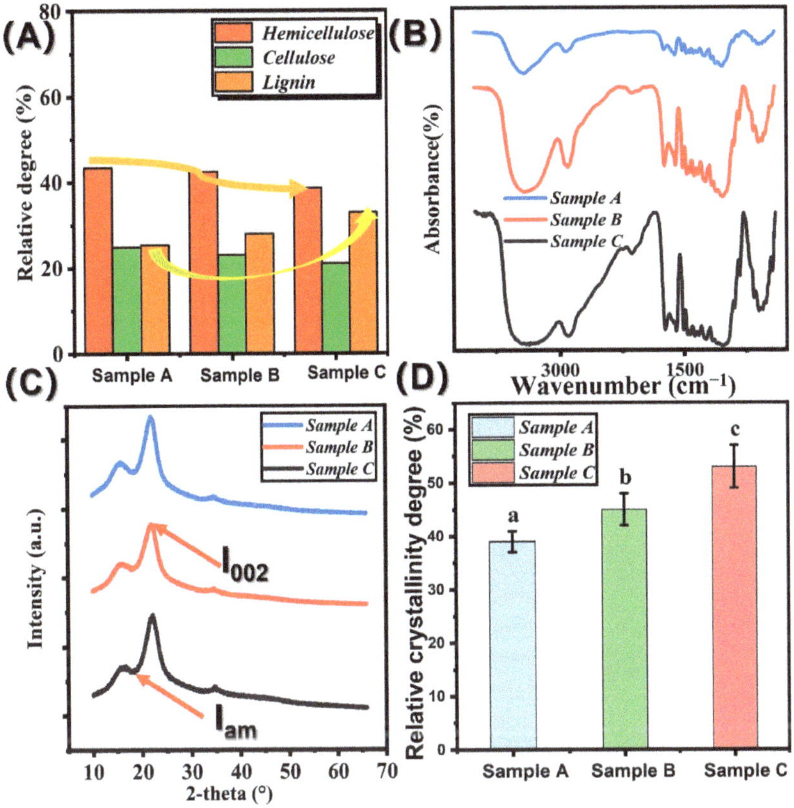

Figure 4. The change in chemical composition, cellulose crystallinity, chemical groups of different bamboo samples: (**A**) hemicellulose, cellulose, and lignin; (**B**): XRD curves; (**C**) Relatively crystallinity degree; (**D**) FTIR curves of different bamboo samples. Different small letters represent the significant difference between heat treatment groups ($p < 0.05$). The error bar in the picture represents the standard deviation.

3.3. Macro-Physical Property of Untreated and Pressure-Steam Treated Bamboo Samples

Figure 5A–D shows the EMC, density, and bending strength of bamboo after pressure-steam heat treatment. As shown in Figure 5A, the EMC of treated bamboo decreased with the increment of treatment temperature. This is because the hygroscopicity of bamboo samples decreased [30]. In addition, the decomposition of the hemicellulose in bamboo cell walls can also contribute to this conclusion. Normally, many free hydroxyl groups exist in hemicellulose. Due to the high-pressure and high-temperature treatment, free hydroxyl groups are greatly decreased. Thus, the EMC of treated bamboo samples decreased. The density of treated bamboo samples decreased with increasing treatment temperature. This can be attributable to the degradation of chemical components in bamboo samples.

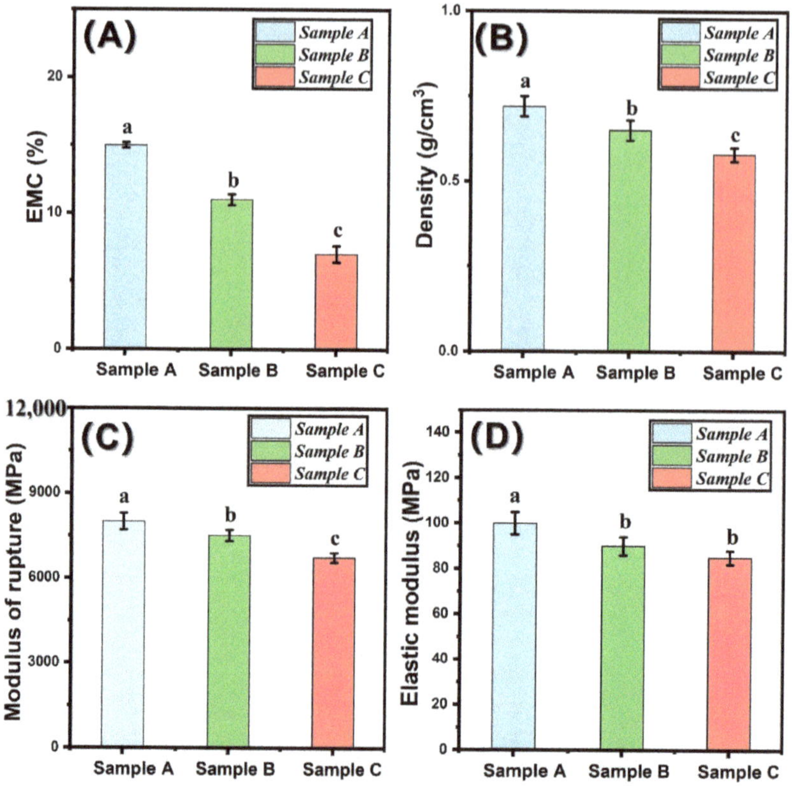

Figure 5. Physical properties of untreated, pressure-steam treated bamboo samples: (**A**) EMC; (**B**) Density; (**C**) modulus of rupture; (**D**) Elastic modulus. Significant difference between heat treatment groups ($p < 0.05$) were presented by different small letters. The error bar in the picture represents the standard deviation.

The macro-/micro-mechanical properties of untreated and pressure-steam-treated bamboo specimens were presented in Figure 5C,D. As shown in Figure 5C, the initial modulus of rupture (MOR) of the control and pressure-steam-treated bamboo specimens were 8700, 8550, and 7200 MPa, respectively. After treatment by pressure-steam, the MOR decreased by 5.8%. With the increment of treatment temperature, the MOR reduced by 16.7%. According to previous literature, firstly, bamboo is softened under the high-pressure and high-temperature condition and then pyrolyzed during the thermal modification process. When the temperature reaches 180 °C, the hemicellulose decreased significantly due to the pressure-steam heat treatment [31]. Pressure-steam heat treatment significantly decreased the modulus of rupture of the bamboo samples.

3.4. Nano-Mechanics of Untreated and Pressure-Steam Treated Bamboo Sample Cell Walls

Nanoindentation (NI) is a useful technology that can analyze the nano-mechanics of bamboo from the cell-wall level. For analyzing the effects of pressure-steam heat treatment on the nano-mechanics of bamboo samples, NI was used to this objective. In Figure 6, we can find that the hardness and modulus of elasticity of the control were 0.52 GPa and 12.7 GPa, respectively. Through ANOVA statistical analysis, we can find that the MOE and hardness of pressure-steam-treated bamboo increased significantly. This is due to the enhanced content of relative crystallinity index and lignin. In addition, according to previous studies, the bending strength and modulus of rupture of bamboo specimens were obtained from the traditional stress–strain relationship. The macro-mechanical properties of bamboo samples showed a decreasing tendency after saturated-steam heat treatment, while the MOE and hardness of bamboo cell walls increased. Normally, the decrement of bending strength is due to the degradation of chemical composition in bamboo cell walls. The micro-mechanical properties of bamboo cell walls are affected for many reasons, such as cellulose crystallinity, lignin content, moisture content, and so on [32–34]. Thus, the mechanical properties of the bamboo cell wall play a subordinate role to macro-mechanical properties. For detail, the Young's modulus values of bamboo obtained from the macroscopic stress–strain relationship increased under elevated compression markedly. In contrast, the mechanical properties of the bamboo cell walls, as expressed by modulus of elasticity and hardness, have shown increasing tendency. The decrement of bamboo density and bamboo's macro-mechanical properties are due to the decomposition of chemical composition, and the micro-mechanical properties of the bamboo cell walls play a subordinate role in this regard [35–38].

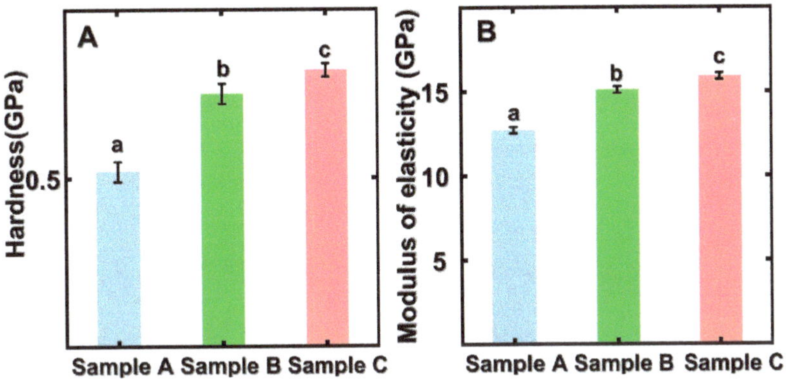

Figure 6. Micro-mechanics of control and treated bamboo samples: (**A**) hardness; (**B**) modulus of elasticity; significant differences between heat treatment groups ($p < 0.05$) were presented by different small letters. The error bar in the picture represents the standard deviation.

3.5. Anti-Mildew Property of Untreated and Pressure-Steam Treated Bamboo Samples

The anti-mildew property of untreated and pressure-steam-treated bamboo samples were shown in Figure 7. As presented in Figure 7, the initial infection ratios of sample A, B, and C were all 0%. The infection ratio of untreated bamboo samples increased from 0% to 100% in the first 10 days, illustrating that the untreated bamboo has poor anti-mildew properties. However, sample B exhibited a better anti-mildew property than that of the control. Unfortunately, after 16 days, the infection ratio of the sample also reached 100%. It is obvious that the infection ratio of sample C increased slowly and kept a 0% infection ratio in the first 8 days. Sample C exhibited the best anti-mildew property in comparison to those of sample A and sample B. It can be attributed to the degradation of hemicellulose and cellulose in bamboo cell walls. The decrement of polysaccharide and starch in bamboo made a positive contribution to the anti-mildew property of bamboo samples [31,39–41]. In addition, the enhanced lignin content in the bamboo surface can inhibit the adhesion between Aspergilus niger and bamboo samples. Thus, pressure-steam treatment can positively enhance the anti-mildew property of bamboo samples. In addition, the summary of the anti-mildew property of treated bamboo in different references are shown in Table 1.

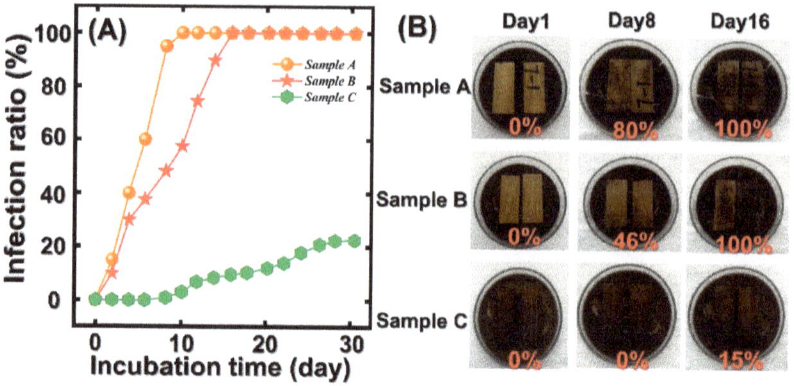

Figure 7. The anti-mildew property of the control and treated bamboo samples in 30 days: (**A**) Infection ratio of untreated bamboo and treated bamboo in 30 days and (**B**) corresponded figures of anti-mildew test.

Table 1. Summary of the anti-mildew property of treated bamboo in different references.

Sample	Treatment Medium	Anti-Mildew Properties in 30 Days	Reference
Arc-shaped Bamboo sheet	Pressure-steam	20%	This work
Bamboo	Linseed oil	25%	[23]
Bamboo	TiO_2	50%	[24]
Bamboo scriber	Fe_3O_4	78.6%	[22]

4. Conclusions and Future Perspectives

Pressure-steam heat treatment, as a newer bamboo thermal modification method, can effectively improve the anti-mildew property of bamboo materials. In this work, we assess the effect of pressure-steam heat treatment on arc-shaped bamboo sheets through analyzing the change in micro-structure, functional groups, chemical components, and so on. Results showed that with the introduction of saturated steam, the content of hemicellulose and cellulose decreased, while the lignin relative content increased significantly. The anti-mildew property of moso bamboo enhanced due to the decomposition of polysaccharide. Last, the modulus of elasticity and hardness of treated moso bamboo cell walls were enhanced after saturated-steam heat treatment. For example, the MOE of treated moso bamboo cell walls increased from 12.7 GPa to 15.7 GPa. This heat-treatment strategy can enhance the anti-mildew property of moso bamboo and can gain more attention from entrepreneurs and scholars. This work introduces pressure-steam heat treatment to entrepreneurs and scholars and analyzes the thermal mechanism for the thermally modified bamboo samples, which can gain more attention from bamboo factories and researchers in the bamboo industry.

Author Contributions: X.L. (Xingyu Liang): Writing & Conceptualization; Y.Y.: data analysis; X.X.: Polished the bamboo samples for SEM analysis and Nanoindentation test; X.L. (Xiaorong Liu): FTIR and XRD analysis; X.W.: methodology; Y.L.: Conceptualization. All authors have read and agreed to the published version of the manuscript.

Funding: The authors acknowledge the funding support from the National Natural Science Foundation of China (Nos. 31971740 and 31901374), Jiangxi Forestry Bureau Forestry Science and Technology Innovation Special Project (No. 202134); Nanping Science and Technology Plan Project (N2020Z001), Zhejiang Provincial Key R&D Program Project (2019C02037); China Postdoctoral Science Foundation

(2021M690531), Nanjing Forestry University Youth Science and Technology Innovation Foundation (CX2019004), and Qing Lan Project.

Conflicts of Interest: The authors declare no conflict of interest.

References

1. Yuan, T.; Wang, X.; Liu, X.; Lou, Z.; Mao, S.; Li, Y. Bamboo Flattening Technology Enables Efficient and Value-Added Utilization of Bamboo in the Manufacture of Furniture and Engineered Composites. *Compos. Part B Eng.* **2022**, *242*, 110097. [CrossRef]
2. Kuai, B.; Tong, J.; Zhang, Y.; Zhan, T.; Lu, J.; Cai, L. Analysis of Micro-Morphology, Mechanical Properties, and Dimensional Stability of Densified Faber Fir Infused with Paraffin. *Holzforschung* **2022**, *76*, 451–462. [CrossRef]
3. Kuai, B.; Wang, Z.; Gao, J.; Tong, J.; Zhan, T.; Zhang, Y.; Lu, J.; Cai, L. Development of Densified Wood with High Strength and Excellent Dimensional Stability by Impregnating Delignified Poplar by Sodium Silicate. *Constr. Build. Mater.* **2022**, *344*, 128282. [CrossRef]
4. Liang, R.; Zhu, Y.-H.; Wen, L.; Zhao, W.-W.; Kuai, B.-B.; Zhang, Y.-L.; Cai, L.-P. Exploration of Effect of Delignification on the Mesopore Structure in Poplar Cell Wall by Nitrogen Absorption Method. *Cellulose* **2020**, *27*, 1921–1932. [CrossRef]
5. Yuan, T.; Xiao, X.; Zhang, T.; Yuan, Z.; Wang, X.; Li, Y. Preparation of Crack-Free, Non-Notched, Flattened Bamboo Board and Its Physical and Mechanical Properties. *Ind. Crops Prod.* **2021**, *174*, 114218. [CrossRef]
6. Guan, M.; Tang, X.; Du, K.; Liu, J.; Li, S. Fluorescence Characterization of the Precuring of Impregnated Fluffed Veneers and Bonding Strength of Scrimber in Relation to Drying Conditions. *Dry. Technol.* **2020**, *40*, 265–272. [CrossRef]
7. Ju, Z.; Zhan, T.; Cui, J.; Brosse, N.; Zhang, H.; Hong, L.; Lu, X. Eco-Friendly Method to Improve the Durability of Different Bamboo (Phyllostachys Pubescens, Moso) Sections by Silver Electrochemical Treatment. *Ind. Crops Prod.* **2021**, *172*, 113994. [CrossRef]
8. Qian, H.; Hou, Q.; Hong, L.; Lu, X.; Ziegler-Devin, I.; Chrusciel, L.; Besserer, A.; Brosse, N. Effect of Highly Efficient Steam Explosion Treatment on Beech, Poplar and Spruce Solid Wood Physicochemical and Permeable Performances. *Ind. Crops Prod.* **2022**, *182*, 114901. [CrossRef]
9. He, Q.; Zhan, T.; Zhang, H.; Ju, Z.; Hong, L.; Brosse, N.; Lu, X. Facile Preparation of High Anti-Fungal Performance Wood by High Voltage Electrostatic Field (HVEF). *J. Clean. Prod.* **2020**, *260*, 120947. [CrossRef]
10. Yang, K.; Li, X.; Wu, Y.; Zheng, X. A Simple, Effective and Inhibitor-free Thermal Treatment for Enhancing Mold-proof Property of Bamboo Scrimber. *Eur. J. Wood Prod.* **2021**, *79*, 1049–1055. [CrossRef]
11. Li, R.; Ekevad, M.; Guo, X.; Cao, P.; Wang, J.; Chen, Q.; Xue, H. Pressure, Feed Rate, and Abrasive Mass Flow Rate Influence on Surface Roughness for Recombinant Bamboo Abrasive Water Jet Cutting. *BioResources* **2015**, *10*, 1998–2008. [CrossRef]
12. Şenol, H.; Erşan, M.; Görgün, E. Optimization of Temperature and Pretreatments for Methane Yield of Hazelnut Shells Using the Response Surface Methodology. *Fuel* **2020**, *271*, 117585. [CrossRef]
13. Campatelli, G.; Lorenzini, L.; Scippa, A. Optimization of Process Parameters Using a Response Surface Method for Minimizing Power Consumption in the Milling of Carbon Steel. *J. Clean. Prod.* **2014**, *66*, 309–316. [CrossRef]
14. Hyeong, S.-K.; Park, M.; Kim, S.-I.; Park, S.; Choi, K.-H.; Im, M.J.; Kim, N.D.; Kim, T.-W.; Lee, S.H.; Park, J.-W.; et al. Compacted Laser-Induced Graphene with Bamboo-Like Carbon Nanotubes for Transformable Capacitive Energy Storage Electrodes. *Adv. Mater. Technol.* **2022**, *7*, 2101105. [CrossRef]
15. Qu, L.; Wang, Z.; Qian, J.; He, Z.; Yi, S. Effects of Aluminum Sulfate Soaking Pretreatment on Dimensional Stability and Thermostability of Heat-Treated Wood. *Eur. J. Wood Prod.* **2020**, *79*, 189–198. [CrossRef]
16. Zhang, Y.M.; Yu, Y.L.; Yu, W.J. Effect of Thermal Treatment on the Physical and Mechanical Properties of Phyllostachys Pubescen Bamboo. *Eur. J. Wood Prod.* **2013**, *71*, 61–67. [CrossRef]
17. Long, Z.; Zhang, L.; Wu, Q.; Tan, Z.; Guo, P. Effect of Temperature on Color and Chemical Composition of Poplar Powder Compacts during Warm-Press Forming. *Eur. J. Wood Prod.* **2021**, *79*, 1461–1468. [CrossRef]
18. Wang, X.; Shan, S.; Shi, S.Q.; Zhang, Y.; Cai, L.; Smith, L.M. Optically Transparent Bamboo with High Strength and Low Thermal Conductivity. *ACS Appl. Mater. Interfaces* **2021**, *13*, 1662–1669. [CrossRef]
19. Forsman, K.; Fredriksson, M.; Serrano, E.; Danielsson, H. Moisture-Dependency of the Fracture Energy of Wood: A Comparison of Unmodified and Acetylated Scots Pine and Birch. *Holzforschung* **2021**, *75*, 731–741. [CrossRef]
20. Wagih, A.; Hasani, M.; Hall, S.A.; Theliander, H. Micro/Nano-Structural Evolution in Spruce Wood during Soda Pulping. *Holzforschung* **2021**, *75*, 754–764. [CrossRef]
21. Takahashi, Y.; Ishiguri, F.; Aiso, H.; Takashima, Y.; Hiraoka, Y.; Iki, T.; Ohshima, J.; Iizuka, K.; Yokota, S. Inheritance of Static Bending Properties and Classification of Load-Deflection Curves in Cryptomeria Japonica. *Holzforschung* **2021**, *75*, 105–113. [CrossRef]
22. Lou, Z.; Han, X.; Liu, J.; Ma, Q.; Yan, H.; Yuan, C.; Yang, L.; Han, H.; Weng, F.; Li, Y. Nano-Fe3O4/Bamboo Bundles/Phenolic Resin Oriented Recombination Ternary Composite with Enhanced Multiple Functions. *Compos. Part B Eng.* **2021**, *226*, 109335. [CrossRef]
23. Wang, Q.; Han, H.; Lou, Z.; Han, X.; Wang, X.; Li, Y. Surface Property Enhancement of Bamboo by Inorganic Materials Coating with Extended Functional Applications. *Compos. Part A Appl. Sci. Manuf.* **2022**, 106848. [CrossRef]

24. Hao, X.; Wang, Q.; Wang, Y.; Han, X.; Yuan, C.; Cao, Y.; Lou, Z.; Li, Y. The Effect of Oil Heat Treatment on Biological, Mechanical and Physical Properties of Bamboo. *J. Wood Sci.* **2021**, *67*, 26. [CrossRef]
25. Chen, C.; Wang, Y.; Wu, Q.; Wan, Z.; Li, D.; Jin, Y. Highly Strong and Flexible Composite Hydrogel Reinforced by Aligned Wood Cellulose Skeleton via Alkali Treatment for Muscle-like Sensors. *Chem. Eng. J.* **2020**, *400*, 125876. [CrossRef]
26. Ge, S.; Ma, N.L.; Jiang, S.; Ok, Y.S.; Lam, S.S.; Li, C.; Shi, S.Q.; Nie, X.; Qiu, Y.; Li, D.; et al. Processed Bamboo as a Novel Formaldehyde-Free High-Performance Furniture Biocomposite. *ACS Appl. Mater. Interfaces* **2020**, *12*, 30824–30832. [CrossRef] [PubMed]
27. Cai, C.; Wei, Z.; Deng, L.; Fu, Y. Temperature-Invariant Superelastic Multifunctional MXene Aerogels for High-Performance Photoresponsive Supercapacitors and Wearable Strain Sensors. *ACS Appl. Mater. Interfaces* **2021**, *13*, 54170–54184. [CrossRef]
28. Cai, C.; Wei, Z.; Ding, C.; Sun, B.; Chen, W.; Gerhard, C.; Nimerovsky, E.; Fu, Y.; Zhang, K. Dynamically Tunable All-Weather Daytime Cellulose Aerogel Radiative Supercooler for Energy-Saving Building. *Nano Lett.* **2022**, *22*, 4106–4114. [CrossRef]
29. Zhou, X.; Han, H.; Wang, Y.; Zhang, C.; Lv, H.; Lou, Z. Silicon-Coated Fibrous Network of Carbon Nanotube/Iron towards Stable and Wideband Electromagnetic Wave Absorption. *J. Mater. Sci. Technol.* **2022**, *121*, 199–206. [CrossRef]
30. Liu, Q.-C.; Liu, T.; Liu, D.-P.; Li, Z.-J.; Zhang, X.-B.; Zhang, Y. A Flexible and Wearable Lithium–Oxygen Battery with Record Energy Density Achieved by the Interlaced Architecture Inspired by Bamboo Slips. *Adv. Mater.* **2016**, *28*, 8413–8418. [CrossRef]
31. Altgen, M.; Hofmann, T.; Militz, H. Wood Moisture Content during the Thermal Modification Process Affects the Improvement in Hygroscopicity of Scots Pine Sapwood. *Wood Sci. Technol.* **2016**, *50*, 1181–1195. [CrossRef]
32. Jiang, S.; Wei, Y.; Shi, S.Q.; Dong, Y.; Xia, C.; Tian, D.; Luo, J.; Li, J.; Fang, Z. Nacre-Inspired Strong and Multifunctional Soy Protein-Based Nanocomposite Materials for Easy Heat-Dissipative Mobile Phone Shell. *Nano Lett.* **2021**, *21*, 3254–3261. [CrossRef] [PubMed]
33. Jiang, S.; Wei, Y.; Tao, L.; Ge, S.; Shi, S.Q.; Li, X.; Li, J.; Le, Q.V.; Xia, C. Microwave Induced Construction of Multiple Networks for Multifunctional Soy Protein-Based Materials. *Prog. Org. Coat.* **2021**, *158*, 106390. [CrossRef]
34. Jiang, S.; Wei, Y.; Li, J.; Li, X.; Wang, K.; Li, K.; Shi, S.Q.; Li, J.; Fang, Z. Development of a Multifunctional Nanocomposite Film with Record-High Ultralow Temperature Toughness and Unprecedented Fatigue-Resistance. *Chem. Eng. J.* **2022**, *432*, 134408. [CrossRef]
35. Lv, H.-F.; Ma, X.-X.; Zhang, B.; Chen, X.-F.; Liu, X.-M.; Fang, C.-H.; Fei, B.-H. Microwave-Vacuum Drying of Round Bamboo: A Study of the Physical Properties. *Constr. Build. Mater.* **2019**, *211*, 44–51. [CrossRef]
36. Meng, F.; Yu, Y.; Zhang, Y.; Yu, W.; Gao, J. Surface Chemical Composition Analysis of Heat-Treated Bamboo. *Appl. Surf. Sci.* **2016**, *371*, 383–390. [CrossRef]
37. Yin, Y.; Berglund, L.; Salmén, L. Effect of Steam Treatment on the Properties of Wood Cell Walls. *Biomacromolecules* **2011**, *12*, 194–202. [CrossRef]
38. Özgenç, Ö.; Durmaz, S.; Boyaci, I.H.; Eksi-Kocak, H. Determination of Chemical Changes in Heat-Treated Wood Using ATR-FTIR and FT Raman Spectrometry. *Spectrochim. Acta Part A Mol. Biomol. Spectrosc.* **2017**, *171*, 395–400. [CrossRef]
39. Kanbayashi, T.; Matsunaga, M.; Kobayashi, M. Cellular-Level Chemical Changes in Japanese Beech (*Fagus Crenata* Blume) during Artificial Weathering. *Holzforschung* **2021**, *75*, 000010151520200229. [CrossRef]
40. Chen, Q.; Wei, P.; Tang, T.; Fang, C.; Fei, B. Quantitative Visualization of Weak Layers in Bamboo at the Cellular and Subcellular Levels. *ACS Appl. Bio. Mater.* **2020**, *3*, 7087–7094. [CrossRef]
41. Gauss, C.; Kadivar, M.; Harries, K.A.; Savastano Jr, H. Chemical Modification of Dendrocalamus Asper Bamboo with Citric Acid and Boron Compounds: Effects on the Physical-Chemical, Mechanical and Thermal Properties. *J. Clean. Prod.* **2021**, *279*, 123871. [CrossRef]

polymers MDPI

Article

Enhancing for Bagasse Enzymolysis via Intercrystalline Swelling of Cellulose Combined with Hydrolysis and Oxidation

Feitian Bai [1,2,†], Tengteng Dong [1,†], Zheng Zhou [1], Wei Chen [1], Chenchen Cai [1] and Xusheng Li [1,*]

1 School of Light Industrial and Food Engineering, Guangxi University, Nanning 530004, China
2 Creating New Greatness Advanced Material Co., Ltd., Changsha 410600, China
* Correspondence: lixusheng@gxu.edu.cn; Tel.: +86-0771-3237-301
† These authors contributed equally to this work.

Abstract: To overcome the biological barriers formed by the lignin–carbohydrate complex for releasing fermentable sugars from cellulose by enzymolysis is both imperative and challenging. In this study, a strategy of intergranular swelling of cellulose combined with hydrolysis and oxidation was demonstrated. Pretreatment of the bagasse was evaluated by one bath treatment with phosphoric acid and hydrogen peroxide. The chemical composition, specific surface area (SSA), and pore size of bagasse before and after pretreatment were investigated, while the experiments on the adsorption equilibrium of cellulose to cellulase and reagent reuse were also performed. Scanning electron microscopy (SEM) and high-performance liquid chromatography (HPLC) were employed for microscopic morphology observations and glucose analysis, respectively. The results showed that pretreated bagasse was deconstructed into cellulose with a nanofibril network, most of the hemicellulose (~100%) and lignin (~98%) were removed, and the SSA and void were enlarged 11- and 5-fold, respectively. This simple, mild preprocessing method enhanced cellulose accessibility and reduced the biological barrier of the noncellulose component to improve the subsequent enzymolysis with a high glucose recovery (98.60%).

Keywords: cellulose; swelling; bagasse; enzymolysis; cellulase

Citation: Bai, F.; Dong, T.; Zhou, Z.; Chen, W.; Cai, C.; Li, X. Enhancing for Bagasse Enzymolysis via Intercrystalline Swelling of Cellulose Combined with Hydrolysis and Oxidation. *Polymers* **2022**, *14*, 3587. https://doi.org/10.3390/polym14173587

Academic Editor: Luis Alves

Received: 5 August 2022
Accepted: 27 August 2022
Published: 30 August 2022

1. Introduction

Lignocellulose biomass is an energy form that plants utilize to store energy gained via photosynthesis [1]. It is known as a carbon-neutral, green resource, with a total annual output of 146 billion tons [2]. Fuel, materials, and chemicals in solid, liquid, and gaseous states are obtained from biomass using various technical means [3–5]. However, biological barriers make it extremely difficult to release fermentable sugars from cellulosic biomass. As a result, large dosages of enzymes are required for hydrolysis, which diminish their cost-efficient features for commercial application [6]. Since the energy crisis in the 1970s, biofuel and biochemical production technologies based on enzymolysis from biomass have been driven in both industry and academia [7]. Although enzymolysis offers potentially higher yields [8], higher selectivity [9], lower energy costs [10], and milder operating conditions [11] compared to chemical processes, the technology still faces significant challenges. Eliminating the biological barriers of lignocellulose biomass to enzymes under mild conditions is vital [12].

Cellulose, as a linear polymer consisting of 300–15,000 D-glucose units, aggregates into 3 to 4 nm-wide elementary fibrils due to intermolecular forces [13–15]. The elementary fibrils are embedded in the hemicellulose matrix and are further aggregated into 10 to 25 nm-wide primary microfibrils [16]. These primary microfibrils are likewise embedded in a lignin–carbohydrate complex (LCC) matrix and are then bonded together, weaving throughout the plant cells in a unique way [17,18]. Limiting the accessibility of polysaccharides and unproductive binding to enzymes are the main mechanisms by which the unique icing that is formed by LCC limits the enzymolysis of lignocellulose biomass [19,20].

To reduce the cellulase required, the accessibility of cellulose through pretreatment by mechanical, chemical, biological, or a combination of these methods has been extensively studied [21,22]. Some early studies show that the dosage of catalyst needed was lower than 20 FPU/g. These pretreatment techniques can be divided into two types according to their mechanism: (1) Based on hydrolysis (dilute acid, hydrothermal, alkali, enzyme) and oxidation (basic hydrogen peroxide) mechanisms to cut off the molecular chain to remove noncellulose components by improving the mass transfer channels [23]. These methods are limited by the resistance of the plant cell wall; it is difficult to completely remove hemicellulose and lignin. (2) Based on mechanical forces (ball milling), molecular forces (ionic liquids, deep eutectic solvents (DESs), inorganic salt hydrates) destroy cellulose aggregation and increase cellulose exposure [24]. The main problem with this kind of method is that cellulose overdisperses or overdissolves and mixes with residual noncellulose components, which may show some deterioration over time [25]. Thus, assuming limited swelling of cellulose combined with hydrolysis and oxidation in one-pot treatment is expected to simultaneously achieve: the removing of the hemicellulose and lignin, and the increasing of cellulose accessibility, although this has not been reported.

H_3PO_4 is an effective cellulose dissolution and swelling agent that can be easily customized for cellulose intercrystalline swelling or dissolution, depending on the properties of cellulose and operating conditions. Walseth [26] first developed a high-reactivity cellulose for cellulase activity analysis by dissolving cellulose using H_3PO_4, which has become one of the most common cellulose substrates for cellulase activity analysis. Previous studies have shown that H_3PO_4 (with the help of H_2O_2) can extract nanofibrils and high-reactivity cellulose suitable for enzymolysis from biomass [27,28]. However, few studies have revealed the influence of intercrystalline swelling of cellulose on bagasse enzymolysis.

In this work, bagasse was pretreated by using an H_3PO_4 and H_2O_2 aqueous solution system under mild conditions, which has three functions, namely, swelling, hydrolysis, and oxidation. Cellulose swelling, hemicellulose hydrolysis, and lignin oxidation degradation occur simultaneously during the pretreatment, and they cooperate and promote each other. Pretreated bagasse was deconstructed into cellulose with a nanofibril network, most of the hemicellulose (~100%) and lignin (~98%) were removed, their pore volume suitable for enzyme entry was magnified 11-fold, and their surface area available for cellulase loading was increased 5-fold. This preprocessing approach enhanced cellulose accessibility and reduced the barrier of noncellulose components to improve the subsequent enzymolysis with a high glucose recovery (98.60%). In addition, the used H_3PO_4 mixture can be reused for subsequent pretreatment or neutralized to produce a fertilizer rich in phosphorus [29]. This study demonstrates a strategy with simple, mild features, which has the potential pretreatment methods for bioethanol processing and a new possible pathway for biomass-refining technology development.

2. Experimental

2.1. Cellulose and Cellulase

The bagasse used in this study was purchased from Guangxi Guitang Group Co., Ltd. (Guitang, China). The bagasse was ground to a 40–60 mesh powder. Analytically pure reagents, phosphoric acid (H_3PO_4, 85% w/v), anhydrous ethanol (98% w/v), and hydrogen peroxide (H_2O_2, 30% w/v) were purchased from Nanning Blue Sky Experimental Equipment Co., Ltd. (Nanning, China).

A total of 30 g bagasse powder, 60 mL H_2O_2, and 240 mL H_3PO_4 were placed in a round-bottomed flask and pretreated at 30 °C with stirring at 300 rpm for 42 h. After the reaction, the solids were recycled by filtration from the suspension and then soaked in 100 mL anhydrous ethanol for 24 h. The pretreated bagasse was recovered by centrifugation at 4000 rpm for 15 min from ethanol suspension.

The chemical composition of raw materials and pretreated bagasse was determined according to the standard method of the US National Renewable Energy Laboratory (NREL) [30]. This involved a two-stage extraction of samples followed by a two-stage acid hydrolysis. Residual solids were quantified as acid-insoluble lignin content. The acid-soluble lignin content was quantified by a UV spectrophotometer (Agilent Cary 3500, Agilent, Santa Clara, CA, USA) in the analytical hydrolysate. The lignin content was the sum of acid-soluble lignin and acid-insoluble lignin content. Structure of cellulose and hemicellulose were quantified as their monomeric forms in the analytical hydrolyzate using high-performance liquid chromatography (HPLC, Agilent 1260 Infinity II, Agilent, Santa Clara, CA, USA) with an HPX-87H column (Agilent, Santa Clara, CA, USA). Cellulase (Novozyme CTec2) was purchased from Sigma-Aldrich (Shanghai, China). The cellulase activity was determined by the filter paper method according to US NREL [31] and protein content was determined using the Bradford method [32].

2.2. *Physicochemical Properties of Cellulose*

2.2.1. X-ray Diffraction

The X-ray diffraction (XRD) pattern was obtained using a MiniFlex 600 advance X-ray diffractometer (Rigaku, Tokyo, Japan) with a Cu Kα radiation source operated at 40 kV and 40 mA. The measurement of 2θ ranged from 10° to 50° at a scanning speed of 5°/min and step size of 0.02°. The crystallinity index (*CrI*) of pretreated cellulose was calculated by subtracting the amorphous contribution from diffraction spectra using an amorphous standard according to a previous study [33]. XRD was calculated by the following formula:

$$CrI\ (\%) = (I_{200} - I_{am})/I_{200} \quad (1)$$

where I_{200} represents the maximum intensity of the lattice diffraction peak at 2θ between 22.5°, and I_{am} represents the intensity scattered by the amorphous component in the sample, which was evaluated as the lowest intensity at 2θ at 18°.

2.2.2. Degree of Polymerization

The intrinsic viscosity degree of polymerization (DP) test [34] was used to calculate the DP of cellulose. The DP was calculated according to the following equation (with an average of three measurements per sample):

$$[\eta]_G = \eta_{sp}/C \times (1 + 0.35\ \eta_{sp}) \quad (2)$$

$$DP = 80\ [\eta]_G \quad (3)$$

where $[\eta]_G$ is the intrinsic viscosity (mL/g), η_{sp} is the specific viscosity, C represents the concentration (g/100 mL), and *DP* is the degree of polymerization.

2.2.3. Specific Surface Area

Nitrogen adsorption (Micromeritics ASAP2460, Norcross, Georgia) was used to measure the specific surface area (SSA) of untreated and pretreated bagasse. The samples were degassed at 90 °C for 12 h prior to analysis to remove moisture and air from the substrate pores. The test was carried out at liquid nitrogen temperature, and the SSA of the sample was calculated using the BET model [35].

2.2.4. Zeta Potential

The surface charge of the pretreated cellulose was evaluated by determining the zeta potential using the zeta potential mode of the Malvern Zetasizer (ZS90X, Melvin, UK) [36]. The pretreated cellulose was uniformly dispersed in a sodium citrate buffer of pH 4.8 to form a 0.5% (*w/v*) suspension, and the suspension was measured and scanned with a cuvette 100 times.

2.2.5. Additional Measurements and Characterization

An X-ray photoelectron spectrometer (XPS) (ULVAC-PHI, Chigasaki-shi, Japan) was used to determine the surface chemical analysis of pretreated cellulose [37]. A Fourier-transform infrared (FTIR) spectrometer (TENSOR II, Brook Technology, Ettlingen, Germany) was used to obtain the FTIR spectra of the untreated bagasse, pretreated bagasse, and enzymolysis residual in the frequency range of 4000–400 cm^{-1} with a resolution of 4 cm^{-1} using the KBr tablet method [38]. Scanning electron microscopy (SEM) SU8220 (Hitachi, Tokyo, Japan) was used to analyze the surface structure of the untreated bagasse and pretreated bagasse. ImageJ software (Version 2.0, National Institutes of Health, Bethesda, MD, USA) was used to determine the diameter of the nanofibers after at least 100 measurements based on SEM images. The bagasse samples were freeze-dried using the Advantage Plus EL-85 freeze-drying system (SP Scientific, Warminster, PA, USA) and the samples were sprayed with gold to improve the conductivity of the samples before observing the samples.

2.3. Adsorption Equilibrium Experiment

Two hundred milligrams of substrate were weighed into a centrifuge tube and a series of concentrations of enzyme solutions were added (0.05 M citrate buffer, refrigerated at 4 °C before use) to form a solid loading of 2% (*w*/*v*). The mixture was shaken at 130 rpm at 4 °C for 2 h. In parallel, a blank control sample was run. After adsorption, the mixture was centrifuged at 10,000 rpm for 5 min, and the supernatant was taken. The protein concentration was determined by the Bradford method and each sample was measured in duplicate. The adsorption capacity was expressed as the difference between the concentration of added enzyme protein and that of supernatant. The adsorption data were fitted using the Langmuir equation [39]:

$$E_b = (E_{bm} \times K_a \times E_f)/(1 + K_a \times E_f) \quad (4)$$

where E_b is the amount of bound cellulase (mg/g substrate), E_{bm} represents the theoretical maximum adsorption capacity of the substrate (mg/g substrate), K_a is the affinity constant (L/mg), and E_f is the free enzyme in the supernatant (mg/mL).

2.4. Enzymatic Hydrolysis

Enzymolysis of pretreated bagasse was carried out in a 50 mM citrate buffer (pH 4.8) with a substrate load of 2% (*w*/*v*; dry matter, DM). Cellulase was introduced at 5, 10, and 20 filter paper unit (FPU)/g cellulose, and 0.1 g/L ampicillin trihydrate was added to avoid microbial interference during hydrolysis. After enzymolysis for 0.5, 2, 4, 8, 16, 48, and 72 h, ~5 mL of solid–liquid mixture was taken out and inactivated at 100 °C for 30 min, passed through a 0.22 µm filter membrane, and stored at 4 °C for further measurement of glucose yield. Enzymolysis of each sample (untreated and pretreated bagasse) was run in parallel. The glucose concentration was measured at 60 °C using an HPLC system equipped with an HPX-87H column (Agilent, Santa Clara, CA, USA). The mobile phase flow rate was at 0.6 mL/min and the detection time was 30 min. The hydrolysis efficiency of the enzyme bound to the cellulose surface was calculated by the hydrolysis rate of the unit bound enzyme in the initial stage of enzymolysis (0.5 h).

3. Results and Discussion

3.1. Physical and Chemical Property Characterization

To assess the efficacy of the pretreatment in the removal of noncellulose components, the chemical composition of untreated and pretreated bagasse is compared in Table 1. Table 1 showed that ~100% of initial hemicellulose in the bagasse was removed during the pretreatment. As shown in the FTIR results (Figure 1), the characteristic peaks at 1737 cm^{-1} (C=O stretching of the acetyl and urate groups of hemicellulose or the ester bond of carboxyl groups in lignin to fragrant acid and ferulic acid) and 1247 cm^{-1} (the alkyl ester of the acetyl group in hemicellulose) of the hemicellulose of pretreated bagasse from untreated bagasse are decreased or completely disappeared [40]. These indicate that hemicellulose removal is complete [41], which is attributed to the fact that cellulose intercrystalline swelling fully exposes the hemicellulose and promotes the hydrolysis of the hemicellulose. Similarly, 98% of the initial lignin was removed during the pretreatment (Table 1). As seen from the FTIR results (Figure 1), the characteristic peaks at 1515 cm^{-1} (C=C stretching of the aromatic skeleton), 1607 cm^{-1} (the aromatic skeletal stretching), and 1458 cm^{-1} (C–H deformation of CH_3 and CH_2) of lignin of pretreated bagasse from untreated bagasse almost disappeared [42]. This demonstrates that lignin was efficiently removed, attributing to the oxidative degradation of lignin by peroxyphosphoric acid (H_3PO_5) formed by H_3PO_4 and H_2O_2 [43]. The cellulose yield possibly reached 96.03% (Table 1), which is due to both the mild reaction conditions and high selectivity of the delignification and hemicellulose removal [43].

Table 1. Composition of bagasse and pretreated bagasse.

	Yield (%) [a]	Cellulose (%)	Hemicellulose (%)	Lingin (%)
Untreated bagasse	100/100	43.26 ± 2.13	22.86 ± 0.97	25.53 ± 1.18
Pretreated bagasse	42.58/96.03	97.56 ± 4.38	0	1.35 ± 0.11

[a] Yield based on the initial amount of biomass/yield based on the initial amount of cellulose in biomass.

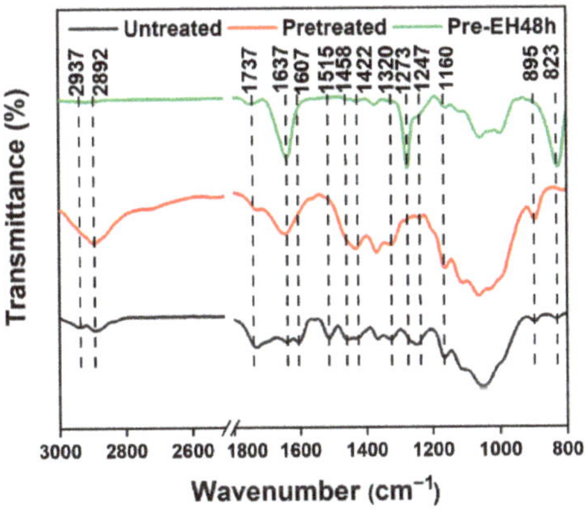

Figure 1. Fourier-transform infrared spectrometer (FTIR) spectrum of the untreated and pretreated bagasse.

As shown in the XPS results (Figure 2a,b), the oxygen-to-carbon (O/C) ratio of untreated bagasse was 0.39. The known theoretical O/C ratios of cellulose, hemicellulose, and lignin are 0.83, 0.81, and 0.33, respectively. The low O/C ratio of natural bagasse can explain the lignin on the surface of the fibrils. The O/C ratio of pretreated bagasse increased to 0.62. The concentrations of C1 (C=C/C-C/C-H), C2 (C-O-C/C-O-H), and C3 (C=O/O-C-O) in untreated bagasse were 41.35%, 47.19%, and 11.46%, respectively. Contributions of cellulose, hemicellulose, and lignin to these peaks have been reported [44,45], with 85% of cellulose signaling to C2 and part of it to C3, 80% of hemicellulose signaling to C2 and the rest to C3, and 50% of lignin signaling to C1 and the rest to C2. The C1 content of pretreated bagasse decreased, while C2 and C3 contents increased. These phenomena suggest that the lignin is removed and the polysaccharides are exposed on the surface of the fibers [46]. This was consistent with the results of the chemical composition and FTIR analysis.

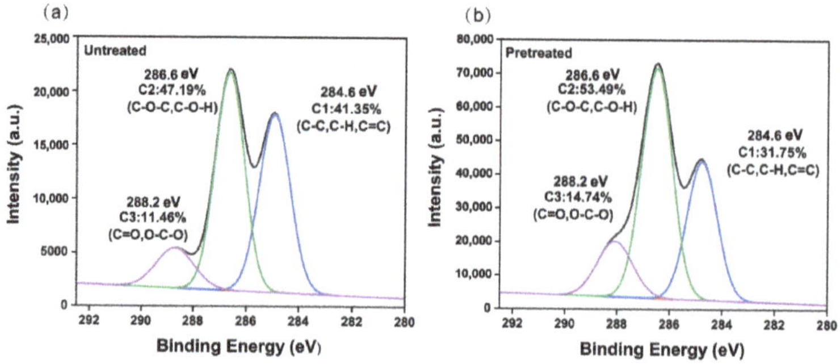

Figure 2. X-ray photoelectron spectrometer (XPS) spectrum of (**a**) untreated bagasse and (**b**) pretreated bagasse.

The structure of untreated bagasse is complete and compact, and the fiber bundles are arranged compactly (Figure 3a). This intact structure greatly impedes the accessibility of the cellulase to the cellulose. Bagasse was pretreated in an aqueous solution of H_3PO_4, and the surface morphology of the pretreated bagasse changed significantly, transforming the dense bagasse into cellulose with a nanofibrils skeleton network structure (Figure 3b,c). The widths of most nanofibers are in the range of 10–60 nm (Figure 3d). This is attributed to the fact that the H_3PO_4 molecules intrude between the fibrils, breaking the hydrogen bonds between adjacent fibrils [47]. The removal of hemicellulose and lignin also increases the number of channels for H_3PO_4 molecules to squeeze into the cell wall, causing the distance between adjacent fibrils to widen.

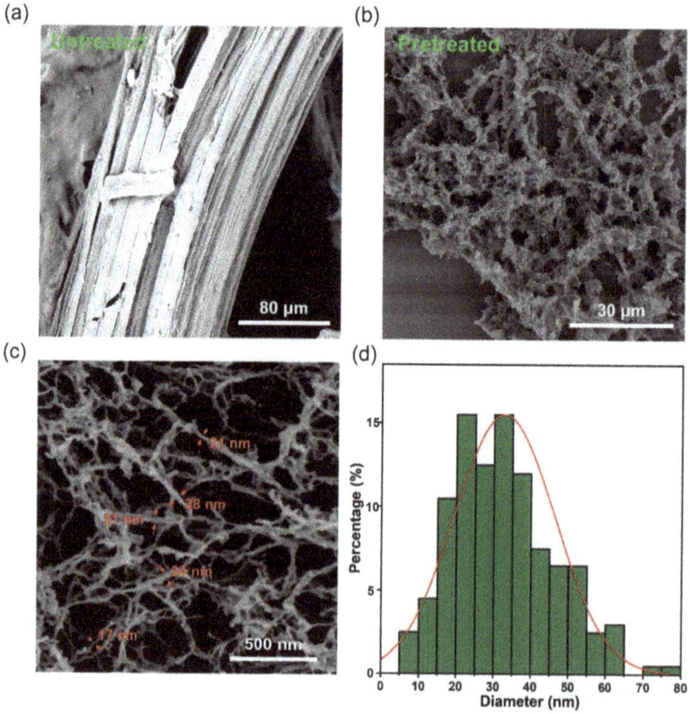

Figure 3. Scanning electron microscopy (SEM) images of (**a**) untreated bagasse and (**b**,**c**) pretreated bagasse; (**d**) diameter distribution of the nanofibers measured based on SEM images.

To evaluate the effect of pretreatment on cellulose aggregation, the XRD patterns of untreated and pretreated bagasse were compared (Figure 4). The peaks [48] at 16° (101), 22° (200), and 34° (004) for cellulose I were significantly strengthened in the XRD patterns of untreated and pretreated bagasse. Similarly, the CrI value of the pretreated bagasse increased from 58.84% to 74.92% (Figure 4). There were no obvious clear peaks at 2θ = 12.1° (110 for cellulose II), and 20.2° (110 for cellulose II) in the XRD patterns of pretreated bagasse as reported in the literature [49]. These indicate that the cellulose crystal structure was unchanged, and the supramolecular structure of cellulose was not visibly broken. This implies that the swelling of H_3PO_4 in cellulose mainly occurs in the intercrystalline spaces rather than the intracrystalline spaces.

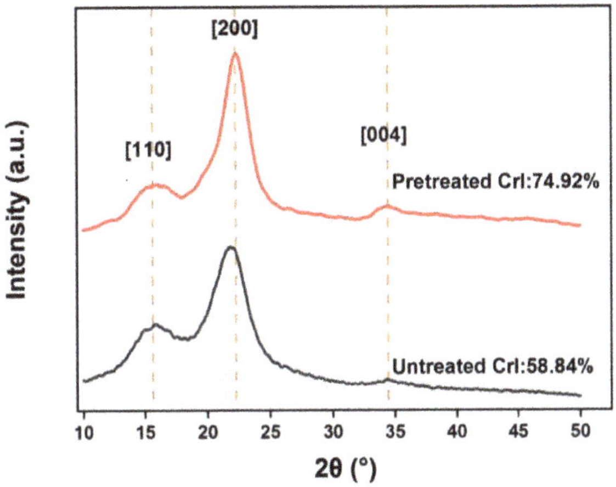

Figure 4. X-ray diffraction (XRD) patterns of the untreated bagasse and pretreated bagasse.

3.2. *Adaptability of Pretreated Bagasse to Cellulase*

To evaluate the effect of cellulose intercrystalline swelling on bagasse enzymolysis, the adaptability of pretreated bagasse to cellulase was analyzed (Figure 5a). As seen from the enzymolysis of pretreated bagasse, the glucose yield (78.19%) achieved at a lower enzyme dosage of 5 FPU/g was 14-fold higher than that achieved with untreated bagasse (5.25%). Further increase in the cellulase dosage to 10 FPU/g resulted in a glucose yield of 95.91% that was five-times higher than that achieved with untreated bagasse (18.07%). However, with 20 FPU/g of cellulase, a glucose yield of 98.60% was obtained: this was two-fold higher than that achieved with untreated bagasse (47.27%). This indicates that pretreated bagasse is highly amenable to cellulase.

Highly selective removal of lignin (~98%) and hemicellulose (~100%) helps to reduce the unproductive adsorption and the physical barrier of bagasse to cellulase (Table 1). These, in addition to the lower noncellulose content of the pretreated bagasse, are also associated with changes in other physicochemical properties including [50,51] pore volume (PV), SSA, degree of polymerization (DP), and CrI, directly and indirectly providing information about enhanced enzymolysis of pretreated bagasse. As seen from the PV results (Figure 5b), a new mesopore (8–23 nm) appeared in the pretreated bagasse and the PV increased to 1.60×10^{-2} cm^3/g from 1.43×10^{-3} cm^3/g (Figure 5b). Pores larger than 5.1 nm allow the enzyme to enter the substrate without being restricted by size [52]. The PV of pretreated bagasse increased 11-fold, meaning that the physical channels through which the enzyme can pass are increased. The SSA of pretreated bagasse significantly increased to 1.9068 m^2/g from 0.3633 m^2/g (Table 2). The increase in the SSA of cellulose means that a larger area is available for enzyme loading [53]. The SSA of pretreated bagasse increased five-fold, meaning that the available surface area of the cellulose for enzyme loading was enhanced. The DP of cellulose dropped to 300 from an initial value of 2877 during the pretreatment

(Table 2). This is attributed to the cleavage of the β-1,4 glycosidic bonds in cellulose by the acid-catalyzed hydrolysis during the pretreatment [54]. It can therefore be inferred that the cellulose was destroyed and depolymerized, meaning that the number of nodes requiring cellulase hydrolysis was reduced.

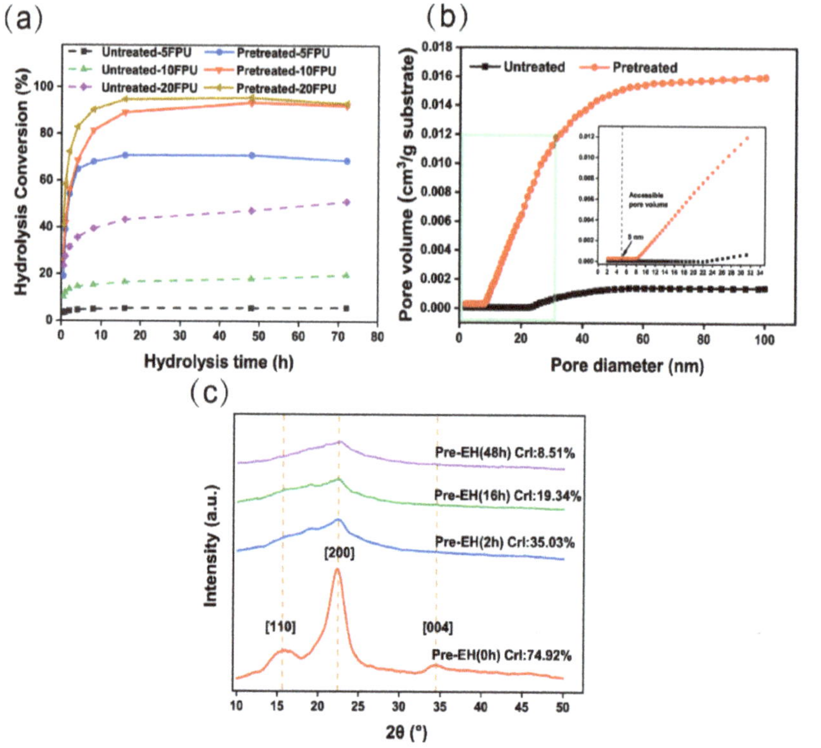

Figure 5. (**a**) Enzymolysis conversion of untreated and pretreated bagasse at 5, 10, and 20 FPU/g; (**b**) cumulative pore volume of untreated and pretreated bagasse; (**c**) X-ray diffraction (XRD) patterns of the pretreated bagasse enzymolysis for 0, 2, 16, and 48 h.

Table 2. Physicochemical properties of bagasse and pretreated bagasse.

	SSA (m^2/g)	DPn	Zeta Potential (mV)
Untreated bagasse	0.3633 ± 0.016	2876.95 ± 26.83	−9.01 ± 0.59
Pretreated bagasse	1.9068 ± 0.207	300.6 ± 7.48	−45.61 ± 1.13

As shown in the XRD results (Figure 5c), the CrI value decreased sharply to 35.03% from 74.92% [55], while the corresponding cellulose conversion to glucose was 55.32% during the 2 h enzymolysis. When the enzymolysis time was extended to 16 h, the percentage of cellulose to glucose increased to 80.02% and the CrI value decreased to 19.34%. The remaining crystalline cellulose was greatly enzymolyzed, the glucose yield reached to 98.6%, and the CrI value of the residue dropped to 8.51%. The cellulase therefore showed a strong preference for the digestion of crystalline cellulose over amorphous cellulose. This may be because amorphous cellulose is mixed with noncellulose components, which hinders the approach of cellulase.

As shown in the XPS results (Table S1), the O/C ratio of the residual from the pretreated bagasse enzymolysis at 20 FPU/g for 48 h decreased from 0.62 to 0.31. In addition, the content of C1 increased (from 31.75% to 58.79%), and the content of C2 decreased (from 53.49% to 27.17%) (Figure S1). As seen from the FTIR results (Figure 1), the characteristic absorption peaks [56,57] at 895 cm^{-1} (the glycosidic bond of cellulose), 2892 cm^{-1} (C–H tensile vibration of methyl and methylene), 1160 cm^{-1} (C–O–C asymmetric stretching of cellulose), and 1066 cm^{-1} (C–O, C–C stretching vibration) of cellulose were weakened. The characteristic absorption peaks [58,59] at 823 cm^{-1} (C–H bending vibration of guaiacyl), 1273 cm^{-1} (C-O stretching vibration of guaiacyl), and 1637 cm^{-1} (C=O conjugated stretching) of lignin were significantly enhanced in the FTIR of the residue (Figure 1). This is attributed to the cellulose being converted to glucose (98.6%) by cellulase and being removed, while the lignin was retained in the enzymolysis residue.

3.3. Enzymolysis Kinetic Behavior of Pretreated Bagasse

The linear correlation coefficient (R^2) was greater than 0.963, indicating that the Langmuir [39,60] equation fits the adsorption isotherm data well (Figure 6a). The affinity constant of pretreated bagasse was 19 L/g, which was three times that of untreated bagasse (6 L/g). This suggests that pretreated bagasse adsorption enzymes require a higher enzyme concentration at saturation than untreated bagasse. The adsorption capacity of cellulase onto the pretreated bagasse decreased to ~29 mg/g from ~40 mg/g (Figure 6a). The adsorption behavior of cellulase onto lignin is well-understood [61] and the hydrophobic lignin enhances the hydrophobic interaction, increasing the adsorption of enzymes onto lignin [62]. The hydrophobic interaction was weakened in pretreated bagasse due to the lower lignin content (Table 1). Cellulase was negatively charged in the buffer at pH 4.8 and demonstrated an electrostatic repulsion of cellulose, which had a negatively charged surface (−45.61 mV). Due to the weakening of the hydrophobic interaction and the enhancement of the electrostatic interaction of pretreated cellulose and cellulase, the adsorption capacity of cellulase onto cellulose decreased, although the SSA of the pretreated bagasse increased (Table 2).

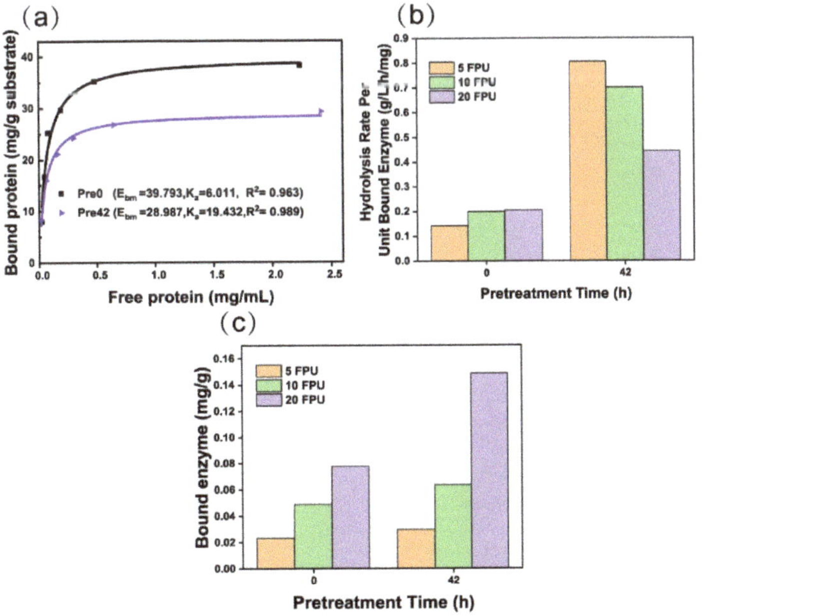

Figure 6. (**a**) Langmuir adsorption isotherms and parameters of cellulase onto pretreated bagasse at 4 °C, (**b**) initial hydrolysis efficiency, and (**c**) bound enzyme in pretreated bagasse with loading of 5, 10, and 20 FPU/g cellulase for 0.5 h.

The unit-bound enzyme efficiency was calculated based on the enzymolysis rate at 0.5 h (Figure 6b). The unit-bound enzyme efficiency of pretreated bagasse was significantly improved (Figure 4b), although the amount of bound enzyme onto bagasse remained unchanged (Figure 6c). This means that productive adsorption increases due to increased cellulose exposure to cellulase after the removal of noncellulose components (Table 1). This implies that the adequate removal of the noncellulose components is necessary for the enzyme to diffuse into or access the cellulose. The unit-bound enzyme efficiency (0.76 g/L/h/mg bound enzyme) of bagasse at low enzyme doses was significantly higher than that of the high enzyme doses of 10 and 20 FPU/g (0.68 and 0.43 g/L/h/mg bound enzyme). This suggests that pretreated bagasse is more conducive to enzyme efficiency at lower enzyme doses.

3.4. Evaluation of H_3PO_4 Recyclability

'Green' and sustainable production is widely recognized by human society [63]. If the final production quality is not disturbed, the reuse of reagents for pretreatment can significantly reduce the cost. The H_3PO_4 mixture was reused five times for subsequent pretreatment of bagasse with the appropriate addition of H_3PO_4, and the resulting glucose yield of pretreated bagasse via enzymolysis was similar to that of the fresh reagent (Figure 7a). Approximately 85% of the H_3PO_4 is recycled directly by filtration, ~10% H_3PO_4 is recovered from ethanol washing solution via rotary evaporation, and only ~5% H_3PO_4 needs to be replenished for reuse (Figure 7b). Residual H_3PO_4 (~5%) in pretreated bagasse almost obviated the need for acid to adjust the pH to 4.8 to meet the requirements of the enzymatic hydrolysis process. Further research will be conducted to recover acid-soluble lignin from the H_3PO_4 mixture to further improve the reuse potential of H_3PO_4.

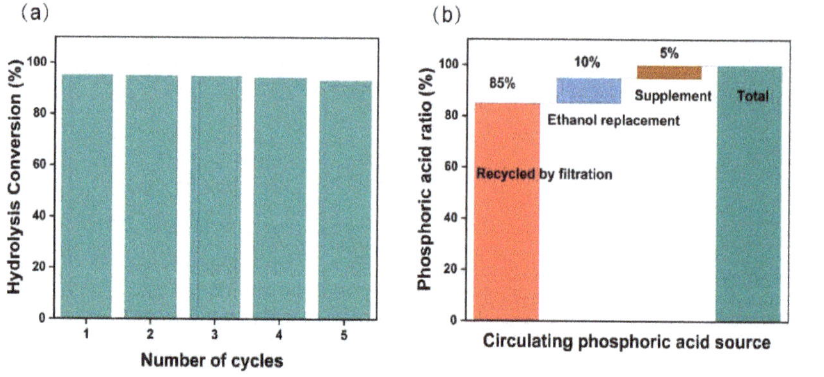

Figure 7. (**a**) Enzymatic hydrolysis conversion at 10 FPU/g for 48 h of bagasse pretreated with five cycles of H_3PO_4; (**b**) source of recycled H_3PO_4.

In addition, phosphorus, in the form of phosphate, is an important nutrient for living things [64]. The treated filtrate rich in H_3PO_4 has the potential to be converted to phosphorus-rich fertilizer by reacting it with calcium hydroxide or ammonia water [65]. These examples indicate biomass pretreatment can be affected with milder environmental consequences.

4. Conclusions

Obtaining green energy and materials from renewable biomass is an indispensable pathway for human society to deal with the energy crisis and environmental issues. To overcome the biological barriers to biomass, a strategy of intergranular swelling of cellulose combined with hydrolysis and oxidation was proposed and demonstrated, which was used for enhancing the release of fermentable sugars by enzymolysis. Due to the fact that cellulose swelling, hemicellulose hydrolysis, and lignin oxidation degradation occur simultaneously during the pretreatment, they cooperate and promote each other. Bagasse was converted into cellulose with a nanoscale size, low DP, high void fraction, and high SSA. The cellulose in pretreated bagasse was sufficiently exposed to cellulase, affording a high glucose yield (98.60%), posing a competitive pretreatment method for enzymatic hydrolysis of biomass. This nanofiber network structure of cellulose provides the possibility for the combined production of fermentable sugars and nanocellulose, which will greatly improve the efficiency of biomass refining and inspire the development of novel cellulose-based materials. In addition, this study found that cellulase preferred crystalline cellulose to amorphous cellulose, which provided new evidence for further understanding the enzymatic hydrolysis mechanism of cellulose.

Supplementary Materials: The following supporting information can be downloaded at: https://www.mdpi.com/article/10.3390/polym14173587/s1, Figure S1: X-ray photoelectron spectrometer (XPS) spectrum of residue from enzymolyzed pretreated bagasse for 48 h, Table S1: oxygen/carbon (O/C) ratio of the X-ray photoelectron spectrometer (XPS) spectrum.

Author Contributions: X.L. conceived and designed the experiments; F.B. and T.D. analyzed the data and wrote the original paper; F.B. and T.D. performed the experiments; W.C., C.C. and Z.Z. edited the manuscript. All authors have read and agreed to the published version of the manuscript.

Funding: This research and the APC were funded by Innovation-driven Development Project of Guangxi Province, grant number AA18118024-1.

Institutional Review Board Statement: Not applicable.

Informed Consent Statement: Not applicable.

Data Availability Statement: The data presented in this study are available on request from the corresponding author.

Acknowledgments: The authors are grateful for the financial support from the Guangxi Key Laboratory of Clean Pulp and Papermaking, and Pollution Control.

Conflicts of Interest: The authors declare no conflict of interest.

References

1. Srivastava, R.K.; Shetti, N.P.; Reddy, K.R.; Kwon, E.E.; Nadagouda, M.N.; Aminabhavi, T.M. Biomass utilization and production of biofuels from carbon neutral materials. *Environ. Pollut.* **2021**, *276*, 116731. [CrossRef] [PubMed]
2. Cheng, B.; Huang, B.; Zhang, R.; Chen, Y.; Jiang, S.; Lu, Y.; Zhang, X.-S.; Jiang, H.; Yu, H. Bio-coal: A renewable and massively producible fuel from lignocellulosic biomass. *Sci. Adv.* **2020**, *6*, eaay0748. [CrossRef] [PubMed]
3. Babin, A.; Vaneeckhaute, C.; Iliuta, M.C. Potential and challenges of bioenergy with carbon capture and storage as a carbon-negative energy source: A review. *Biomass Bioenergy* **2021**, *146*, 105968. [CrossRef]
4. Cherubini, F. The biorefinery concept: Using biomass instead of oil for producing energy and chemicals. *Energy Convers. Manag.* **2010**, *51*, 1412–1421. [CrossRef]
5. Kim, H.; Lee, S.; Lee, B.; Park, J.; Lim, H.; Won, W. Improving revenue from lignocellulosic biofuels: An integrated strategy for coproducing liquid transportation fuels and high value-added chemicals. *Fuel* **2021**, *287*, 119369. [CrossRef]
6. Pihlajaniemi, V.; Kallioinen, A.; Sipponen, M.H.; Nyyssölä, A. Modeling and optimization of polyethylene glycol (PEG) addition for cost-efficient enzymatic hydrolysis of lignocellulose. *Biochem. Eng. J.* **2021**, *167*, 107894. [CrossRef]
7. Meng, D.; Wei, X.; Zhang, Y.-H.P.J.; Zhu, Z.; You, C.; Ma, Y. Stoichiometric Conversion of Cellulosic Biomass by in Vitro Synthetic Enzymatic Biosystems for Biomanufacturing. *ACS Catal.* **2018**, *8*, 9550–9559. [CrossRef]
8. Tang, C.; Shan, J.; Chen, Y.; Zhong, L.; Shen, T.; Zhu, C.; Ying, H. Organic amine catalytic organosolv pretreatment of corn stover for enzymatic saccharification and high-quality lignin. *Bioresour. Technol.* **2017**, *232*, 222–228. [CrossRef]
9. Singh, J.K.; Vyas, P.; Dubey, A.; Upadhyaya, C.P.; Kothari, R.; Tyagi, V.V.; Kumar, A. Assessment of different pretreatment technologies for efficient bioconversion of lignocellulose to ethanol. *Front. Biosci. Sch.* **2018**, *10*, 350–371.
10. Reiniati, I.; Hrymak, A.N.; Margaritis, A. Recent developments in the production and applications of bacterial cellulose fibers and nanocrystals. *Crit. Rev. Biotechnol.* **2017**, *37*, 510–524. [CrossRef]
11. Buntić, A.V.; Pavlović, M.D.; Antonović, D.G.; Šiler-Marinković, S.S.; Dimitrijević-Branković, S.I. Utilization of spent coffee grounds for isolation and stabilization of Paenibacillus chitinolyticus CKS1 cellulase by immobilization. *Heliyon* **2016**, *2*, e00146. [PubMed]

12. Aui, A.; Wang, Y.; Mba-Wright, M. Evaluating the economic feasibility of cellulosic ethanol: A meta-analysis of techno-economic analysis studies. *Renew. Sustain. Energy Rev.* **2021**, *145*, 111098.
13. Zugenmaier, P. *Crystalline Cellulose and Derivatives: Characterization and Structures*; Springer: Berlin/Heidelberg, Germany, 2008.
14. Rosén, T.; He, H.; Wang, R.; Zhan, C.; Chodankar, S.; Fall, A.; Aulin, C.; Larsson, P.T.; Lindström, T.; Hsiao, B.S. Cross-sections of nanocellulose from wood analyzed by quantized polydispersity of elementary microfibrils. *ACS Nano* **2020**, *14*, 16743–16754. [CrossRef] [PubMed]
15. Zhang, S.; Chi, M.; Mo, J.; Liu, T.; Liu, Y.; Fu, Q.; Wang, J.; Luo, B.; Qin, Y.; Wang, S.; et al. Bioinspired asymmetric amphiphilic surface for triboelectric enhanced efficient water harvesting. *Nat. Commun.* **2022**, *13*, 4168. [CrossRef] [PubMed]
16. Song, B.; Zhao, S.; Shen, W.; Collings, C.; Ding, S.-Y. Direct measurement of plant cellulose microfibril and bundles in native cell walls. *Front. Plant Sci.* **2020**, *11*, 479. [CrossRef]
17. Zhao, H.-K.; Wei, X.-Y.; Xie, Y.-M.; Feng, Q.-H. Preparation of nanocellulose and lignin-carbohydrate complex composite biological carriers and culture of heart coronary artery endothelial cells. *Int. J. Biol. Macromol.* **2019**, *137*, 1161–1168. [CrossRef]
18. Chen, W.; Dong, T.; Bai, F.; Wang, J.; Li, X. Lignin–carbohydrate complexes, their fractionation, and application to healthcare materials: A review. *Int. J. Biol. Macromol.* **2022**, *203*, 29–39.
19. Várnai, A.; Siika-aho, M.; Viikari, L. Restriction of the enzymatic hydrolysis of steam-pretreated spruce by lignin and hemicellulose. *Enzym. Microb. Technol.* **2010**, *46*, 185–193.
20. Lin, L.; Yan, R.; Liu, Y.; Jiang, W. In-depth investigation of enzymatic hydrolysis of biomass wastes based on three major components: Cellulose, hemicellulose and lignin. *Bioresour. Technol.* **2010**, *101*, 8217–8223. [CrossRef]
21. Meenakshisundaram, S.; Fayeulle, A.; Leonard, E.; Ceballos, C.; Pauss, A. Fiber degradation and carbohydrate production by combined biological and chemical/physicochemical pretreatment methods of lignocellulosic biomass—A review. *Bioresour. Technol.* **2021**, *331*, 125053.
22. Zhang, J.; Zhou, H.; Liu, D.; Zhao, X. Chapter 2—Pretreatment of lignocellulosic biomass for efficient enzymatic saccharification of cellulose. In *Lignocellulosic Biomass to Liquid Biofuels*; Yousuf, A., Pirozzi, D., Sannino, F., Eds.; Academic Press: Cambridge, MA, USA, 2020; pp. 17–65.
23. Chu, Q.; Tong, W.; Wu, S.; Jin, Y.; Hu, J.; Song, K. Eco-friendly additives in acidic pretreatment to boost enzymatic saccharification of hardwood for sustainable biorefinery applications. *Green Chem.* **2021**, *23*, 4074–4086.
24. BenYahmed, N.; Jmel, M.A.; Smaali, I. Impact of Pretreatment Technology on Cellulosic Availability for Fuel Production. In *Substrate Analysis for Effective Biofuels Production*; Springer: Berlin/Heidelberg, Germany, 2020; pp. 217–242.
25. Satlewal, A.; Agrawal, R.; Bhagia, S.; Sangoro, J.; Ragauskas, A.J. Natural deep eutectic solvents for lignocellulosic biomass pretreatment: Recent developments, challenges and novel opportunities. *Biotechnol. Adv.* **2018**, *36*, 2032–2050. [PubMed]
26. Walseth, C.S. *Enzymatic Hydrolysis of Cellulose*; Georgia Institute of Technology: Atlanta, GA, USA, 1948.
27. Wan, X.; Yao, F.; Tian, D.; Shen, F.; Hu, J.; Zeng, Y.; Yang, G.; Zhang, Y.; Deng, S. Pretreatment of wheat straw with phosphoric acid and hydrogen peroxide to simultaneously facilitate cellulose digestibility and modify lignin as adsorbents. *Biomolecules* **2019**, *9*, 844. [CrossRef] [PubMed]
28. Bai, F.; Dong, T.; Chen, W.; Wang, J.; Li, X. Nanocellulose hybrid lignin complex reinforces cellulose to form a strong, water-stable lignin-cellulose composite usable as a plastic replacement. *Nanomaterials* **2021**, *11*, 3426.
29. Qiu, J.; Ma, L.; Shen, F.; Yang, G.; Zhang, Y.; Deng, S.; Zhang, J.; Zeng, Y.; Hu, Y. Pretreating wheat straw by phosphoric acid plus hydrogen peroxide for enzymatic saccharification and ethanol production at high solid loading. *Bioresour. Technol.* **2017**, *238*, 174–181.
30. Stefaniak, T.R.; Dahlberg, J.A.; Bean, B.W.; Dighe, N.; Wolfrum, E.J.; Rooney, W.L. Variation in Biomass Composition Components among Forage, Biomass, Sorghum-Sudangrass, and Sweet Sorghum Types. *Crop Sci.* **2012**, *52*, 1949–1954.
31. Adney, B.; Baker, J. Measurement of Cellulase Activities. 2008. Available online: https://www.researchgate.net/publication/306200684_Measurement_of_Cellulase_Activities (accessed on 23 August 2022).
32. Kruger, N.J. The Bradford method for protein quantitation. In *The Protein Protocols Handbook*; Springer: Berlin/Heidelberg, Germany, 2009; pp. 17–24.
33. Mazlita, Y. *Catalytic Synthesis of Nanocellulose from Oil Palm Empty Fruit Bunch Fibres, Mazlita Yahya*; University of Malaya: Kuala Lumpur, Malaysia, 2016.
34. ASTM International. *Standard Test Method for Intrinsic Viscosity of Cellulose*; ASTM: West Conshohocken, PA, USA, 2013.
35. Brunauer, S.; Emmett, P.H.; Teller, E. Adsorption of gases in multimolecular layers. *J. Am. Chem. Soc.* **1938**, *60*, 309–319.
36. Lu, M.; Li, J.; Han, L.; Xiao, W. An aggregated understanding of cellulase adsorption and hydrolysis for ball-milled cellulose. *Bioresour. Technol.* **2019**, *273*, 1–7.
37. Hu, Y.; Hu, F.; Gan, M.; Xie, Y.; Feng, Q. Facile one-step fabrication of all cellulose composites with unique optical performance from wood and bamboo pulp. *Carbohydr. Polym.* **2021**, *274*, 118630.
38. Treichel, H.; Fongaro, G.; Scapini, T.; Frumi Camargo, A.; Spitza Stefanski, F.; Venturin, B. *Utilising Biomass in Biotechnology*; Springer: Berlin/Heidelberg, Germany, 2020.
39. Bolster, C.H.; Hornberger, G.M. On the use of linearized langmuir rquations. *Soil Sci. Soc. Am. J.* **2007**, *71*, 1796–1806.
40. Yao, S.; Nie, S.; Yuan, Y.; Wang, S.; Qin, C. Efficient extraction of bagasse hemicelluloses and characterization of solid remainder. *Bioresour. Technol.* **2015**, *185*, 21–27. [PubMed]

41. Sluiter, A.; Hames, B.; Ruiz, R.; Scarlata, C.; Sluiter, J.; Templeton, D.; Crocker, D. *Determination of Structural Carbohydrates and Lignin in Biomass*; Laboratory Analytical Procedure 2010, (TP-510-42618); National Renewable Energy Laboratory: Golden, CO, USA, 2010.
42. Huang, Y.; Wang, L.; Chao, Y.; Nawawi, D.S.; Akiyama, T.; Yokoyama, T.; Matsumoto, Y. Analysis of Lignin Aromatic Structure in Wood Based on the IR Spectrum. *J. Wood Chem. Technol.* **2012**, *32*, 294–303.
43. Wang, J.; Wang, Q.; Wu, Y.; Bai, F.; Wang, H.; Si, S.; Lu, Y.; Li, X.; Wang, S. Preparation of cellulose nanofibers from bagasse by phosphoric acid and hydrogen peroxide enables fibrillation via a swelling, hydrolysis, and oxidation cooperative mechanism. *Nanomaterials* **2020**, *10*, 2227.
44. Chundawat, S.P.S.; Venkatesh, B.; Dale, B.E. Effect of particle size based separation of milled corn stover on AFEX pretreatment and enzymatic digestibility. *Biotechnol. Bioeng.* **2007**, *96*, 219–231. [PubMed]
45. Li, J.; Lu, M.; Guo, X.; Zhang, H.; Li, Y.; Han, L. Insights into the improvement of alkaline hydrogen peroxide (AHP) pretreatment on the enzymatic hydrolysis of corn stover: Chemical and microstructural analyses. *Bioresour. Technol.* **2018**, *265*, 1–7. [PubMed]
46. Martinsson, A.; Hasani, M.; Potthast, A.; Theliander, H. Modification of softwood kraft pulp fibres using hydrogen peroxide at acidic conditions. *Cellulose* **2020**, *27*, 7191–7202.
47. Zhang, Y.-H.P.; Cui, J.; Lynd, L.R.; Kuang, L.R. A transition from cellulose swelling to cellulose dissolution by o-phosphoric acid: Evidence from enzymatic hydrolysis and supramolecular structure. *Biomacromolecules* **2006**, *7*, 644–648.
48. Sun, J.; Sun, X.; Zhao, H.; Sun, R. Isolation and characterization of cellulose from sugarcane bagasse. *Polym. Degrad. Stab.* **2004**, *84*, 331–339.
49. Kolpak, F.; Blackwell, J. Determination of the structure of cellulose II. *Macromolecules* **1976**, *9*, 273–278. [CrossRef]
50. Xiao, K.; Zhou, W.; Geng, M.; Feng, W.; Wang, Y.; Xiao, N.; Zhu, D.; Zhu, F.; Liu, G. Comparative evaluation of enzymatic hydrolysis potential of eichhornia crassipes and sugarcane bagasse for fermentable sugar production. *BioResources* **2018**, *13*, 4897–4915.
51. Fu, H.; Mo, W.; Shen, X.; Li, B. Impact of centrifugation treatment on enzymatic hydrolysis of cellulose and xylan in poplar fibers with high lignin content. *Bioresour. Technol.* **2020**, *316*, 123866. [PubMed]
52. Grethlein, H.E. The effect of pore size distribution on the rate of enzymatic hydrolysis of cellulosic substrates. *Nat. Biotechnol.* **1985**, *3*, 155–160.
53. Zhao, X.; Wen, J.; Chen, H.; Liu, D. The fate of lignin during atmospheric acetic acid pretreatment of sugarcane bagasse and the impacts on cellulose enzymatic hydrolyzability for bioethanol production. *Renew. Energy* **2018**, *128*, 200–209.
54. Dussán, K.J.; Silva, D.; Moraes, E.; Arruda, P.V.; Felipe, M. Dilute-acid hydrolysis of cellulose to glucose from sugarcane bagasse. *Chem. Eng. Trans.* **2014**, *38*, 433–438.
55. Guo, J.-M.; Wang, Y.-T.; Cheng, J.-R.; Zhu, M.-J. Enhancing enzymatic hydrolysis and fermentation efficiency of rice straw by pretreatment of sodium perborate. *Biomass Convers. Biorefin.* **2020**, *12*, 361–370.
56. Wiley, J.H. *Raman Spectra of Celluloses*; Georgia Institute of Technology: Atlanta, GA, USA, 1986.
57. Kumar, A.; Negi, Y.S.; Choudhary, V.; Bhardwaj, N.K. Characterization of cellulose nanocrystals produced by acid-hydrolysis from sugarcane bagasse as agro-waste. *J. Mater. Phys. Chem.* **2014**, *2*, 1–8.
58. Shukry, N.; Fadel, S.; Agblevor, F.; El-Kalyoubi, S. Some physical properties of acetosolv lignins from bagasse. *J. Appl. Polym. Sci.* **2008**, *109*, 434–444.
59. Mancera, A.; Fierro, V.; Pizzi, A.; Dumarçay, S.; Gérardin, P.; Velásquez, J.; Quintana, G.; Celzard, A. Physicochemical characterisation of sugar cane bagasse lignin oxidized by hydrogen peroxide. *Polym. Degrad. Stab.* **2010**, *95*, 470–476.
60. Kinniburgh, D.G. General purpose adsorption isotherms. *Environ. Sci. Technol.* **1986**, *20*, 895–904.
61. Lu, X.; Zheng, X.; Li, X.; Zhao, J. Adsorption and mechanism of cellulase enzymes onto lignin isolated from corn stover pretreated with liquid hot water. *Biotechnol. Biofuels* **2016**, *9*, 118.
62. Nakagame, S.; Chandra, R.P.; Kadla, J.F.; Saddler, J.N. The isolation, characterization and effect of lignin isolated from steam pretreated Douglas-fir on the enzymatic hydrolysis of cellulose. *Bioresour. Technol.* **2011**, *102*, 4507–4517. [PubMed]
63. Jiang, J.; Xiao, F.; He, W.-M.; Wang, L. The application of clean production in organic synthesis. *Chin. Chem. Lett.* **2021**, *32*, 1637–1644.
64. Anand, K.; Kumari, B.; Mallick, M. Phosphate solubilizing microbes: An effective and alternative approach as biofertilizers. *Int. J. Pharm. Sci.* **2016**, *8*, 37–40.
65. Gilmour, R. *Phosphoric Acid: Purification, Uses, Technology, and Economics*; CRC Press: Boca Raton, FL, USA, 2013.

polymers MDPI

Article

Effects of Fluorine-Based Modification on Triboelectric Properties of Cellulose

Qiuxiao Zhu †, Tingting Wang †, Xiaoping Sun, Yuhe Wei, Sheng Zhang, Xuchong Wang and Lianxin Luo *

Guangxi Key Laboratory of Clean Pulp and Papermaking and Pollution Control, School of Light Industrial and Food Engineering, Guangxi University, Nanning 530004, China
* Correspondence: luolianxin@gxu.edu.cn
† These authors contribute equally to this work.

Abstract: The hydroxyl groups on the cellulose macromolecular chain cause the cellulose surface to have strong reactivity. In this study, 1H, 1H, 2H, 2H-perfluorodecyltriethoxysilane (PDOTES) was used to modify cellulose to improve its triboelectric properties, and a triboelectric nanogenerator (TENG) was assembled. The introduction of fluorine groups reduced the surface potential of cellulose and turned it into a negative phase, which enhanced the ability to capture electrons. The electrical properties increased by 30% compared with unmodified cellulose. According to the principles of TENGs, a self-powered human-wearable device was designed using PDOTES-paper, which could detect movements of the human body, such as walking and running, and facilitated a practical method for the preparation of efficient wearable sensors.

Keywords: cellulose; triboelectric nanogenerator; paper; contact electrification; wearable devices

Citation: Zhu, Q.; Wang, T.; Sun, X.; Wei, Y.; Zhang, S.; Wang, X.; Luo, L. Effects of Fluorine-Based Modification on Triboelectric Properties of Cellulose. *Polymers* **2022**, *14*, 3536. https://doi.org/10.3390/polym14173536

Academic Editor: Agnieszka Tercjak

Received: 21 July 2022
Accepted: 20 August 2022
Published: 28 August 2022

1. Introduction

With the rapid growth of global energy demand, the overexploitation of non-renewable fossil energy resources such as oil, coal, and natural gas has led to a serious energy crisis and environmental and ecological problems [1]. Therefore, how to obtain sustainable and environmentally friendly energy from the surrounding environment has attracted much attention. In 2012, Wang et al. proposed the concept of a triboelectric nanogenerator (TENG) [2,3], which is a technology based on the coupling effect of friction-generated electricity and electrostatic induction, with the advantages of low cost, small size, portability, and applicability to a variety of scenarios [4,5]. However, the relatively low output power density is still a limitation of the friction nanogenerator and is one of the main obstacles to the wide application of TENGs. Therefore, researchers have also made many attempts to enhance the electrical properties, such as the selection of suitable frictional electric materials, the introduction of functional groups [6,7], the doping of high dielectric materials [8], and the design of micropatterns [9,10]. The selection of suitable frictional materials is the most effective way to fundamentally improve the performance of TENGs.

Cellulose is the most abundant natural polymer compound on Earth, offering the advantages of low cost, good processability, good mechanical flexibility, and the ability to exhibit a unique combination of chemical, structural, dielectric, and optical properties [11–13]. These advantages and properties can make cellulose-based functional materials attractive as TENG substrates or components, a highly promising green material in the field of constituent materials for electronic devices. The polarity of frictional electrical materials is determined by the chemical properties of the material itself and is closely related to the functional groups present on the surface of the material [14]. The surface of cellulose is rich in hydroxyl groups, and by using hydroxyl groups in chemical reactions, some functional groups, chain segments, or molecules can be introduced to the cellulose molecules to synthesize cellulose derivatives. In recent years, many researchers have used chemical methods to modify various groups that promote electron transfer to cellulose to improve the frictional electrical properties. Yao et al. [15] reported a method to modify the frictional polarity of strong CNFs by chemically introducing nitro and methyl groups on the cellulose surface to enhance the frictional electrical output of TENGs, and after methylation and nitroxylation, the charge density of CNFs, respectively. Roy et al. [16] found that allicin from garlic juice enhanced the frictional electrical properties of cellulose nanofibers (CNFs) with strong efficacy, and the grafting of "thiol-ene" onto CNFs increased the TENG of pristine cellulose by about 6.5 times. Nie et al. [17] found that aminosilane modification of CNF films resulted in a significant enhancement of the positive charge on the CNF surface, allowing for excellent frictional charge density and hydrophobicity on CNFs. More studies have been conducted to modify cellulose in the positive direction, and fewer studies have been conducted to enhance the frictional negative polarity of cellulose. Fluorine is the most electronegative element with a small radius and a high ionization energy, and the introduction of the fluorine group is a reasonable choice to enhance the friction-negative polarity of cellulose.

In this study, 1H, 1H, 2H, 2H-perfluorodecyltriethoxysilane (PDOTES) was used to introduce fluorine groups with high electronegativity into cellulose, which improved the negative friction polarity of cellulose. FTIR, XRD, and XPS were used to characterize the modification of the cellulose, while SEM and AFM were used to examine the surface morphology and surface potential distribution of the modified cellulose. A TENG was assembled with modified cellulose as anode material and nylon (PA) film, and the triboelectric properties of this TENG were investigated. In addition, a TENG-based human wearable device is designed on this basis. At the same time, the prepared TENG can charge and power external devices, showing great potential in self-powered sensing systems and providing a factual basis for further expanding the application of friction materials.

2. Materials and Methods

2.1. Raw Materials and Reagents

Fast growing Eucalyptus cellulose fibers (Nanning, China). 1H, 1H, 2H, 2H-perfluorodecyltriethoxysilane (PDOTES) was purchased from Aladdin Biochemical Technology Co., Ltd. (Shanghai, China). A polymethyl methacrylate (PMMA) plate was purchased from Deyao Building Materials Co., Ltd. (Guangzhou, China), a conductive tape (copper) was purchased from Sitejie Office Store (Shenzhen, China), and a nylon (PA) film was purchased from Aofa Plastics Co., Ltd. (Suzhou, China).

2.2. Fluoro Modification of Cellulose

One gram of PDOTES was added to a 250 mL conical flask, 100 mL of ethanol/water solution at a volume ratio of 8:2 was added, and the mixture was stirred at room temperature for 2 h to obtain the hydrolysate of PDOTES. One gram of cellulose was added to the solution, and the mixture was stirred vigorously at room temperature for 5 min followed by sonication at 800 W for 10 min. After that, the solution was transferred to a 250 mL hydrothermal reactor and stirred in a constant temperature water bath at 80 °C for 6 h. After the reaction, the mixture was filtered and washed repeatedly with acetone and distilled water. The obtained product was PDOTES-cellulose.

2.3. Preparation of Paper and Fluorinated Paper (PDOTES-Paper)

The vacuum filtration method was employed to fabricate paper and PDOTES-paper films with a thickness of 50 μm. The paper-making process is shown in Figure 1. First, 0.6 g of cellulose or PDOTES-cellulose was placed in a 500 mL conical flask; then, 300 mL of distilled water was added. Next, the cellulose suspension was sonicated at room temperature for 10 min and then stirred for 2 h to disperse the cellulose evenly. A microporous filter membrane was placed at the bottom of a G5 sand core funnel, and the cellulose suspension was poured into it, forming a sheet of cellulose paper by vacuum suction filtration. After the sheet was formed, it was removed and placed in an automatic dryer. The temperature was set to 75 °C, and the drying time was set to 15 min. In this way, the paper and PDOTES-paper required for the experiments were obtained.

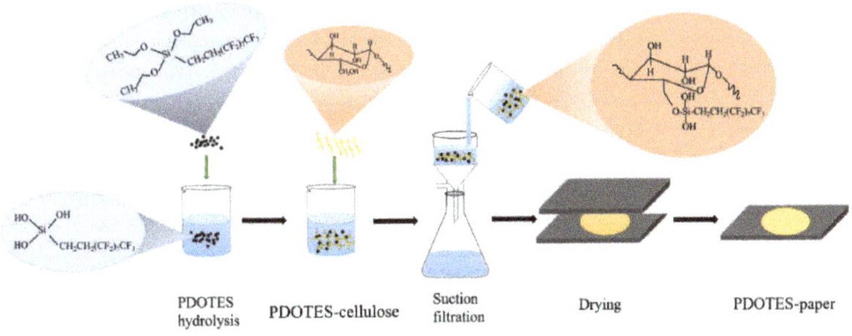

Figure 1. Preparation of PDOTES-paper.

2.4. Structure of TENG and Wearable Device

The structures of the TENG and the wearable device are shown in Figure 2. An acrylic plate was cut into a 7 × 7 cm square as the TENG support material, and conductive double-sided tape, paper, PDOTES-paper, and PA film were cut into 4 × 4 cm squares. The conductive double-sided adhesive was attached to the middle of the acrylic plate, and then the paper, PDOTES-paper, and PA film were adhered to the conductive double-sided adhesive. The structure of the human-wearable device was similar to that of the TENG. It used xerographic paper as a soft support material, PDOTES-paper as a negative friction electrode material, and PA film as a positive electrode friction material, which were cut into 4 × 4 cm squares and fixed onto electrostatic copy paper with conductive double-sided tape. The two electrodes were connected with a sponge, and a layer of polyethylene film was coated on the outer layer of the wearable equipment to prevent erosion.

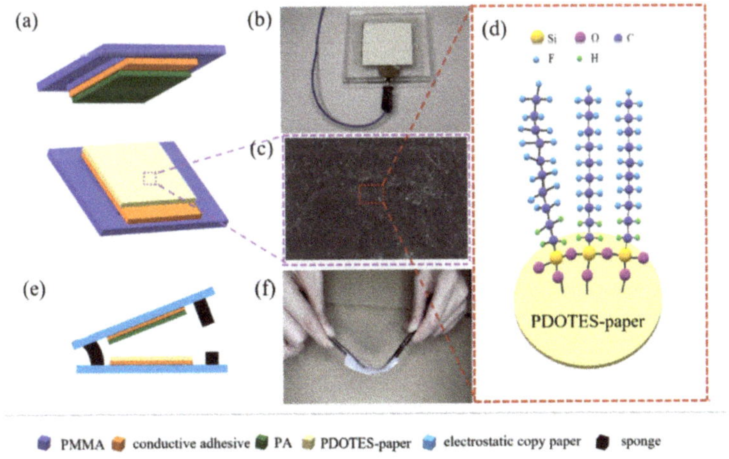

Figure 2. (**a**) Structural diagram of TENG; (**b**) TENG physical drawing; (**c**) SEM image of PDOTES-paper; (**d**) schematic diagram of TENG electrode; (**e**) diagram of wearable device structure; (**f**) physical drawing of wearable device.

2.5. Characterization

After spraying gold on cellulose, scanning electron microscopy (SEM, Hitachi su8020, Tokyo, Japan) was used to observe the surface morphology and the elemental content, and atomic force microscopy (AFM, Dimension Edge, Bruker Co., Ltd., Berlin, Germany) was used to observe the surface roughness. KPFM images were taken by atomic force microscopy (AFM) at room temperature under dark conditions using an AFM system (XE-100, Park Systems, Bruker Co., Ltd., Berlin, Germany). The bias applied to the KPFM tip was 2.0 V for all samples. Changes in the functional groups of cellulose before and after fluoro modification were determined by Fourier transform infrared spectroscopy (FTIR, Vertes 70, Berlin, Germany) in the range 4000–400 cm^{-1}. Changes in the surface elemental content of cellulose before and after fluorine modification were analyzed by X-ray photoelectron spectroscopy (XPS, Kratos Axis Ultra DLD, London, UK), and the change in crystallinity was analyzed by X-ray diffraction (XRD, Smartlab 3 kW, Tokyo, Japan). The crystallinity index (*CrI*) was calculated according to the Segal Formula (1):

$$CrI(\%) = \frac{(I_{002} - I_{am})}{I_{002}} \times 100\% \quad (1)$$

where I_{am} is the intensity of the diffraction peak at 2θ = 18°, and I_{002} is the intensity of the diffraction peak at 2θ = 22°.

When testing the TENG, the periodic movement of the electrode was controlled by a tubular linear motor (Linmot H10–70 × 240/210, CA, USA) and a vibration exciter (JZK-10, Shenzhen, China), in which the tubular linear electrode controlled the motion frequency and the vibration exciter controlled the working pressure. The electrical signal generated by the TENG was collected using an electrometer (Keithley 6514, CA, USA) and output was sent to a Ni USB-6259, CA, USA, acquisition card.

3. Results

3.1. Characterization of Cellulose

3.1.1. FTIR Analysis

FTIR can quantitatively and semi-quantitatively analyze samples. The increase in and loss of functional groups of cellulose before and after amino modification can be determined by FTIR reaction, and changes in peak strength can also indicate changes in functional group content [18]. As shown in Figure 3a, PDOTES-cellulose retains its absorption peak for cellulose. After the cellulose was treated with PDOTES, a new functional group absorption peak appeared in the infrared spectrum. As shown in Figure 3b, the absorption peak at 1245 cm^{-1} originates from the stretching vibration of -CF_2 [19]. The absorption peaks at 1100–1000 cm^{-1} are attributed to Si-O-Si and Si-O-C [20]. These results preliminarily confirm that cellulose was successfully modified by PDOTES.

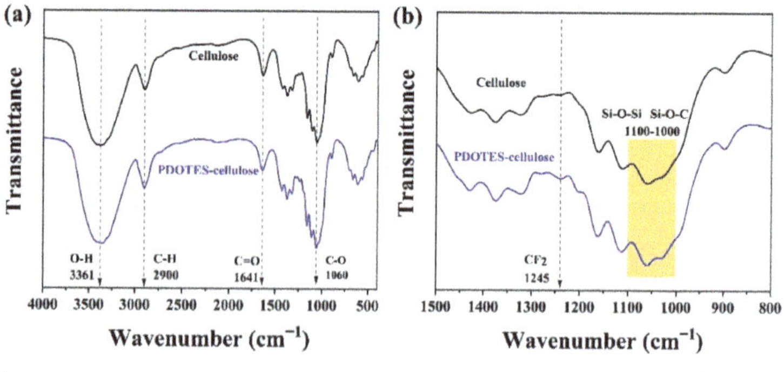

Figure 3. FTIR spectra of cellulose before and after modification: (**a**) 4000–500 cm^{-1}; (**b**) 1500–800 cm^{-1}.

3.1.2. XRD Analysis

XRD can be used to determine the structure and crystal morphology of cellulose and to measure changes in cellulose crystallinity [18]. As shown in Figure 4, the peaks at 2θ = 16° and 22° are diffraction absorption peaks of cellulose *I* [21], and indicate that fluorine-based modification did not destroy the crystalline region of cellulose, i.e., the cellulose was still a typical cellulose *I* after modification. It can be seen from Figure 4 that the diffraction peak intensity at I_{002} is significantly weakened after cellulose modification. After modification, the calculated crystallinity of cellulose decreased from 84.36% to 70.14%, owing to the introduction of amorphous PDOTES on the cellulose surface.

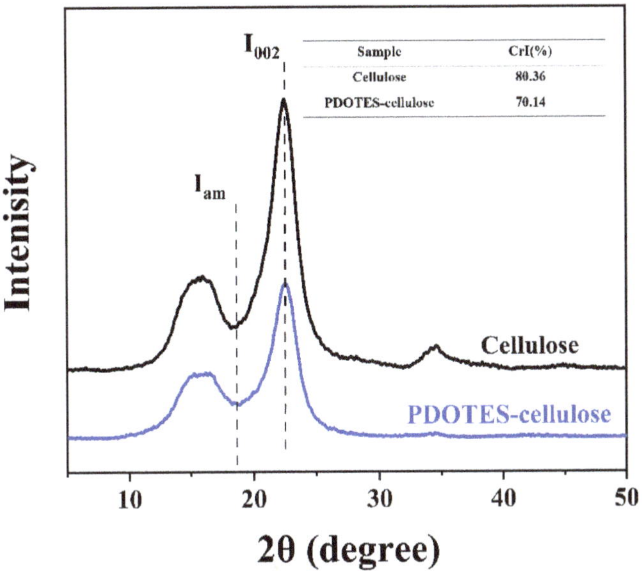

Figure 4. XRD spectrum of cellulose before and after modification.

3.1.3. XPS Analysis

XPS can be used for quantitative and semi-quantitative analysis of the surface elements of samples [22]. As shown in Figure 5a,b, peaks for both cellulose and PDOTES-cellulose appear at 532.4 and 286.5 eV [23], corresponding to O1s and C1s, respectively, and additional peaks for PDOTES-cellulose appear at 688.2, 153.4, and 102.2 eV, corresponding to F1s, Si2s, and Si2p, respectively. The proportions of elements on the surface of the sample analyzed by XPS are listed in Table 1. Compared with cellulose, F1s is present on the surface of PDOTES-cellulose, proving that the fluoro modification of cellulose was successful. From the XPS data, the F element is 37.64%, and the substitution degree of cellulose is calculated to be 0.77 [24]. At the same time, the C1s content of PDOTES-cellulose increased, while the O1s content and O/C decreased. This is because PDOTES was introduced into cellulose, and the amount of C1s on the cellulose surface increased significantly.

Table 1. Cellulose surface element ratios.

Samples	C1s (%)	O1s (%)	F1s (%)	O/C (%)
Cellulose	55.72	44.28	-	79.47
PDOTES-cellulose	43.77	18.59	37.64	42.47

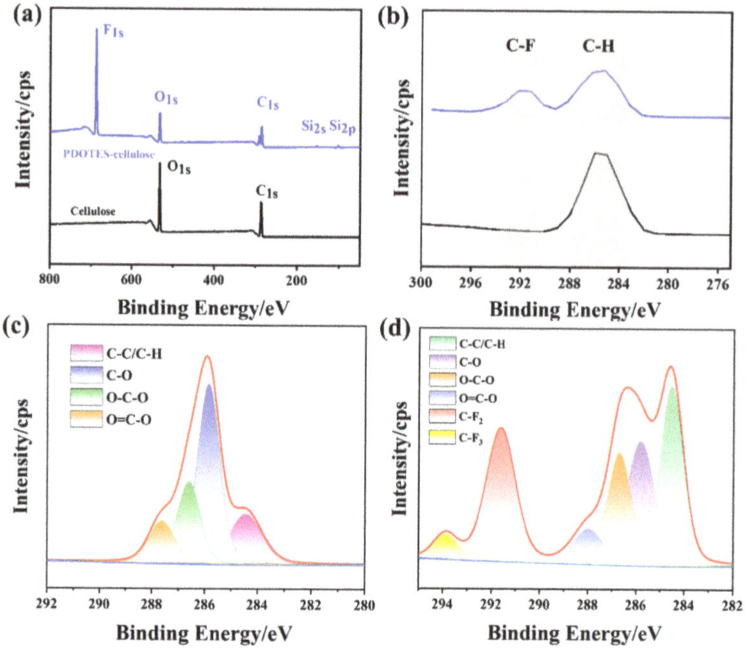

Figure 5. (**a**,**b**) Full XPS spectrum before and after cellulose modification; (**c**) C1s peaks on the surface of cellulose; (**d**) C1s peak deconvolution on PDOTES-cellulose surface.

3.2. Surface Morphology and Elemental Distribution

Figure 6a–d show SEM images of cellulose at different magnifications. Before modification, the surface of cellulose is smooth, the structure is dense, and there are no small particles on the surface. After modification, the surface of cellulose becomes rougher, holes and small particles appear, and fracturing of the cellulose can be observed. This shows that after modification, the crystalline area of cellulose was destroyed and glucoside bonds began to break, which is consistent with outcomes from XRD. Figure 6e–g depicts the surface element distribution of PDOTES-cellulose, where Si and F elements are present.

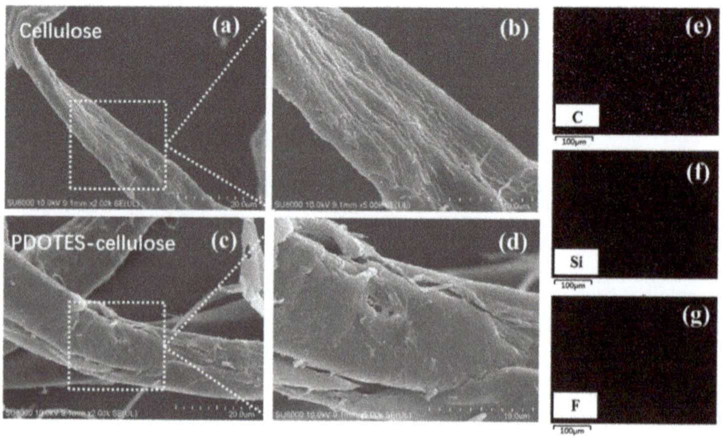

Figure 6. SEM images at different magnifications: (**a**,**b**) Cellulose; (**c**,**d**) PDOTES-cellulose; (**e**–**g**) surface elemental distribution of PDOTES-cellulose.

3.3. Surface Roughness and Surface Potential Distribution

Figure 7 shows that the surface roughness Ra of pure cellulose paper is 127 nm and that of PDOTES-paper is 195 nm. The surface roughness increases after modification, which corresponds to the situation of voids on the fiber surface after modification as observed by SEM. In order to determine the impact of chemical alteration on the material's surface potential from a microscopic perspective and to investigate its triboelectric properties, the surface potential of cellulose was measured in the KPFM mode of atomic force microscopy (AFM). Figure 7c,f correspond to the surface potential of cellulose paper before and after modification. The surface potential of cellulose paper is 78.5 mV, and the surface potential of PDOTES-paper is −44.37 mV. The surface potential after fluorosilane modification not only decreased but also appeared in the opposite direction, which is due to the strong electronegativity of F, which made it have a strong electron-withdrawing ability and improved the negative frictional polarity of cellulose.

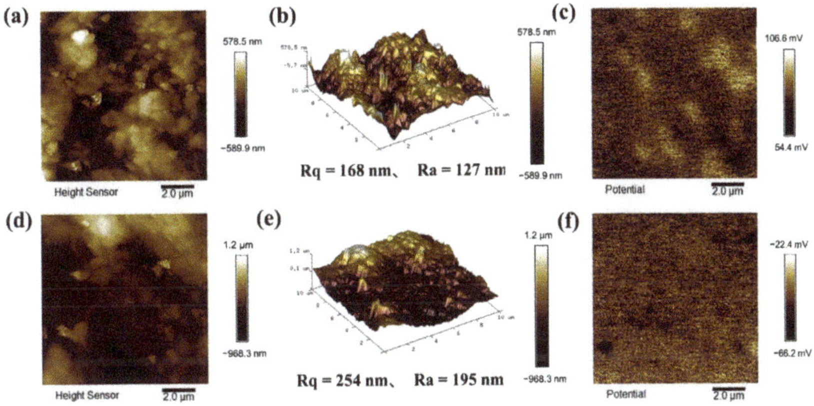

Figure 7. AFM images before and after modification: (**a**) 2D surface topography of paper; (**b**) 3D surface topography of paper; (**c**) surface potential of paper; (**d**) 2D surface topography of PDOTES-paper; (**e**) 3D surface topography of PDOTES-paper; (**f**) surface potential of PDOTES-paper.

3.4. Working Principle of Triboelectric Nanogenerator

Figure 8a–d show the working principle of a vertical contact separated TENG. In the initial state, as shown in Figure 8a, friction layers are separated from each other, and no charge is generated. As shown in Figure 8b, when the friction layers are pressed together, due to the different electron gain and loss abilities of the two friction layers, electrons from PA film are transferred to the surface of PDOTES-paper, so that the surface of PA film is positively charged, and the surface of PDOTES-paper is negatively charged. As shown in Figure 8c, when the pressure decreases and the electrodes are separated again, electrons flow from PDOTES-paper to PA film owing to the potential difference between the two friction layers. As shown in Figure 8d, when the friction layers stop separating, the circuit tends to balance, and no electrons move. As shown in Figure 8e, when the electrodes start to approach each other again, the potential of PDOTES-paper is higher than that of PA film, and the electrons move from PDOTES-paper to PA film. When the TENG electrodes carry out the above movements continuously, a stable alternating current is generated in the circuit [25].

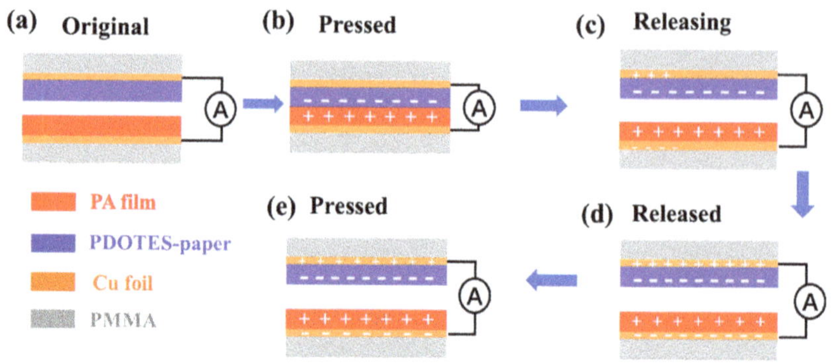

Figure 8. Structure and working principle of TENG. (**a**) PDOTES-paper-based TENG original structure; (**b**)TENG friction layers are in contact with each other by external force; (**c**) Remove the external force to separate the TENG friction layer; (**d**) The TENG friction layer is separated to the maximum distance; (**e**) Cyclic pair reapplying external forces.

3.5. Output Performance of Triboelectric Nanogenerator

Figure 9 shows the effect of fluoro modification on contact electrification of the cellulose. Using paper-based and PDOTES-paper-based TENGs as the experimental objects, a linear motor was used to move the two TENG friction layers through periodic contact separation movements. The maximum distance between the electrodes was 7 mm, and the acceleration was 0.5 m/s^2. As shown in Figure 9, after fluorine-based modification of cellulose, the open circuit voltage of TENG increased from 9.3 to 12.6 V (a 35.48% increase), the short circuit current increased from 79.3 to 108.6 nA (a 36.71% increase), and the charge density increased from 109.5 to 141.1 pC·cm^{-2} (a 28.86% increase). This is because the fluorine group has a strong electronic function, which improves the negative polarity of cellulose and the output performance of the TENG, showing that the introduction of fluorine groups improves the contact electrification performance of cellulose.

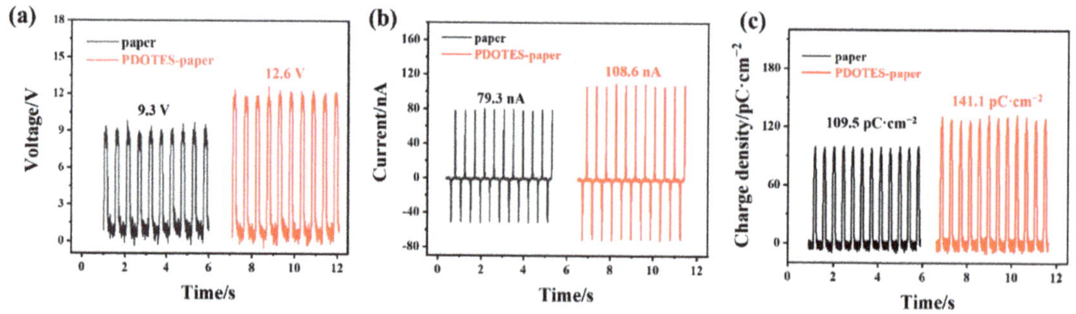

Figure 9. Output performance of paper-based TENG before and after cellulose modification: (**a**) Open circuit voltage; (**b**) short circuit current; (**c**) charge density.

A change in working conditions greatly affects the output performance of the TENG. In the next experiment, a PDOTES-paper-based TENG was used as the experimental object to study the effect of working conditions on the output performance of the TENG. As shown in Figure 10a–c, when the contact pressure of the TENG was increased from 10 to 50 N using a vibration exciter, the open-circuit voltage, short-circuit current, and charge density were increased by 73.75%, 24.27%, and 57.20%, respectively. According to the analysis of the paper surface shown in Figure 7, the surface of the PDOTES-paper was rougher. With an increase in working pressure, PDOTES-paper gradually deforms and fills the gap with the PA film, so that the contact area between electrodes gradually increases, resulting in an improvement in the TENG output performance [26]. A linear motor was used to control the working acceleration of the TENG. As shown in Figure 10d–f, when the acceleration of indirect contact movement of the electrode controlled by the linear motor was increased from 0.5 m/s^2 to 0.9 m/s^2, the open-circuit voltage, short-circuit current, and charge density were increased by 26.42%, 60.51%, and 31.21%, respectively. Compared with the open-circuit voltage and charge density, the short-circuit current increased significantly more because the higher the acceleration of contact separation between electrodes, the shorter the time for electrons in the circuit to reach flow equilibrium, and the shorter the duration of the corresponding current peak, thus increasing the current peak.

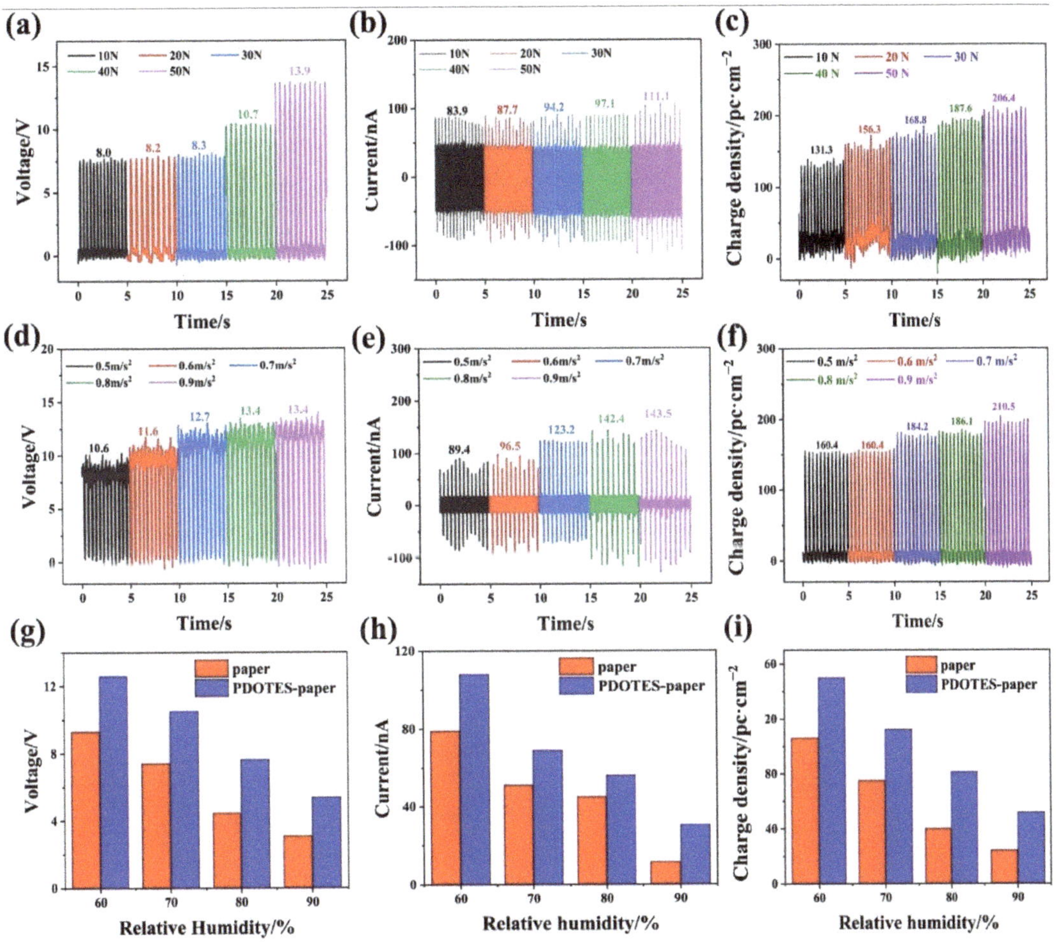

Figure 10. Effect of working conditions on PDOTES-paper TENG output performance. (**a**) Voltage, (**b**) current, and (**c**) charge density at various working pressures; (**d**) voltage, (**e**) current, and (**f**) charge density at various accelerations. The output performance of cellulose-based TENG before and after modification under different ambient humidity: Voltage (**g**), current (**h**), and charge density (**i**).

In addition to the working pressure and working acceleration, the relative humidity of the environment is also an important factor affecting TENG output performance [27,28]. As shown in Figure 10g–i, the motion acceleration between the TENG electrodes was

0.5 m/s^2, and as the air humidity in the control environment increased from 60% to 90%, the output performance of the paper-based and PDOTES-paper-based TENGs was reduced to varying degrees. The open-circuit voltage, short-circuit current, and charge density of the paper-based TENG decreased by 66.67%, 85.37%, and 77.05%, respectively, while the output performance loss of the PDOTES-paper-based TENG was less, with decreases of 57.14%, 71.72%, and 65.40%, respectively. This is because the introduction of a fluorine group into cellulose reduces the surface energy of PDOTES-paper and improves its negative friction polarity. On the other hand, taken in combination with the results in Figure 7, because the surface of PDOTES-paper is rougher than that of the paper, PDOTES-paper is less likely to be penetrated by water droplets in the air, so the PDOTES-paper-based TENG has stronger moisture resistance.

TENGs can be used to supply power for small pieces of electronic equipment, so to study their output power we created a TENG-powered circuit [29,30]. As shown in Figure 11a,b, when the resistance of the variable resistance box in the TENG circuit increased from 10^3 Ω to 10^8 Ω, the output voltage of the TENG gradually increased, the output current gradually decreased, and the output power first increased and then decreased. When the resistance was 4×10^7 Ω, the instantaneous power reached its maximum value of 9.9 nW·cm^{-2}. When the PDOTES-paper-based TENG was connected to the LED circuit board, as shown in Figure 11c, more than 23 LED bulbs could be lit. The output stability of the TENG is an important index for evaluating its practical application [31]. As shown in Figure 12d,e, two stability tests were conducted on the PDOTES-paper-based TENG. First, the voltage was measured for 2000 continuous cycle operations, and was stable at approximately 12 V from the beginning to the end of the test. The other test was to measure the voltage of the TENG three times in one month. After 15 days, the voltage remained stable, and the voltage decreased slightly after one month.

3.6. Applications of FG-TENGs in Self-Powered Sensing

As a wearable device, a paper-based TENG is lightweight, thin, and flexible. It has potential for a wide number of applications as self-powered sensors [32,33]. As shown in Figure 12, a wearable device was manufactured with PDOTES-paper as a negative friction electrode adhered to socks. By simulating human movement, electrical signals were generated during the process of heel–ground contact separation. We performed two typical movements: walking and running. As shown in Figure 12, different electrical signals were detected. It can be seen from the electrical signal that the frequency of the electrical signal reflects the frequency of human movement. Moreover, when the movement state changes from walking to running, the peak value of the electrical signal also increases, because not only the working frequency of the wearable device increases, but the pressure exerted by the foot on the wearable device also increases. This shows that a paper-based TENG, as a self-powered sensor, can effectively monitor human motion.

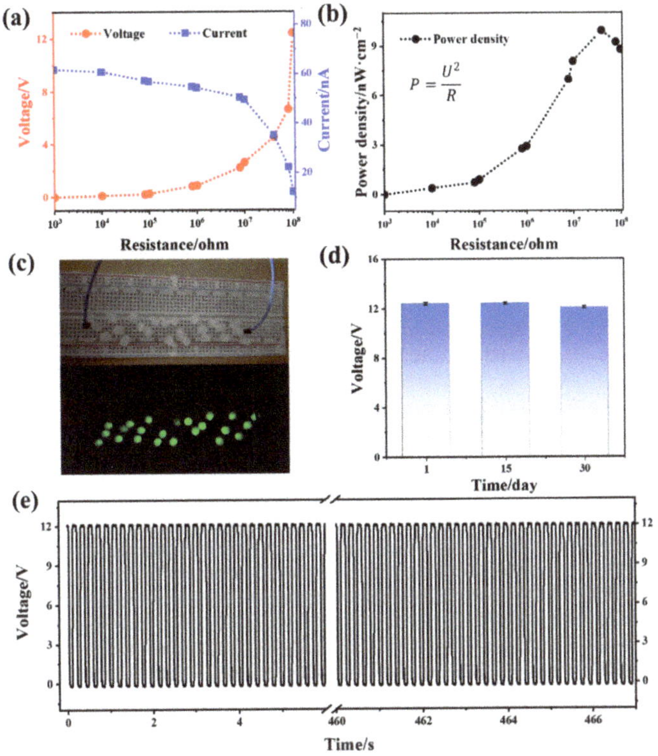

Figure 11. PDOTES-paper-based TENG: (**a**) Output voltage and current changes under external resistance; (**b**) power output curve under external resistance; (**c**) demonstrating supply of current to an LED lamp; (**d**) output performance vs. time; (**e**) output performance after 2000 cycles.

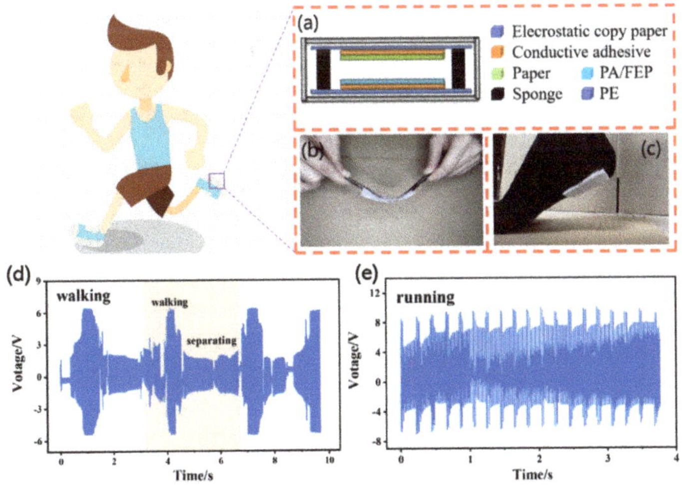

Figure 12. Structure diagram of the wearable device (**a**–**c**); open circuit voltage during walking (**d**); open circuit voltage during running (**e**).

4. Conclusions

The cellulose is chemically modified by 1H, 1H, 2H, 2H-perfluorodecyltriethoxysilane. The access of the fluorine group increases the surface roughness of cellulose, and the surface potential changes from a positive to a negative phase, so that the modified cellulose has negative triboelectricity. A TENG was prepared with PDOTES-paper as a friction-negative material. The output electrical performance was approximately 30% higher than that of a paper-based TENG. When the resistance in the working circuit was 4×10^7 Ω, the output power reached 9.9 nW·cm^{-2}. When connected to an LED circuit board, more than 20 LED bulbs could be lit at the same time. Using a PDOTES-paper-based TENG, a self-powered human-wearable device was designed to detect human movement. This shows that the contact electrification of cellulose has the potential for broad applications.

Author Contributions: Conceptualization, L.L.; methodology, Q.Z., S.Z. and T.W.; validation, X.W., X.S. and Y.W.; formal analysis, S.Z., Q.Z. and T.W.; investigation, S.Z.; resources, L.L. and S.Z.; data curation, S.Z.; writing original draft preparation, S.Z. and Q.Z.; writing—review and editing, L.L. and Q.Z.; visualization, S.Z.; supervision, L.L.; project administration, L.L.; funding acquisition, L.L. All authors have read and agreed to the published version of the manuscript.

Funding: This project was funded by the Natural Science Foundation Project of Guangxi (2018GXNSFAA281336).

Institutional Review Board Statement: Not applicable.

Informed Consent Statement: Not applicable.

Data Availability Statement: Data available in a publicly accessible repository.

Acknowledgments: Thanks to the Guangxi key laboratory of clean pulp and papermaking and pollution control for providing technology and financial support. The authors would like to thank shiyanjia lab for the support of AFM analysis.

Conflicts of Interest: The authors declare no conflict of interest.

Nomenclature

PDOTES	1H, 1H, 2H, 2H-perfluorodecyltriethoxysilane
TENG	Triboelectric nanogenerator
CNFs	Cellulose nanofibers
PMMA	Polymethyl methacrylate
PA	Nylon

References

1. Wei, X.; Zhao, Z.; Zhang, C.; Yuan, W.; Wu, Z.; Wang, J.; Wang, Z.L. All-Weather Droplet-Based Triboelectric Nanogenerator for Wave Energy Harvesting. *ACS Nano* **2021**, *15*, 13200–13208. [CrossRef]
2. Wang, Z.L.; Wu, W. Nanotechnology-Enabled Energy Harvesting for Self-Powered Micro-/Nanosystems. *Angew. Chem. Int. Ed.* **2012**, *51*, 11700–11721. [CrossRef]
3. Niu, S.; Wang, S.; Lin, L.; Liu, Y.; Zhou, Y.S.; Hu, Y.; Wang, Z.L. Theoretical study of contact-mode triboelectric nanogenerators as an effective power source. *Energy Environ. Sci.* **2013**, *6*, 3576–3583. [CrossRef]
4. Chen, C.; Chen, L.; Wu, Z.; Guo, H.; Yu, W.; Du, Z.; Wang, Z.L. 3D double-faced interlock fabric triboelectric nanogenerator for bio-motion energy harvesting and as self-powered stretching and 3D tactile sensors. *Mater. Today* **2020**, *32*, 84–93. [CrossRef]
5. Chandrasekhar, A.; Vivekananthan, V.; Khandelwal, G.; Kim, S.J. A fully packed water-proof, humidity resistant triboelectric nanogenerator for transmitting Morse code. *Nano Energy* **2019**, *60*, 850–856. [CrossRef]
6. Rajabi-Abhari, A.; Lee, J.; Tabassian, R.; Kim, J.; Lee, H.; Oh, I. Antagonistically Functionalized Diatom Biosilica for Bio-Triboelectric Generators. *Small* **2022**, *18*, 2107638. [CrossRef]
7. Yao, L.; Zhou, Z.; Zhang, Z.; Du, X.; Zhang, Q.-L.; Yang, H. Dyeing-Inspired Sustainable and Low-Cost Modified Cellulose-Based TENG for Energy Harvesting and Sensing. *ACS Sustain. Chem. Eng.* **2022**, *10*, 3909–3919. [CrossRef]
8. Shi, K.; Zou, H.; Sun, B.; Jiang, P.; He, J.; Huang, X. Dielectric Modulated Cellulose Paper/PDMS-Based Triboelectric Nanogenerators for Wireless Transmission and Electro polymerization Applications. *Adv. Funct. Mater.* **2019**, *30*, 1904536. [CrossRef]
9. Kim, I.; Roh, H.; Choi, W.; Kim, D. Air-gap embedded triboelectric nanogenerator via surface modification of non-contact layer using sandpapers. *Nanoscale* **2021**, *13*, 8837–8847. [CrossRef]

10. Mule, A.R.; Dudem, B.; Yu, J.S. High-performance and cost-effective triboelectric nanogenerators by sandpaper-assisted micropatterned polytetrafluoroethylene. *Energy* **2018**, *165*, 677–684. [CrossRef]
11. Zhao, D.; Zhu, Y.; Cheng, W.; Chen, W.; Wu, Y.; Yu, H. Cellulose-Based Flexible Functional Materials for Emerging Intelligent Electronics. *Adv. Mater.* **2021**, *33*, 2000619. [CrossRef]
12. Zhang, R.; Dahlstrom, C.; Zou, H.; Jonzon, J.; Hummelgard, M.; Ortegren, J.; Blomquist, N.; Yang, Y.; Andersson, H.; Olsen, M.; et al. Cellulose-Based Fully Green Triboelectric Nanogenerators with Output Power Density of 300 W m^{-2}. *Adv. Mater.* **2020**, *32*, 2002824. [CrossRef]
13. Qin, Y.; Mo, J.; Liu, Y.; Zhang, S.; Wang, J.; Fu, Q.; Wang, S.; Nie, S. Stretchable Triboelectric Self-Powered Sweat Sensor Fabricated from Self-Healing Nanocellulose Hydrogels. *Adv. Funct. Mater.* **2022**, *32*, 2201846. [CrossRef]
14. Wang, N.; Liu, Y.; Ye, E.; Li, Z.; Wang, D. Control methods and applications of interface contact electrification of triboelectric nanogenerators: A review. *Mater. Res. Lett.* **2022**, *10*, 97–123. [CrossRef]
15. Yao, C.; Yin, X.; Yu, Y.; Cai, Z.; Wang, X. Chemically Functionalized Natural Cellulose Materials for Effective Triboelectric Nanogenerator Development. *Adv. Funct. Mater.* **2017**, *27*, 1700794. [CrossRef]
16. Roy, S.; Ko, H.-U.; Maji, P.K.; Van Hai, L.; Kim, J. Large amplification of triboelectric property by allicin to develop high performance cellulosic triboelectric nanogenerator. *Chem. Eng. J.* **2019**, *385*, 123723. [CrossRef]
17. Nie, S.; Cai, C.; Lin, X.; Zhang, C.; Lu, Y.; Mo, J.; Wang, S. Chemically Functionalized Cellulose Nanofibrils for Improving Triboelectric Charge Density of a Triboelectric Nanogenerator. *ACS Sustain. Chem. Eng.* **2020**, *8*, 18678–18685. [CrossRef]
18. Tabaght, F.E.; El Idrissi, A.; Aqil, M.; Elbachiri, A.; Tahani, A.; Maaroufi, A. Grafting method of fluorinated compounds to cellulose and cellulose acetate: Characterization and biodegradation stund. *Cellul. Chem. Technol.* **2021**, *55*, 511–528. [CrossRef]
19. Nie, S.; Fu, Q.; Lin, X.; Zhang, C.; Lu, Y.; Wang, S. Enhanced performance of a cellulose nanofibrils-based triboelectric nanogenerator by tuning the surface polarizability and hydrophobicity. *Chem. Eng. J.* **2021**, *404*, 126512. [CrossRef]
20. Kaynak, B.; Alpan, C.; Kratzer, M.; Ganser, C.; Teichert, C.; Kern, W. Anti-adhesive layers on stainless steel using thermally stable dipodal perfluoroalkyl silanes. *Appl. Surf. Sci.* **2017**, *416*, 824–833. [CrossRef]
21. French, A.D. Idealized powder diffraction patterns for cellulose polymorphs. *Cellulose* **2014**, *21*, 885–896. [CrossRef]
22. Xu, S.; Wu, L.; Lu, S.; Wu, H.; Huang, L.; Chen, L. Preparation and Characterization of Cellulose-g-PFOEMA. *J. Cellul. Sci. Technol.* **2016**, *24*, 7–11.
23. Jing-qiang, Z.; Lu, L.I.N.; Bei-hai, H.E.; Shi-jie, L.; Ping-kai, O. X-Ray Photoelectron Spectroscopic Analysis of Celluloses with Different Crystallization Index. *Chem. Ind. For. Prod.* **2009**, *29*, 30–34.
24. Levdanskya, V.A.; Kondracenkoa, A.S.; Levdanskya, A.V.; Kuznetsova, B.N.; Djakovitchc, L.; Pinelc, C. Sulfation of Microcrystalline Cellulose with Sulfamic Acid in N, N-Dimethylformamide and Diglyme. *J. Sib. Fed. Univ.* **2014**, *7*, 162–167.
25. Mi, H.Y.; Jing, X.; Zheng, Q.; Fang, L.; Huang, H.-X.; Turng, L.-S.; Gong, S. High-performance flexible triboelectric nanogenerator based on porous aerogels and electrospun nanofibers for energy harvesting and sensitive self-powered sensing. *Nano Energy* **2018**, *48*, 327–336. [CrossRef]
26. Xu, Y.; Min, G.; Gadegaard, N.; Dahiya, R.; Mulvihill, D.M. A unified contact force-dependent model for triboelectric nanogenerators accounting for surface roughness. *Nano Energy* **2020**, *76*, 105067. [CrossRef]
27. Wen, R.; Guo, J.; Yu, A.; Zhai, J.; Wang, Z.L. Humidity-Resistive Triboelectric Nanogenerator Fabricated Using Metal Organic Framework Composite. *Adv. Funct. Mater.* **2019**, *29*, 1807655. [CrossRef]
28. Wang, N.; Zheng, Y.; Feng, Y.; Zhou, F.; Wang, D. Biofilm material based triboelectric nanogenerator with high output performance in 95% humidity environment. *Nano Energy* **2020**, *77*, 105088. [CrossRef]
29. Li, Y.; Cheng, G.; Lin, Z.-H.; Yang, J.; Lin, L.; Wang, Z.L. Single-electrode-based rotationary triboelectric nanogenerator and its applications as self-powered contact area and eccentric angle sensors. *Nano Energy* **2015**, *11*, 323–332. [CrossRef]
30. Mule, A.R.; Dudem, B.; Patnam, H.; Graham, S.A.; Yu, J.S. Wearable Single-Electrode-Mode Triboelectric Nanogenerator via Conductive Polymer-Coated Textiles for Self-Power Electronics. *ACS Sustain. Chem. Eng.* **2019**, *7*, 16450–16458. [CrossRef]
31. Wu, J.; Xi, Y.; Shi, Y. Toward wear-resistive, highly durable and high performance triboelectric nanogenerator through interface liquid lubrication. *Nano Energy* **2020**, *72*, 104659. [CrossRef]
32. He, X.; Zou, H.; Geng, Z.; Wang, X.; Ding, W.; Hu, F.; Zi, Y.; Xu, C.; Zhang, S.L.; Yu, H.; et al. A Hierarchically Nanostructured Cellulose Fiber-Based Triboelectric Nanogenerator for Self-Powered Healthcare Products. *Adv. Funct. Mater.* **2018**, *28*, 1805540. [CrossRef]
33. Sriphan, S.; Charoonsuk, T.; Maluangnont, T.; Pakawanit, P.; Rojviriya, C.; Vittayakorn, N. Multifunctional Nanomaterials Modification of Cellulose Paper for Efficient Triboelectric Nanogenerators. *Adv. Mater. Technol.* **2020**, *5*, 2000001. [CrossRef]

polymers

MDPI

Article

Synthesis of Covalent Organic Frameworks (COFs)-Nanocellulose Composite and Its Thermal Degradation Studied by TGA/FTIR

Chunxia Zhu [1], Shuyu Pang [1], Zhaoxia Chen [1], Lehua Bi [2], Shuangfei Wang [1], Chen Liang [1,*] and Chengrong Qin [1]

1 Guangxi Key Laboratory of Clean Pulp & Papermaking and Pollution Control, School of Light Industrial and Food Engineering, Guangxi University, Nanning 530004, China; 2016301052@st.gxu.edu.cn (C.Z.); 2116301042@st.gxu.edu.cn (S.P.); 2116391003@st.gxu.edu.cn (Z.C.); wangsf@gxu.edu.cn (S.W.); qin_chengrong@163.com (C.Q.)

2 Xingjian College of Science and Liberal Arts, Guangxi University, Nanning 530004, China; bilele@163.com

* Correspondence: liangchen@gxu.edu.cn

Abstract: At present, the synthesis methods of crystalline porous materials often involve powder products, which not only affects the practical application but also has complex synthesis operations and limited scale. Based on the mechanochemical method, we choose COF-TpPa-1, preparing TpPa-1-DANC composites. Covalent organic frameworks (COFs) are a kind of crystalline material formed by covalent bonds of light elements. COFs possess well pore structure and high thermal stability. However, the state of synthesized powders limits their application. Cellulose nanocrystals (CNCs) are promising renewable micron materials with abundant hydroxyl groups on their surface. It is possible to prepare high-strength materials such as film, water, and aerogel. Firstly, the nanocellulose was oxidized by the sodium periodate method to obtain aldehyde cellulose nanocrystals (DANC). TpPa-1-DANC not only had the crystal characteristic peak of COFs at $2\theta \approx 5^{\circ}$ but also had a BET surface area of 247 m^2/g. The chemical bonds between COFs and DANC formed by Schiff base reaction appeared in FTIR and XPS. The pyrolysis behavior of the composite was characterized by TG-IR, which showed that the composite had good thermal stability. With the advantages of nanocellulose as a material in every dimension, we believe that this method can be conducive to the large-scale synthesis of COFs composites, and has the possibility of multi-form synthesis of COFs.

Keywords: covalent organic frameworks; aldehyde cellulose nanocrystals; mechanochemistry; TGA/FTIR

Citation: Zhu, C.; Pang, S.; Chen, Z.; Bi, L.; Wang, S.; Liang, C.; Qin, C. Synthesis of Covalent Organic Frameworks (COFs)-Nanocellulose Composite and Its Thermal Degradation Studied by TGA/FTIR. *Polymers* **2022**, *14*, 3158. https://doi.org/10.3390/polym14153158

Academic Editor: Luis Alves

Received: 8 July 2022
Accepted: 30 July 2022
Published: 2 August 2022

1. Introduction

Porous organic materials have received much attention due to their excellent performance in absorption, catalysis, biomedicine, energy storage, and other valuable applications [1–3]. Covalent Organic Frameworks (COFs) are crystalline polymers fabricated by covalent bonds of light elements (C, N, B, H, etc.) [4]. COFs were first synthesized in the form of boron oxygen bonds by Yaghi [5] using a hydrothermal method in 2005, opening a new door to porous materials. COFs materials possess advantages including robust and ordered porous structure, modifiability, low density, and high thermal stability [6]. It has become a research hotspot in the fields of gas adsorption and separation [7], catalysis [8], and energy storage [9]. Nevertheless, due to the crystallization characteristics and rigid structure of COFs, most of the as-synthesized COFs are insoluble powders. Firstly, COFs materials are difficult to process and shape [10]. Secondly, powder makes it difficult for COFs to be recycled after use, which limits the application of COFs materials. Thirdly, it is not conducive to the development of COFs composites. In this current situation, it would be highly desirable to form a new method.

Cellulose is an ample natural polymeric material on the earth and is derived from plants, marine creatures, and bacteria [11–13], etc. Cellulose nanofibers (CNFs) possess advantages including tensile strength and easy membrane-forming [14] that can be used as composite skeleton or substrate. With outstanding biocompatibility [15,16], nanocellulose currently has achieved a variety of nano-adopting forms, including carbon nanotubes [17] and metal organic frameworks (MOFs) [18], etc. Qian et al. [19] found that adding CNF can significantly improve the mechanical properties of MOFs/cellulose composites. Wan et al. [20] synthesized Pd@COF/NFC composite membranes in situ on modified nanocellulose membranes based on a hydrothermal method. However, considering the compatibility of the two-dimensional (2D) layered structure of COFs with substrates, it remains a challenge to fabricate defect-free COFs layers on substrates by in situ hydrothermal methods [21]. Therefore, Abdul et al. [22] developed a strategy of in-situ solid phase doping of carbon nanofibers (CNFs) to prepare COF-CNF hybrid films by mechanochemical methods. The common preparation methods of COFs include the hydrothermal method [23,24] and the microwave-assisted method [25,26], which usually have complicated operation, harsh conditions, and long reaction times and accumulate the organic solvent waste. It is necessary to explore suitable solvent systems and catalysts and conditions such as the ratio of monomers [27]. From the perspective of ecology and energy consumption, it is exigent to explore an approachable and workable strategy of COFs with simpler and faster operation.

Comparatively, the mechanochemical synthesis of COFs is an energy-efficient and low-cost synthesis method [28]. The mechanochemical method can be run under a shorter cycle and milder condition of no or less solvent, which has the feasibility of environmental friendliness and the possibility of expanding the scale of material production [29]. In mechanochemical reactions, such as ball milling, the size of solid particles decreases, and the accumulated potential energy leads to the chemical reaction of substances [30]. The mechanochemical method is currently used in the synthesis of nanomaterials or compounds, such as nanoparticles, zeolites, porous carbons, metal complexes, metal organic frameworks (MOFs), etc. [31,32]. Khayumc et al. [33] used synthetic flexible COF flakes as electrode materials, and the use of mechanochemistry could allow one to avoid the use of binders or additives. Due to the outstanding capacity of COFs, considering the plasticity and scalability of mechanochemical methods and the multidimensional nature of nanocellulose, a simple solvent-free method is explored to synthesize covalent organic framework-nanocellulose composites.

To explore efficient, environmentally friendly, and easy-operate synthetic methods, mechanochemical methods were selected to prepare covalent organic framework (COF-TpPa-1) and covalent organic framework-nanocellulose composite (COF-DANC). Aldehyde group modified nanocellulose crystals were selected to synthesize composites with a certain degree of crystallinity and specific surface area, and the preparation and characterization of covalent organic framework-nanocellulose composites were explored. At the same time, the thermogravimetric analysis combined with infrared spectroscopy was used to analyze the pyrolysis products and yield analysis, and to explore the feasibility of mechanochemical synthesis of composite materials.

2. Materials and Methods

2.1. Materials

P-phenylenediamine (Pa-1) was purchased from Sigma, 98%; Trialdehyde phloroglucinol (Tp) was purchased from McLean, Shanghai, China, 97%; Acetone was purchased from Chengdu Kelong Chemical Co., Ltd., Chengdu, China, AR. Microcrystalline cellulose (MCC) was purchased from Sinopharm Chemical Reagent Co., Ltd. Shanghai, China, column chromatography.

P-toluenesulfonic acid (PTSA) was purchased from Tianjin Damao Chemical Reagent Factory, Tianjin, China, AR; N,N-dimethylformamide (DMF) was purchased from Tianjin Zhiyuan Chemical Reagent Co., Ltd., Tianjin, China, AR; Sulfuric acid (H_2SO_4) was purchased from Ningbo Xinzhi Biotechnology, Ningbo, China, analytically pure; sodium periodate ($NaIO_4$) was purchased from Aladdin reagent, Shanghai, China, AR; Ethylene glycol was purchased from Thermo Fisher Scientific, Waltham, MA, USA, AR. All reagents and solvents are commercially available and used without further purification.

Dialysis bag MD77 was purchased from United Carbon, MWCO 3500. The portable grinder was purchased from Shanghai Wanbo Bio, Shanghai, China, Mini-2-5.

2.2. Preparation of Dialdehyde Cellulose Nanocrystals (DANC) by Periodic Acid Oxidation

Cellulose nanocrystals (CNCs) were prepared from microcrystalline cellulose (MCC) by sulfuric acid hydrolysis [34]. Oxidation of CNCs suspension was conducted using sodium periodate to obtain 2,3-dialdehyde nanocrystalline cellulose (DANC) [35]. 30 g CNCs suspension (0.5 g absolute dry) and a certain amount of $NaIO_4$ (molar ratio 1:4) were added to a 100 mL flask. The flask was covered with aluminum foil to avoid light to avoid the decomposition of $NaIO_4$ in light. The oxidation reaction was carried out by stirring at 70 °C for 180 min. Then, 10 mL of ethylene glycol was added and allowed to react for another 30 min to remove unreacted $NaIO_4$ to terminate the oxidation reaction. The suspension was dialyzed (MWCO 3500) for one week in deionized water, and was stopped when the conductivity of the filtrate was less than 2 μs/cm on a conductivity meter. The yield was calculated, and the suspension was stored at 4 °C for standby.

2.3. Preparation of TpPa-1-DANC

Covalent organic framework nanocellulose composite (COF-DANC) was prepared in a portable grinder. Then, 475 mg of P-toluenesulfonic acid (PTSA), 47.5 mg of p-phenylenediamine (Pa-1), and 100 mg of 2,3-dialdehyde nanocrystalline cellulose (DANC) were added into the centrifuge tube, and two 3 cm diameter zirconia balls were ground in a portable grinder for 5 min. To this 0.3 mmol (63 mg) of trialdehyde phloroglucinol (Tp) was added and ground for 10 min, then grinding was continued with 5.5 mmol (100 μL) deionized water for 5 min. The mixture was removed from the centrifuge tube, then heated in the oven at 170 °C for 60 s. After cooling to room temperature, it was filled with water to wash away PTSA, and then ultrasonic washed with DMF and acetone 2–3 times to remove unreacted monomer and oligomer impurities. After methanol soxhlet extraction for 12 h and vacuum drying at 60 °C for one night, the powder was COF-DANC. When the addition amount of DANC was 50 mg, the material was named COF-DANC-A.

2.4. Characterization

Fourier transform infrared (FT-IR) spectra were recorded with Nicolet iS50 (Thermo Fisher Scientific) with a Diamond ATR (Golden Gate, MA, USA) accessory in the 600–4000 cm^{-1} region, the sample was vacuum dried at 60 °C overnight before testing. X-ray photoelectron spectrometer (XPS) was tested with ESCALAB 250XI+ (Thermo Fisher Scientific). Solid state NMR (SSNMR) was taken in Agilent 600M (Agilent, Santa Clara, CA, USA), chemical shifts were expressed in parts per million (δ scale). The X-ray Diffraction (XRD) was performed with a SmartLab 3KW (Rigaku, Tokyo, Japan) for Cu Kα radiation (λ = 1.5406 Å). The operating voltage and current were 40 kV and 40 mA in the range of 3°–40° with the scanning speed of 2°/min and the step of 0.01°. The samples were dried and gently ground into suitable powder for the XRD test. The crystallinity index (C_{r1}) of nanocellulose samples can be calculated by Equation (1). [36,37]. In the Segal method, the C_{r1} of nanocellulose can be interpreted by main crystalline peak I_{002} ($2\theta \approx 22°$) and amorphous peak I_{am} ($2\theta \approx 18°$). Measurements were repeated at least twice.

$$C_{r1} = \frac{I_{002} - I_{am}}{I_{002}}, \tag{1}$$

Scanning electron microscopy (SEM) images were recorded using a Hitachi SU8220 (HITACHI, Tokyo, Japan) with tungsten filament as an electron source operated at 10 kV. Then, 1 mg of sample was dispersed in 10 mL of isopropanol and sonicated for 60 min, then the dispersion was dropped casting multiple times on the silicon bottom specimen stage, and Au sputtering was conducted before analysis. Transmission scanning electron microscopy (TEM) analysis was performed using HITACHI HT7700 at an accelerating voltage of 200 kV. For TEM sample preparation, the sample was dispersed in isopropanol, and the supernatant was dropped on the copper grid (200 mesh). After air-drying at room temperature for several minutes, the sample was negatively colored with phosphotungstic acid dye under dark conditions. After 30 min, the excess dye was absorbed with filter paper.

The specific surface area of the samples was measured by nitrogen adsorption–desorption experiments at 77 K performed on the ASAP 2460 (Micromeritics, Atlanta, GR, USA). The samples were outgassed under vacuum (150 °C, −0.1 MPa) overnight prior to the N_2 adsorption studies. The surface areas were evaluated using the Brunauer–Emmett–Teller (BET) model. Thermogravimetric infrared spectroscopy (TGA-FTIR) experiments were performed using a TGA55 thermogravimetric analyzer (TA, Everett, MA, USA) combined with a Nicolet iS50 Fourier transform infrared spectrometer (Thermo Fisher). The heating rate of the experiment was 15 °C/min, the flow rate of the nitrogen atmosphere was 40 mL/min, and about 10 mg of the sample was taken. The thermogravimetric overflow gas was detected synchronously in the infrared equipment through the connected insulation pipe.

3. Results and Discussion

Here, according to the previous report by Karak [38], COF-TpPa-1 can be prepared more easily and on a larger scale by the mechanochemical method (Figure 1a). Microcrystalline cellulose (MCC) introduced aldehyde groups on the surface by the sodium periodate method to obtain 2,3-dialdehyde nanocrystalline cellulose (Figure 1b). Aldehyde-based nanocellulose and organic monomer were added step by step to prepare the TpPa-1-DANC composite. Figure 1c provides a schematic of the synthesis procedures of COF-TpPa-1 and composite with the mechanochemical method.

Ground P-toluenesulfonic acid (PTSA), with DANC and Pa-1, was gradually added to Tp-1 and a small amount of water, and grinding was continued until evenly mixed and then the mixture was heated. The aldehyde group on DANC reacted with the amine group on Pa-1 to form Schiff base C=N, which helps DANC connect to the COF nanosheet. PTSA can promote Schiff base reaction and contribute to the formation of imino groups in COF.

3.1. Morphology Analysis of CNCs and DANC

As shown in Figure 2, the SEM pictures showed that microcrystalline cellulose (MCC) (g) was acidolysis to obtain cellulose nanocrystals (CNCs) (h), and then the 2,3-dialdehyde nanocrystalline cellulose (DANC) was modified by sodium periodate method. As can be seen in the following (i), a wide needle DANC with a width of 20 ± 7 nm (average ± one standard deviation) and a length of about 200 ± 50 nm (average ± one standard deviation), and the dispersion degree was better than NCC. According to previous studies, the surface modification of nanocellulose [39] can reduce the agglomeration phenomenon, so as to improve the dispersion of CNCs and the interface interaction of the substrate [40]. By introducing aldehyde groups to the surface of CNCs, a well-dispersed DANC can be obtained. In Figure 3d the nanocellulose in XRD presented the main peaks at around 14, 16, and 22°, belonging to the typical characteristics of cellulose I-type structure, corresponding to 101, $1\bar{0}1$, and 002 crystal planes, respectively. The absorption peak near 34° came from the 004 crystal plane of cellulose I-type structure. It proved that the crystalline structure of DANC does not rearrange after modification. According to Equation (1), the crystallinity indexes (C_{r1}) of CNCs and DANC were 0.73 and 0.056. This was because, with the oxidation of sodium periodate, glucopyranose opened the ring, resulting in the change from the crystal surface to the internal structure, so the crystallinity index decreased. Referring to the literature [41], based on the oxidation reaction between the aldehyde group and hydroxylamine hydrochloride, aldehyde content of 7.1 mmol/g was obtained by calculation, which is conducive to the synthesis of nanocellulose matrix composites.

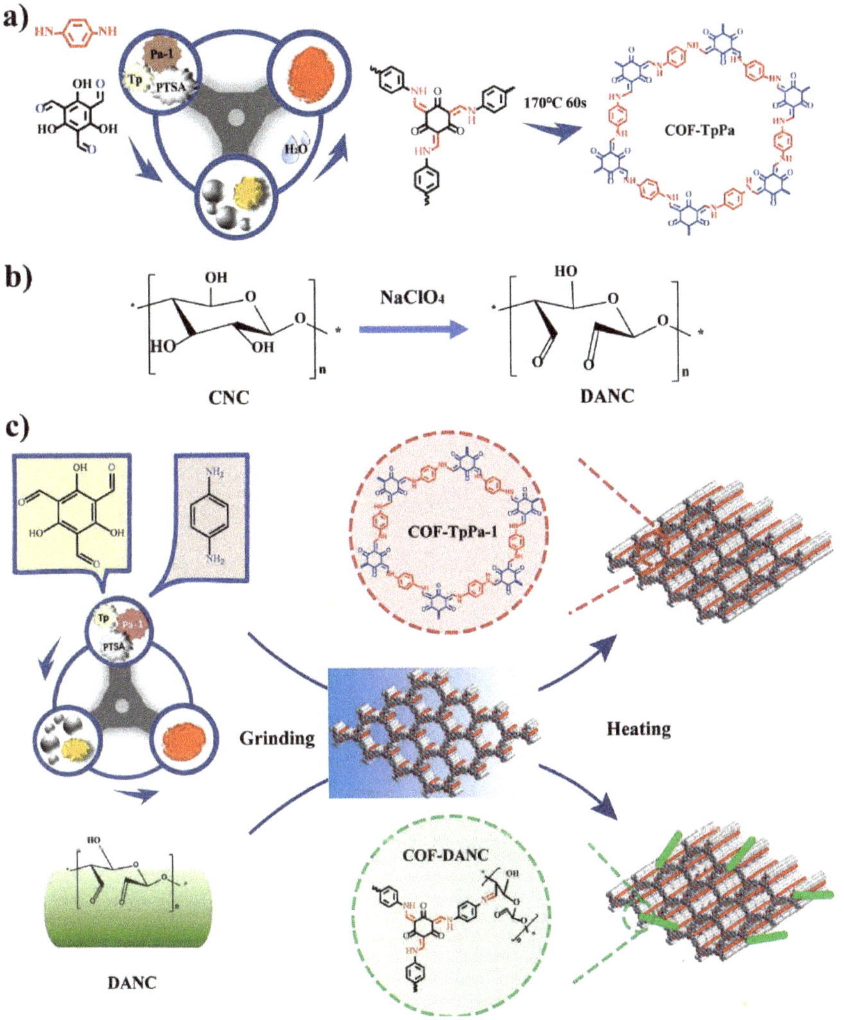

Figure 1. (**a**) Preparation of covalent organic framework COF-TpPa-1 by grinding method; (**b**) Preparation of 2,3-dialdehyde nanocrystalline cellulose (DANC) by sodium periodate method; (**c**) Preparation process of TpPa-1-DANC composite.

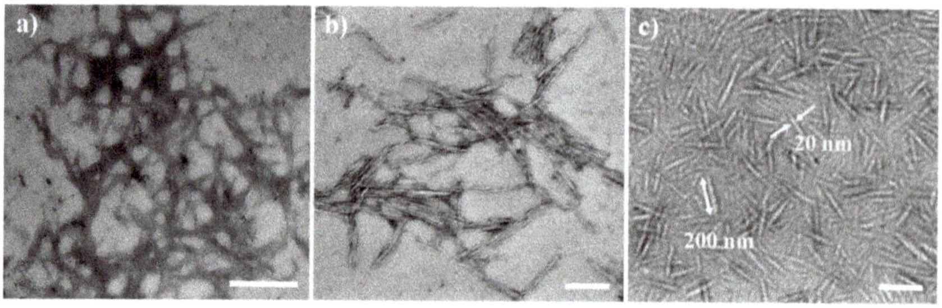

Figure 2. Transmission electron microscope (TEM) of (**a**) MCC, (**b**) NCC and (**c**) DANC; scales of (**a**) is 1 μm, (**b**,**c**) are 200 nm.

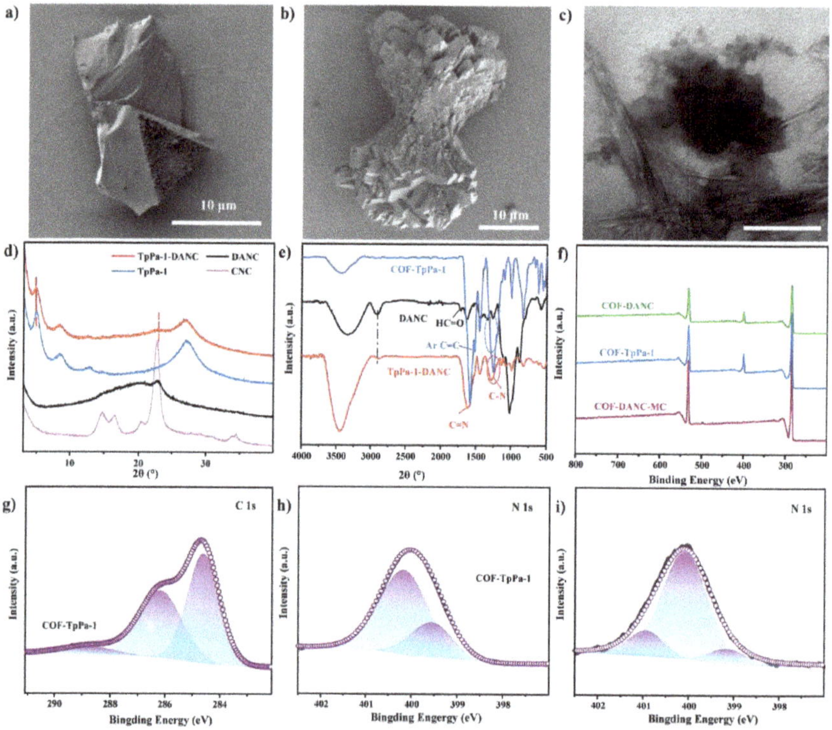

Figure 3. SEM (10 μm) (**a**,**b**) and TEM images (200 nm) (**c**) of TpPa-1-DANC; (**d**) XRD and (**e**) FTIR of TpPa-1, DANC, CNC and TpPa-1-DANC; (**f**) XPS full spectrum of TpPa-1 and TpPa-1-DANC; (**g**) C 1s image of COF-TpPa-1; (**h**) N 1s maps of TpPa-1 and (**i**) COF-DANC.

3.2. Characterization of COF-TpPa-1 by Mechanochemical Method

From Figure 4a,b, in COFs synthesized by the mechanochemical method, it can be seen that the transverse dimension of COF-TpPa-1 was 10–15 μm. Its lamellar shape corresponds to the layer stacking seen in the TEM of Figure 4c, which is consistent with the results in the literature [38].

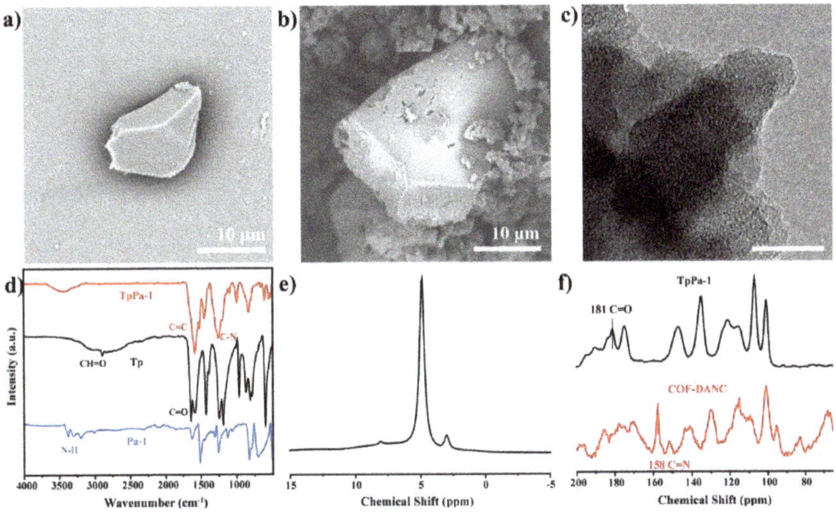

Figure 4. Scanning electron microscope (**a**,**b**) and TEM image (**c**) of TpPa-1; FTIR image (**d**) of TpPa-1 and its monomer; ^{1}H NMR (**e**) of COF-TpPa-1; ^{13}C NMR (**f**) of COF-TpPa-1 and TpPa-1-DANC.

The structure of the covalent organic framework is an organic framework with a repeated unit structure, so it has certain predictability and regularity. The crystal structure of COFs and their composites were analyzed by X-ray diffraction (XRD), and the structural information such as the regularity of the framework or the stacking degree of layers of the samples could be known. The XRD spectrum of the TpPa-1 sample prepared by the mechanochemical method was shown in Figure 3d. It can be seen that there was an obvious characteristic peak at 4.9°, corresponding to the (100) plane of covalent organic framework~8.7° and 11.9° are attributed to (200) and (210) reflection planes, respectively. The 2θ = 27° angle belonged to the amorphous peak, which is attributed to the (001) plane π-π Stacking, and layered stacking distance was 3.3 Å. These results were consistent with previous reports on pure covalent organic frameworks [42].

Figure 4d showed the infrared (FTIR) diagram of Tp, Pa-1, and COF-TpPa-1. It can be seen that the C–H characteristic peak 2894 cm^{-1} of Tp and the C=O characteristic peak 1640 cm^{-1} of carbonyl stretching band disappear. The peak of COF-TpPa-1 at 3400–3500 cm^{-1} was the stretching vibration peak of –OH, and there was no N–H characteristic peak (3296 and 3367 cm^{-1}), indicating that Pa-1 was consumed after the reaction. The C–N stretching vibration peak appeared at 1256 cm^{-1}, and the secondary amine (N–H) bending vibration peak appeared at 1517 cm^{-1}. The absorption peak of the benzene ring skeleton appeared at 1454 cm^{-1}. A new peak of C=C appeared at 1585 cm^{-1}. The C–H characteristic peak of aldehyde –CHO disappeared at 2820 cm^{-1}, indicating that the aldehyde group in the molecular structure reacted, and the characteristic peak of ketone C=O appeared at 1604 cm^{-1}, which is in good agreement with previous reports [38]. In conclusion, the synthesized COF-TpPa-1 had the structure of aldehyde C=O structure conjugated with C=C, and the structure of the para-substituted benzene ring and imine bond (C–N). This indicates that the condensation reaction and imine bond formation have occurred, that is, β-ketene amine skeleton formation.

Solid nuclear magnetism (NMR) reflects the chemical environment within the nucleus by combining the internal and surrounding structures of the nucleus. The structural network of COF-TpPa-1 was analyzed by ^{1}H NMR and ^{13}C NMR, which provided a basis for thermogravimetric infrared analysis and a reference for further optimizing the synthesis process. In the ^{1}H NMR diagram of Figure 4e, it can be seen that a strong peak of ~5 ppm was assigned to water molecules from COF-TpPa-1. An obvious signal centered at

~8.0 ppm was assigned to the protons of amino groups, indicating that amino groups and adjacent carbonyls may form strong intramolecular hydrogen bonds and mutual shielding generated by aromatic rings [43]. As shown in Figure 4f, it was the solid-state ^{13}C NMR spectrum of COF-TpPa-1. The chemical shift at 146 ppm can be considered as carbon (=CNH–) on enamine, and the signal at 181 ppm came from carbonyl carbon (–C=O). And the 157 ppm of TpPa-1-DANC came from the C=N bond produced by the aldehyde amine condensation reaction [44]. The small peak at 191 ppm was attributed to the carbon atom of the aldehyde group at the end of COF-TpPa-1 in the composite, which is consistent with the FTIR results. It was confirmed that in the covalent organic framework, β-ketene amine bond formation occurred.

3.3. *Characterization of TpPa-1-DANC Composites by Mechanochemical Method*

3.3.1. Physical Properties

TpPa-1 and DANC were compounded by the mechanochemical method. The morphology of TpPa-1-DANC was analyzed by SEM and TEM. In Figure 3d,e of TpPa-1-DANC, the average particle size scale increased to 15 μm. It can be seen by transmission electron microscope that the contact between flake stacked COF and rod-needle-shaped DANC was a transitional blur, rather than a clear interface, which indicates that the two are not going through simple physical blending, but covalent grafting [22].

Compared with TpPa-1, the XRD of TpPa-1-DANC not only showed the characteristic peak of COF but also showed the characteristic peak of (110) crystal plane of cellulose at about 22°, indicating that DANC is successfully incorporated into the TpPa-1 network. The peak at $2\theta = 27.22°$ decreased, which may be due to the fact that the COF grown on the (001) crystal plane was parallel to the direction of cellulose [20].

3.3.2. Chemical Properties

TpPa-1-DANC had an obvious characteristic peak at about 1400 cm^{-1}, and about 2900 cm^{-1} came from the C–H bond of 2,3-dialdehyde nanocrystalline cellulose (DANC). The stretching vibration of the C=N bond at 1629 cm^{-1} [45] shows the formation of the imine bond, the weakening of the C–N bond of COFs, the disappearance of C=O of DANC, and the weakening of C–N, which proves that they are successfully grafted through reaction [46].

X-ray photoelectron spectroscopy (XPS) characterization technology irradiates the sample surface with X-ray photons. According to the analysis of the electronic energy distribution of the elements on the sample surface, judge whether the relevant chain chemistry is formed, and investigate the content and grafting of COFs composite. XPS analysis of TpPa-1 and its composites are shown in Figure 3. It can be seen from the full spectrum f) that there is an obvious N peak in TpPa-1. With the addition of DANC, the proportion of the N peak decreases slightly. Measuring the content ratio of the C 1s atom to the N 1s atom of TpPa-1 (C/N = 8.66), it was lower than that of COF-DANC-A (C/N = 9.54). And this was consistent with the experimental data of materials by EDS in Table 1. This proved the existence of 2,3-dialdehyde nanocrystalline cellulose in the obtained material. Under the conditions of the pure grinding method, the N content was low, which may be due to the agglomeration of COF-TpPa-1 powder and DANC powder without catalyst, and the N peak was not characterized when the material surface is scanned by XPS.

Table 1. Summary of XPS atomic mass ratio of different materials.

Element	Atomic %	
	COF	**COF-DANC**
C 1s	73.15	72.18
O 1s	18.40	20.25
N 1s	8.45	7.57

Further high-resolution C spectrum analysis of the sample showed that 284.50, 286.08, and 288.74 eV belong to (sp_2) C=C, C–O/C–H, and C=O, respectively in the C 1s high resolution of TpPa-1 [47]. N 1s appeared at 399.47 eV and 400.06 eV, and the peaks belonging to Ar–NH_2 and Ar–NH–C were divided. After loading, TpPa-1-DANC had a new peak of 400.09 eV C=N [48]. It was proved that the reaction of TpPa-1 and DANC produces a C=N bond through covalent grafting.

3.3.3. Thermal Analysis

The thermogravimetric-infrared combined technology connects the thermogravimetric analyzer and infrared analysis through the insulation pipe and analyzes the combined spectrum, including a thermogravimetric diagram, infrared diagram of each time (temperature), thermogravimetric infrared three-dimensional spectrum, and intensity change diagram with time. Thermogravimetric infrared technology obtains the thermal stability and purity of the material by synchronously analyzing the infrared spectrum of the polymer at the weight loss temperature, which is an effective analysis method for the current situation of low synthesis of COFs. In the experiment, the infrared three-dimensional spectrum of pyrolysis products obtained by TGA-FTIR technology at the heating rate of 15°/min shows the corresponding characteristic absorption peaks in the fixed wave number range under the pyrolysis temperature range.

The thermogravimetric curve and thermogravimetric infrared three-dimensional diagrams were shown in Figure 5. The TpPa-1 in Figure 5a had a pyrolysis temperature of 330 and 444 °C with a pyrolysis residue of about 40%. The temperature of the maximum decomposition rate decreased with the increase of DANC, similar to the literature [49]. In Figure 5c, the first pyrolysis peak of TpPa-1-DANC was larger than the area of COF because the addition of DANC increased the total carbon content of the material. The results showed that the composites with high COFs content had advantages in thermal stability. As the proportion of COFs increased, the ash component of composites increased, which is due to the gradual aromatization of COFs residues during carbonization. Therefore, the smaller the proportion of COF in the complex, the residue will decrease with the redduction of aromatics. It has been reported in the literature [18] that the ash content of pure MOF powder TGA is 40.79%, increasing by 20–50% with the loading of MOF in CNC–CMC aerogel powder, thus increasing the thermal stability of composites with the loading of MOF in aerogel. Compared to the thermogravimetric curves of TpPa-1-DANC and DANC, the maximum weightlessness rate of composites was higher than that of DANC, suggesting that the stability of DANC was influenced by the combination of COFs and the residual rate of the composites at high temperature was higher than 20% of DANC. This was because the COFs have a stable aromatic ring, which is more stable than the ring structure of hemiacetal at high temperatures.

The thermogravimetric curves of the composite TpPa-1-DANC still showed two major pyrolysis peaks similar to those of COFs. With the addition of DANC, the initial decomposition temperature of the composites was lower than that of COFs, close to that of nanocellulose, and the maximum decomposition rate remained at about 300 °C, but the decomposition range became increased, thus effectively increasing the thermal stability of the nanocellulose composites. DANC had a significant mass loss at 100 °C due to its strong absorptivity and a significant breakdown of the substance at 200 °C. At 11.6 min, it was about 150 °C, at which point the product has partial water and organic solvents, such as DMF, with characteristic peaks around 2900–3000, 1700, and 1000 cm^{-1} corresponding to C–H stretching vibrations, carbonyl C=O double bond telescopic vibrations, and C–O and C–C skeletal vibrations within the C–H in-plane bending vibration of DANC [49].

As shown in Figure 5c, this meant that TpPa-1-DANC has higher thermal stability than DANC. In the temperature range of 300–450 °C, glycosidic bonds, partial C–O, and C–C bonds are broken. The maximum weight loss rate and 600 °C residue rate of TpPa-1-DANC were higher than those of DANC. This was because COFs have stable aromaticity at high temperatures, which is more stable than the ring structure of semi acetals and plays a positive role in the wide application of COFs materials.

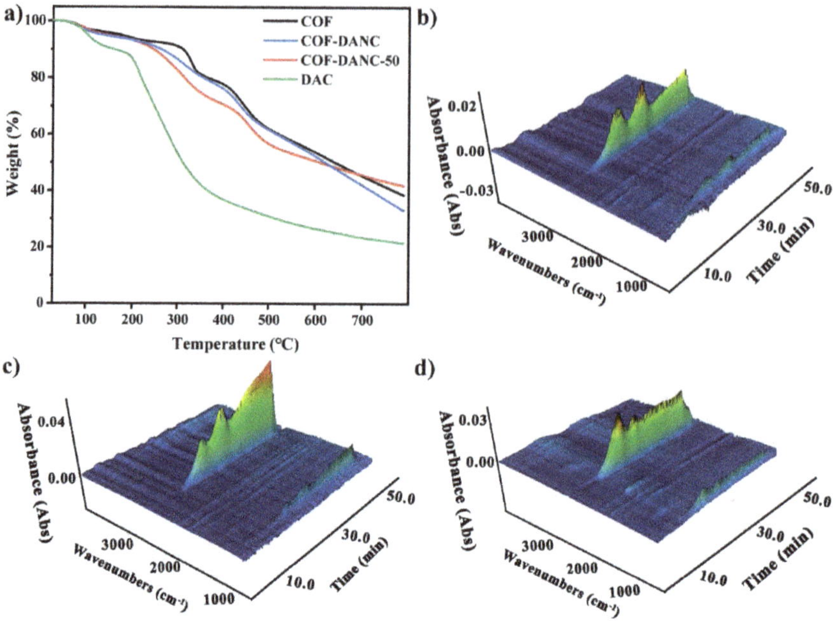

Figure 5. (**a**) the thermogravimetric curve of different materials; (**b**–**d**) the thermogravimetric infrared three-dimensional diagrams of (**b**) COF-TpPa-1, (**c**) TpPa-DANC-50 composites and (**d**) COF monomer mixing.

According to Beer's law, the intensity of characteristic absorption peak larger, that is, the higher the absorbance, and the higher relative content of gas components in the total gas. Figure 6 showed the thermogravimetric curve of COF-TpPa-1 at a heating rate of 15 °C/min and the FTIR spectrum of pyrolysis products at different temperatures. It can be seen from the infrared image that pyrolysis is divided into three steps:

1. At about 100 °C, it can be seen in Figure 6 that at 100 (5 min) and 137 °C (7 min), the characteristic peak of water increased at 3400–3700 and 1500–1700 cm^{-1}, which may come from the water on the material surface. In addition, there was a certain time error in the delay of gas transmission.
2. At about 19 min (and 300 °C), the absorbance of CO increased slightly. The appearance at 1380 cm^{-1} indicated the vibration of the C–C skeleton. By 30.6 min, a large amount of CO_2 was produced.
3. The third stage was about 400 °C, which shows the large-scale cracking of the frame. With the increase in temperature, the peak value of 2300 cm^{-1} increased. The maximum weight loss peak was about 40.2 min. The absorption peak at 3500–3700 cm^{-1} showed that the gas still contains water. This means that when the temperature reached 400 °C, the composite began to decompose.

The absorbance of the corresponding CO_2 reached the maximum, indicating that the material has good thermal stability at about 300 °C.

Generally speaking, pyrolysis at low temperatures comes from untreated solvents or oligomers in COFs. This is consistent with the excellent thermal stability of COFs, so there is no detection of branch chain fracture or frame cracking of COFs at low temperatures. Therefore, the thermogravimetric infrared images of COFs after methanol extraction are compared to judge whether the COFs material is clean and whether the pores are clean. In contrast, the extraction of COF-TpPa-1 in advance showed similar thermogravimetric performance in three cracking stages.

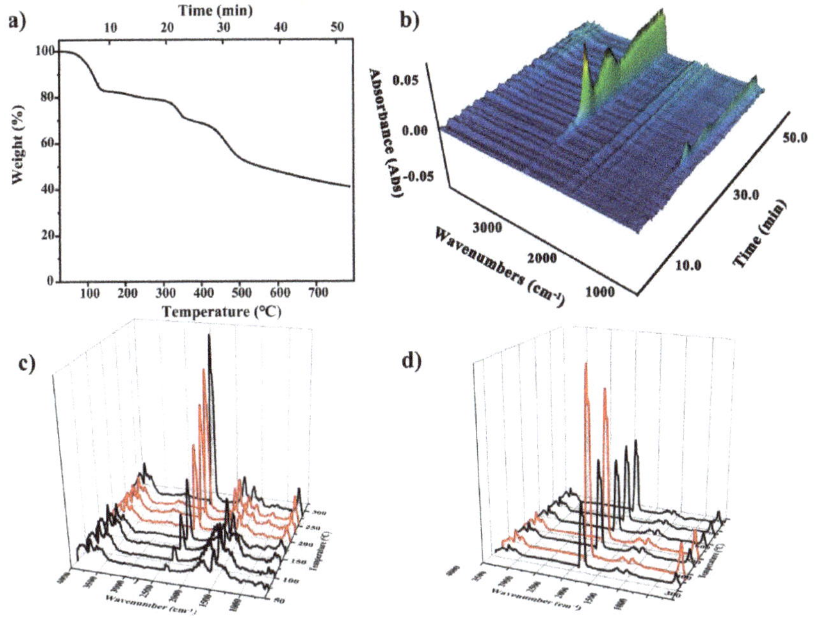

Figure 6. (**a**) Thermogravimetric diagram and (**b**) the thermogravimetric infrared three-dimensional diagram of COF-TpPa-1; Infrared images in the temperature ranges (**c**) 0–300 °C and (**d**) 300–800 °C of COF-TpPa-1.

The difference was that the characteristic peak of CH_4 appears in the decomposition process. As shown in Figure 7, alkanes begin to be produced at 4.86 min, and the CH_4 characteristic peak appears at 2970–2900 cm^{-1}. With the increase in temperature, the absorption peak of methane increased, reaches the absorption peak at 7.3 min, and disappeared at about 19 min. The change in the CH_4 peak was similar to that of 1390 and 1032 cm^{-1} characteristic peaks. The gas produced in this temperature range was considered to be coming from DMF (boiling point is 157 °C), and characteristic peaks appeared around 2900–3000, 1390, 1700, and 1000 cm^{-1}, corresponding to C–H stretching vibration, carbonyl C=O double bond stretching vibration, C–H plane bending vibration and C–O and C–C skeleton vibration of DMF [49].

XRD and BET analysis showed that the TpPa-1-DANC composite maintained good crystallinity and specific surface area of TpPa-1. Considering various harsh environmental requirements in practical application, a thermal stability test is very important. The thermal stability of COFs was obtained by thermogravimetric analysis, which basically had good thermal stability at 300–400 °C.

3.3.4. Structure Analysis of Composites

The structural stability and permanent porosity of the materials were studied by a nitrogen adsorption-desorption experiment at 77 K. As shown in Figure 8a, COF-TpPa-1 showed a steep absorption curve under the low-temperature line ($P/P_0 < 0.5$), showing the characteristics of microporous materials. The specific surface area decreases with the increase of the amount of DANC. The BET specific surface area of TpPa-1 prepared by the mechanical grinding method is 359 m^2/g, which is bigger than 247 m^2/g when DANC was added. However, they were less than the specific surface area of COFs prepared by the hydrothermal method [50]. It was speculated that the long-range order may be limited due to certain stripping of COFs in the mechanical method. Also the residue of oligomers in the formation process may be caused by insufficient pores for nitrogen adsorption [51]. Pore size distribution analysis of COF-TpPa-1 from the adsorption isotherms indicated that the samples contain micropores with diameters of 1.1 nm (Figure 8b).

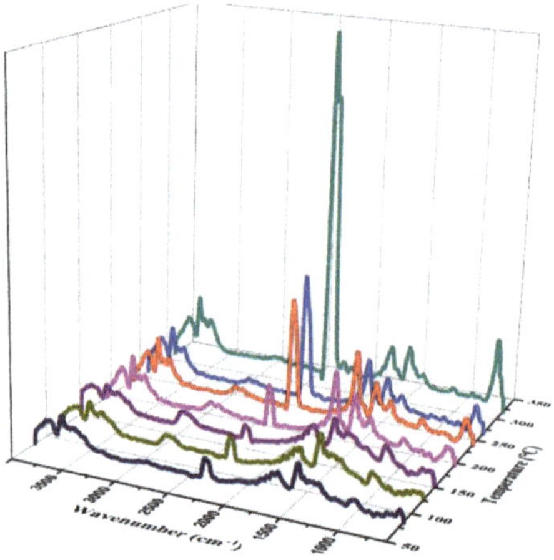

Figure 7. Thermogravimetric infrared three-dimensional diagram of COF-TpPa-1 without Soxhlet extraction in the temperature range of 0–350 °C.

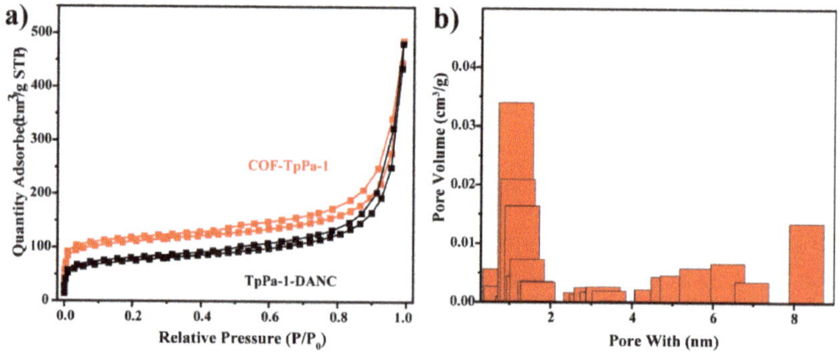

Figure 8. (**a**) Nitrogen desorption curves of TpPa-1 and TpPa-1-DANC; (**b**) Pore size distributions of the pure COF-TpPa-1.

4. Conclusions

COF-TpPa-1 and TpPa-1-DANC were prepared by the mechanochemical method from bottom to top. The successful synthesis of the material was proved by various characterizations. The obtained TpPa-1-DANC composite had a certain specific surface area (BET, 247 m^2/g). The pyrolysis behavior of the material was analyzed combined with thermogravimetric infrared technology. The results showed that TpPa-1-DANC not only improves the specific surface area but also improves the thermal stability. Combined with the film-forming and gel properties of nanocellulose, our experiment provides a basis for the preparation of morphological covalent organic framework composites based on nano cellulose and improves the application feasibility of COFs materials.

Author Contributions: Conceptualization, C.L. and C.Z.; methodology, C.Z. and S.P.; software, S.P.; resources, Z.C.; data curation, L.B.; writing—original draft preparation, C.Z.; writing—review and editing, C.L.; project administration, S.W. and C.Q. All authors have read and agreed to the published version of the manuscript.

Funding: National Natural Science Foundation of China (No.21968004); Guangxi Natural Science Foundation of China (2020GXNSFAA159027).

Institutional Review Board Statement: Not applicable.

Informed Consent Statement: Informed consent was obtained from all subjects involved in the study.

Data Availability Statement: The data presented in this study are available in the manuscript's figure.

Conflicts of Interest: The authors declare no conflict of interest. The funders had no role in the design of the study; in the collection, analyses, or interpretation of data; in the writing of the manuscript, or in the decision to publish the results.

References

1. Hasell, T.; Miklitz, M.; Stephenson, A.; Little, M.A.; Chong, S.Y.; Clowes, R.; Chen, L.; Holden, D.; Tribello, G.A.; Jelfs, K.E.; et al. Porous Organic Cages for Sulfur Hexafluoride Separation. *J. Am. Chem. Soc.* **2016**, *138*, 1653–1659. [CrossRef]
2. Lee, J.-S.M.; Wu, T.-H.; Alston, B.M.; Briggs, M.E.; Hasell, T.; Hu, C.-C.; Cooper, A.I. Porosity-engineered carbons for supercapacitive energy storage using conjugated microporous polymer precursors. *J. Mater. Chem. A* **2016**, *4*, 7665–7673. [CrossRef]
3. Benyettou, F.; Kaddour, N.; Prakasam, T.; Das, G.; Sharma, S.K.; Thomas, S.A.; Bekhti-Sari, F.; Whelan, J.; Alkhalifah, M.A.; Khair, M.; et al. In vivo oral insulin delivery via covalent organic frameworks. *Chem. Sci.* **2021**, *12*, 6037–6047. [CrossRef] [PubMed]
4. Diercks, C.S.; Yaghi, O.M. The atom, the molecule, and the covalent organic framework. *Science* **2017**, *355*, eaal1585. [CrossRef] [PubMed]
5. Côté, A.P.; Benin, A.I.; Ockwig, N.W.; O'Keeffe, M.; Matzger, A.J.; Yaghi, O.M. Porous, Crystalline, Covalent Organic Frameworks. *Science* **2005**, *310*, 1166–1170. [CrossRef] [PubMed]
6. Geng, K.; He, T.; Liu, R.; Dalapati, S.; Tan, K.T.; Li, Z.; Tao, S.; Gong, Y.; Jiang, Q.; Jiang, D. Covalent Organic Frameworks: Design, Synthesis, and Functions. *Chem. Rev.* **2020**, *120*, 8814–8933. [CrossRef]
7. Huang, N.; Chen, X.; Krishna, R.; Jiang, D. Two-Dimensional Covalent Organic Frameworks for Carbon Dioxide Capture through Channel-Wall Functionalization. *Angew. Chem. Int. Ed.* **2015**, *54*, 2986–2990. [CrossRef]
8. Chen, R.; Wang, Y.; Ma, Y.; Mal, A.; Gao, X.-Y.; Gao, L.; Qiao, L.; Li, X.-B.; Wu, L.-Z.; Wang, C. Rational design of isostructural 2D porphyrin-based covalent organic frameworks for tunable photocatalytic hydrogen evolution. *Nat. Commun.* **2021**, *12*, 1354. [CrossRef]
9. Shinde, D.B.; Aiyappa, H.B.; Bhadra, M.; Biswal, B.P.; Wadge, P.; Kandambeth, S.; Garai, B.; Kundu, T.; Kurungot, S.; Banerjee, R. A mechanochemically synthesized covalent organic framework as a proton-conducting solid electrolyte. *J. Mater. Chem. A* **2016**, *4*, 2682–2690. [CrossRef]
10. Wang, H.; Zeng, Z.; Xu, P.; Li, L.; Zeng, G.; Xiao, R.; Tang, Z.; Huang, D.; Tang, L.; Lai, C.; et al. Recent progress in covalent organic framework thin films: Fabrications, applications and perspectives. *Chem. Soc. Rev.* **2019**, *48*, 488–516. [CrossRef]
11. Bondancia, T.J.; Mattoso, L.H.C.; Marconcini, J.M.; Farinas, C.S. A new approach to obtain cellulose nanocrystals and ethanol from eucalyptus cellulose pulp via the biochemical pathway. *Biotechnol. Prog.* **2017**, *33*, 1085–1095. [CrossRef] [PubMed]
12. Bondancia, T.J.; Florencio, C.; Baccarin, G.S.; Farinas, C.S. Cellulose nanostructures obtained using enzymatic cocktails with different compositions. *Int. J. Biol. Macromol.* **2022**, *207*, 299–307. [CrossRef] [PubMed]
13. Kamperidou, V.; Terzopoulou, P. Anaerobic Digestion of Lignocellulosic Waste Materials. *Sustainability* **2021**, *13*, 12810. [CrossRef]
14. Lagerwall, J.P.F.; Schütz, C.; Salajkova, M.; Noh, J.; Hyun Park, J.; Scalia, G.; Bergström, L. Cellulose nanocrystal-based materials: From liquid crystal self-assembly and glass formation to multifunctional thin films. *NPG Asia Mater.* **2014**, *6*, e80. [CrossRef]
15. Joly, F.-X.; Coulis, M. Comparison of cellulose vs. plastic cigarette filter decomposition under distinct disposal environments. *Waste Manag.* **2018**, *72*, 349–353. [CrossRef]
16. Schiavi, D.; Francesconi, S.; Taddei, A.R.; Fortunati, E.; Balestra, G.M. Exploring cellulose nanocrystals obtained from olive tree wastes as sustainable crop protection tool against bacterial diseases. *Sci. Rep.* **2022**, *12*, 6149. [CrossRef]
17. Zhang, Y.; Liu, G.; Yao, X.; Gao, S.; Xie, J.; Xu, H.; Lin, N. Electrochemical chiral sensor based on cellulose nanocrystals and multiwall carbon nanotubes for discrimination of tryptophan enantiomers. *Cellulose* **2018**, *25*, 3861–3871. [CrossRef]
18. Zhu, H.; Yang, X.; Cranston, E.D.; Zhu, S. Flexible and Porous Nanocellulose Aerogels with High Loadings of Metal-Organic-Framework Particles for Separations Applications. *Adv. Mater.* **2016**, *28*, 7652–7657. [CrossRef]
19. Qian, L.; Lei, D.; Duan, X.; Zhang, S.; Song, W.; Hou, C.; Tang, R. Design and preparation of metal-organic framework papers with enhanced mechanical properties and good antibacterial capacity. *Carbohydr. Polym.* **2018**, *192*, 44–51. [CrossRef]
20. Wan, X.; Wang, X.; Chen, G.; Guo, C.; Zhang, B. Covalent organic framework/nanofibrillated cellulose composite membrane loaded with Pd nanoparticles for dechlorination of dichlorobenzene. *Mater. Chem. Phys.* **2020**, *246*, 122574. [CrossRef]

21. Zhang, Y.; Guo, J.; Han, G.; Bai, Y.; Ge, Q.; Ma, J.; Lau, C.H.; Shao, L. Molecularly soldered covalent organic frameworks for ultrafast precision sieving. *Sci. Adv.* **2021**, *7*, eabe8706. [CrossRef] [PubMed]
22. Mohammed, A.K.; Vijayakumar, V.; Halder, A.; Ghosh, M.; Addicoat, M.; Bansode, U.; Kurungot, S.; Banerjee, R. Weak Intermolecular Interactions in Covalent Organic Framework-Carbon Nanofiber Based Crystalline yet Flexible Devices. *ACS Appl Mater Interfaces* **2019**, *11*, 30828–30837. [CrossRef] [PubMed]
23. Maschita, J.; Banerjee, T.; Lotsch, B.V. Direct and Linker-Exchange Alcohol-Assisted Hydrothermal Synthesis of Imide-Linked Covalent Organic Frameworks. *Chem. Mater.* **2022**, *34*, 2249–2258. [CrossRef] [PubMed]
24. Furukawa, H.; Yaghi, O.M. Storage of Hydrogen, Methane, and Carbon Dioxide in Highly Porous Covalent Organic Frameworks for Clean Energy Applications. *J. Am. Chem. Soc.* **2009**, *131*, 8875–8883. [CrossRef]
25. Díaz de Greñu, B.; Torres, J.; García-González, J.; Muñoz-Pina, S.; de los Reyes, R.; Costero, A.M.; Amorós, P.; Ros-Lis, J.V. Microwave-Assisted Synthesis of Covalent Organic Frameworks: A Review. *ChemSusChem* **2021**, *14*, 208–233. [CrossRef]
26. Głowniak, S.; Szczęśniak, B.; Choma, J.; Jaroniec, M. Advances in Microwave Synthesis of Nanoporous Materials. *Adv. Mater.* **2021**, *33*, 2103477. [CrossRef]
27. Li, Y.; Chen, W.; Xing, G.; Jiang, D.; Chen, L. New synthetic strategies toward covalent organic frameworks. *Chem. Soc. Rev.* **2020**, *49*, 2852–2868. [CrossRef]
28. Das, G.; Balaji Shinde, D.; Kandambeth, S.; Biswal, B.P.; Banerjee, R. Mechanosynthesis of imine, beta-ketoenamine, and hydrogen-bonded imine-linked covalent organic frameworks using liquid-assisted grinding. *Chem. Commun. (Camb)* **2014**, *50*, 12615–12618. [CrossRef]
29. Do, J.L.; Friscic, T. Mechanochemistry: A Force of Synthesis. *ACS Cent. Sci.* **2017**, *3*, 13–19. [CrossRef]
30. Szczęśniak, B.; Borysiuk, S.; Choma, J.; Jaroniec, M. Mechanochemical synthesis of highly porous materials. *Mater. Horiz.* **2020**, *7*, 1457–1473. [CrossRef]
31. Casco, M.E.; Kirchhoff, S.; Leistenschneider, D.; Rauche, M.; Brunner, E.; Borchardt, L. Mechanochemical synthesis of N-doped porous carbon at room temperature. *Nanoscale* **2019**, *11*, 4712–4718. [CrossRef] [PubMed]
32. Cheng, H.; Tan, J.; Ren, Y.; Zhao, M.; Liu, J.; Wang, H.; Liu, J.; Zhao, Z. Mechanochemical Synthesis of Highly Porous CeMnOx Catalyst for the Removal of NOx. *Ind. Eng. Chem. Res.* **2019**, *58*, 16472–16478. [CrossRef]
33. Khayum, M.A.; Vijayakumar, V.; Karak, S.; Kandambeth, S.; Bhadra, M.; Suresh, K.; Acharambath, N.; Kurungot, S.; Banerjee, R. Convergent Covalent Organic Framework Thin Sheets as Flexible Supercapacitor Electrodes. *ACS Appl. Mater. Interfaces* **2018**, *10*, 28139–28146. [CrossRef] [PubMed]
34. Bondeson, D.; Mathew, A.; Oksman, K. Optimization of the isolation of nanocrystals from microcrystalline celluloseby acid hydrolysis. *Cellulose* **2006**, *13*, 171. [CrossRef]
35. Liimatainen, H.; Suopajärvi, T.; Sirviö, J.; Hormi, O.; Niinimäki, J. Fabrication of cationic cellulosic nanofibrils through aqueous quaternization pretreatment and their use in colloid aggregation. *Carbohydr. Polym.* **2014**, *103*, 187–192. [CrossRef]
36. Ahvenainen, P.; Kontro, I.; Svedström, K. Comparison of sample crystallinity determination methods by X-ray diffraction for challenging cellulose I materials. *Cellulose* **2016**, *23*, 1073–1086. [CrossRef]
37. Segal, L.; Creely, J.J.; Martin, A.E.; Conrad, C.M. An Empirical Method for Estimating the Degree of Crystallinity of Native Cellulose Using the X-Ray Diffractometer. *Text. Res. J.* **1959**, *29*, 786–794. [CrossRef]
38. Karak, S.; Kandambeth, S.; Biswal, B.P.; Sasmal, H.S.; Kumar, S.; Pachfule, P.; Banerjee, R. Constructing Ultraporous Covalent Organic Frameworks in Seconds via an Organic Terracotta Process. *J. Am. Chem. Soc.* **2017**, *139*, 1856–1862. [CrossRef]
39. Ansari, F.; Salajková, M.; Zhou, Q.; Berglund, L.A. Strong Surface Treatment Effects on Reinforcement Efficiency in Biocomposites Based on Cellulose Nanocrystals in Poly(vinyl acetate) Matrix. *Biomacromolecules* **2015**, *16*, 3916–3924. [CrossRef]
40. Zheng, T.; Clemons, C.M.; Pilla, S. Comparative Study of Direct Compounding, Coupling Agent-Aided and Initiator-Aided Reactive Extrusion to Prepare Cellulose Nanocrystal/PHBV (CNC/PHBV) Nanocomposite. *ACS Sustain. Chem. Eng.* **2020**, *8*, 814–822. [CrossRef]
41. Li, H.; Wu, B.; Mu, C.; Lin, W. Concomitant degradation in periodate oxidation of carboxymethyl cellulose. *Carbohydr. Polym.* **2011**, *84*, 881–886. [CrossRef]
42. Kandambeth, S.; Mallick, A.; Lukose, B.; Mane, M.V.; Heine, T.; Banerjee, R. Construction of crystalline 2D covalent organic frameworks with remarkable chemical (acid/base) stability via a combined reversible and irreversible route. *J. Am. Chem. Soc.* **2012**, *134*, 19524–19527. [CrossRef] [PubMed]
43. Ma, J.; Fu, X.-B.; Li, Y.; Xia, T.; Pan, L.; Yao, Y.-F. Solid-state NMR study of adsorbed water molecules in covalent organic framework materials. *Microporous Mesoporous Mater.* **2020**, *305*, 110287. [CrossRef]
44. Halder, A.; Ghosh, M.; Khayum, M.A.; Bera, S.; Addicoat, M.; Sasmal, H.S.; Karak, S.; Kurungot, S.; Banerjee, R. Interlayer Hydrogen-Bonded Covalent Organic Frameworks as High-Performance Supercapacitors. *J. Am. Chem. Soc.* **2018**, *140*, 10941–10945. [CrossRef]
45. Wang, P.; Wu, Q.; Han, L.; Wang, S.; Fang, S.; Zhang, Z.; Sun, S. Synthesis of conjugated covalent organic frameworks/graphene composite for supercapacitor electrodes. *RSC Adv.* **2015**, *5*, 27290–27294. [CrossRef]
46. Çalışkan, M.; Baran, T. Design of a palladium nanocatalyst produced from Schiff base modified dialdehyde cellulose and its application in aryl halide cyanation and reduction of nitroarenes. *Cellulose* **2022**, *29*, 4475–4493. [CrossRef]
47. Gao, R.; Xiao, S.; Gan, W.; Liu, Q.; Amer, H.; Rosenau, T.; Li, J.; Lu, Y. Mussel Adhesive-Inspired Design of Superhydrophobic Nanofibrillated Cellulose Aerogels for Oil/Water Separation. *ACS Sustain. Chem. Eng.* **2018**, *6*, 9047–9055. [CrossRef]

48. Zhong, X.; Liang, W.; Lu, Z.; Qiu, M.; Hu, B. Ultra-high capacity of graphene oxide conjugated covalent organic framework nanohybrid for U(VI) and Eu(III) adsorption removal. *J. Mol. Liq.* **2021**, *323*, 114603. [CrossRef]
49. Cui, S. Thermal Degradation of As-Synthesized MOFs Studied by TG-FTIR. *Sepectroscopy Spectr. Anal.* **2012**, *32*, 131–132. [CrossRef]
50. Pérez-Carvajal, J.; Boix, G.; Imaz, I.; Maspoch, D. The Imine-Based COF TpPa-1 as an Efficient Cooling Adsorbent That Can Be Regenerated by Heat or Light. *Adv. Energy Mater.* **2019**, *9*, 1901535. [CrossRef]
51. Gopalakrishnan, V.N.; Nguyen, D.-T.; Becerra, J.; Sakar, M.; Ahad, J.M.E.; Jautzy, J.J.; Mindorff, L.M.; Béland, F.; Do, T.-O. Manifestation of an Enhanced Photoreduction of CO2 to CO over the In Situ Synthesized rGO–Covalent Organic Framework under Visible Light Irradiation. *ACS Appl. Energy Mater.* **2021**, *4*, 6005–6014. [CrossRef]

polymers

MDPI

Article

Chemical and Structural Elucidation of Lignin and Cellulose Isolated Using DES from Bagasse Based on Alkaline and Hydrothermal Pretreatment

Na Wang [1], Baoming Xu [1], Xinhui Wang [1], Jinyan Lang [1] and Heng Zhang [1,2,*]

[1] College of Marine Science and Biological Engineering, Qingdao University of Science and Technology, Qingdao 260412, China; wlalala21@163.com (N.W.); 14763738886@163.com (B.X.); wxhameq@163.com (X.W.); ljy17806248212@163.com (J.L.)

[2] Guangdong Provincial Key Lab of Green Chemical Product Technology, Guangzhou 510640, China

* Correspondence: hgzhang@sina.com

Abstract: The separation of cellulose, hemicellulose, and lignin components using deep eutectic solvent, which is a green solvent, to obtain corresponding chemicals can realize the effective separation and high-value utilization of these components at low cost. In this study, we used waste biomass sugarcane bagasse as the raw material, choline chloride as the hydrogen bond acceptor, and lactic acid as the hydrogen bond donor to synthesize a deep eutectic solvent of choline chloride/lactic acid (L-DES) and treated sugarcane bagasse pretreated by alkali or hydrothermal methods to separate cellulose, hemicellulose, and lignin. In addition, we comparatively studied the effect of different pretreatment methods on lignin removal by DES and found that the lignin removal rate by L-DES after alkaline pretreatment was significantly higher than that after hydrothermal pretreatment, and the mechanism of action causing this difference is discussed.

Keywords: bagasse; DESs; separation; lignin

Citation: Wang, N.; Xu, B.; Wang, X.; Lang, J.; Zhang, H. Chemical and Structural Elucidation of Lignin and Cellulose Isolated Using DES from Bagasse Based on Alkaline and Hydrothermal Pretreatment. *Polymers* **2022**, *14*, 2756. https://doi.org/10.3390/polym14142756

Academic Editors: Luis Alves and Shuangquan Yao

Received: 25 May 2022
Accepted: 2 July 2022
Published: 6 July 2022

1. Introduction

Finding an economical and efficient development pathway is important to achieve the high-value utilization of biomass resources. Data crop residues account for up to 54% of waste biomass resources [1]. Among the residues, sugarcane bagasse, which is produced abundantly, yields approximately 540 million tons worldwide every year [2].

Sugarcane bagasse is rich in cellulose, which can be used as a raw material for pulp and paper making, and its rich hemicellulose content is also conducive to the realization of various industries, such as xylose and xylitol; more than 200 value-added chemicals can be produced by refining lignocellulosic biomass [3], such as ethanol, levulinic acid, furan, sorbitol, xylitol, glycerol, and their derivatives [4]. Therefore, the prospects for the utilization of sugarcane bagasse as a waste biomass resource are very promising.

Deep eutectic solvent (DES) is an environmentally friendly green solvent that can solubilize lignin and increase the hydrophilicity of lignocellulose, thus increasing the glycation rate of lignocellulose. Hou et al. [5] found that the ester bond of the linkage between lignin and hemicellulose was broken from the infrared of lignocellulose treated with DESs, the hydrogen bond of the linkage between hemicellulose and cellulose was broken, and hemicellulose was hydrolyzed into oligosaccharides dissolved in DESs. Many studies have shown that DESs can be used as an effective solvent for the lignin separation of bagasse lignin, cellulose, and hemicellulose fractions, with a lignin removal of up to 81% [6–10].

As shown in Figure 1, hemicellulose acts as a binder to closely interpenetrate lignin and cellulose. Hemicellulose and lignin have chemical connections, and hydrogen bonding connections and Van der Waals forces exist between them and cellulose. Therefore, separating the three kinds of material while removing the hemicellulose is difficult and complicated. In addition, the glycosides bond of hemicellulose breaks under acidic conditions, leading to the degradation of hemicellulose into oligosaccharides [11]. Given that DES is acidic, the separation efficiency and purity of lignin can be further improved if the hemicellulose is pre-separated to improve the implementability of the DES separation of lignin and avoid the difficulty of separating hemicellulose and lignin by turning them into small molecules after acid hydrolysis.

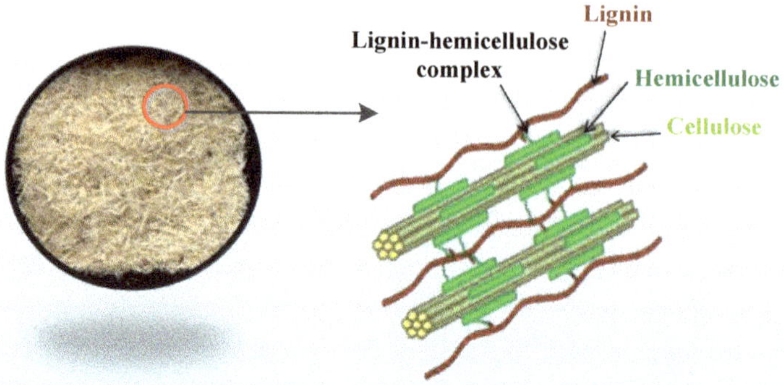

Figure 1. Three-dimensional structure of cellulose, hemicellulose, and lignin in bagasse.

Hemicellulose glycans can be dissolved in alkaline solution or hydrolyzed by dilute acid. The commonly used methods for hemicellulose separation include dilute acid hydrolysis, alkali extraction, hydrothermal extraction, and organic solvent extraction. Considering the concept of green process flow, the alkaline and hydrothermal hemicellulose extraction methods were selected and achieved highly satisfactory results [12,13].

However, the current study used DESs to remove lignin from sugarcane bagasse, and the products still retained hemicellulose, which did not report the effect of the pretreatment by alkali and hydrothermal methods on the subsequent lignin separation, regarding whether it would favor the lignin solubilization, or on the lignin structure.

In order to further improve the lignin removal rate and achieve effective separation of the components, therefore, a two-step treatment method was used in this study. Cellulase lignin (CEL), which has few structural alterations of lignin [14], was extracted from sugarcane bagasse feedstock as a representative of the total lignin in lignocellulosic feedstock, and choline chloride was used as the hydrogen bond acceptor and lactic acid as the hydrogen bond donor to synthesize the deep eutectic solvent of choline chloride/lactic acid (L-DES), which had good lignin removal and was essentially insoluble in cellulose in the previous experiments. The sugarcane bagasse pretreated by hydrothermal process and then treated by L-DES was hydrothermal lignin (HL), and after alkali pretreatment and L-DES treatment, the crude lignocellulosic was alkali lignin (AL). The structural changes of crude lignin HL and AL were compared and analyzed, and the differences in the mechanisms of lignin separation between the alkaline and hydrothermal pretreatments were also compared.

2. Materials and Method

2.1. Materials

Choline chloride, sodium acetate and hydrochloric acid were purchased from Sinopharm Chemical Reagent Co., Ltd. (Shanghai, China). Lactic acid was obtained from Shanghai Rinn Technology Development Co., Ltd. (Shanghai, China). Cellulase and dioxane were purchased from Shanghai Rinn Reagent Co., Ltd. (Shanghai, China). All the above reagents are analytically pure.

2.2. Experimental Methods

(1) Synthesis of L-DES

The sugarcane bagasse feedstock was dried and de-watered for 24 h before the experiment, and a molar ratio of choline chloride:lactic acid (1:9) was placed in a flask, heated, and stirred with N_2 at 80 °C. The reaction was carried out for 1–2 h. When a uniform and transparent solution was obtained, the reaction was allowed to go on for another 20 min, further stabilizing the formed solution, and then the reaction was stopped. Finally, L-DES was obtained.

(2) Separation method of hemicellulose, cellulose, and lignin

1. Alkali separation of sugarcane bagasse hemicellulose

After benzene-alcohol extraction, bagasse was weighed, reacted in the alkali solution at a high temperature for a period of time, and then kept warm. After holding, the extract was poured into a beaker, the flask was washed three times with deionized water, and the washing solution was poured into the beaker.

The filtrate and the residue were separated by suction filtration. The residue was washed with deionized water until neutral, and the cake residue (bagasse after alkaline extraction, residue AB for short) was dried to a constant weight and used for the subsequent separation experiments. The filtrate (filtrate and the washing solution for cleaning the residue) was adjusted to a pH of approximately 5.5 using glacial acetic acid, left to stand, and then centrifuged. The precipitate was dried to obtain hemicellulose A. After the supernatant was concentrated by rotary evaporation, thrice the volume of 95% ethanol was added and allowed to stand for 2 h. The precipitate was washed with 70% ethanol after suction filtration and dried to obtain hemicellulose B (crude xylan sample). The supernatant was concentrated and dried to obtain alcohol-soluble lignin.

2. Hydrothermal separation of sugarcane bagasse hemicellulose

The bagasse, after benzene-alcohol extraction, was weighed and insulated for a period of time in a hydrothermal autoclave. After the insulation reaction, the hydrothermal autoclave was left to stand overnight under natural conditions and cooled to room temperature. The extract was poured into a beaker, the hydrothermal autoclave was washed three times with deionized water, and the washing solution was poured into the beaker.

The filtrate and residue were separated by suction filtration. After washing the residue several times with deionized water, the filter cake residue (bagasse after hydrothermal extraction, residue HB for short) was dried to a constant weight and used for the subsequent separation experiments. The filtrate (filtrate and the washing solution for cleaning the residue) was concentrated by distillation, and the hemicellulose was precipitated with thrice the volume of 95% ethanol. The experimental study showed that the hemicellulose yield was the highest when three times the volume of ethanol was used. Finally, the precipitate was centrifuged and freeze-dried to obtain a crude hemicellulose product, which was weighed to calculate the hemicellulose removal rate.

3. DES separation of sugarcane bagasse lignin

The dried residue obtained from the above two pretreatment methods was wetted with water (to reduce the lignin softening temperature) and reacted with L-DES at a high temperature for a certain time. After the reaction, the reaction solution was poured into a beaker, the four-necked flask was washed three times with anhydrous ethanol, and the washing solution was poured into the beaker.

The filtrate and the residue were separated by suction filtration, and the residue was washed with anhydrous ethanol for neutralization and dried in an oven to a constant weight to obtain crude cellulose. The filtrate was diluted with a large amount of deionized water and left to stratify, filtered to obtain a crude lignin solid, and dried to a constant weight and weighed to record the mass of crude cellulose and lignin. The filtrate was distilled from a large amount of anhydrous ethanol and deionized water using a rotary evaporator to obtain a small amount of viscous liquid, which was collected and recovered to obtain the DES for reuse. A flow chart is shown in Figure 2.

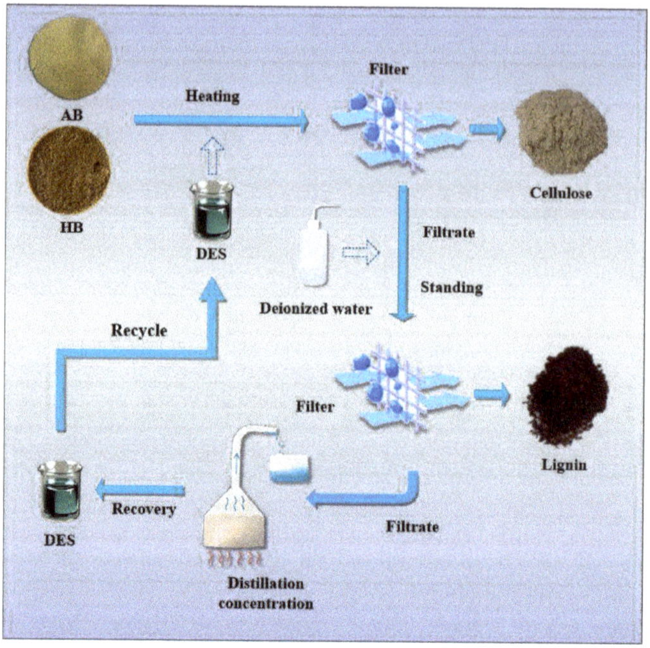

Figure 2. Flow chart of DES separation of lignin and cellulose.

(3) Separation and extraction of cellulolytic enzyme lignin (CEL)

An enzymatic hydrolysis reaction was carried out by adding cellulase and acetic acid-sodium acetate buffer to the ground bagasse powder [15,16]. The precipitated residue after the enzymatic treatment was washed several times with acetic acid-sodium acetate buffer and deionized water and then freeze-dried. The residue was extracted using 96% aqueous dioxane solution, and the supernatant was concentrated by evaporation [16]. The concentrated solution was left to stand for precipitation with a hydrochloric acid solution, and the CEL precipitated out.

2.3. *Characterization of Cellulose and Lignin*

A small amount of dried bagasse, bagasse AB and HB after the respective hydrothermal and alkaline pre-extractions, cellulose AC and HC, and lignin AL and HL, AC which were separated after L-DES treatment, were compressed into tablets using KBr. Infrared measurements were taken using a VECTOR22 Fourier transform infrared spectrometer (Bruker Corporation, Ettlingen, Germany), with a scanning range of 4000–400 cm^{-1}, 60 scan times, a resolution of 4 cm^{-1}, and a signal-to-noise ratio of 55,000:1.

Crude cellulose AC and HC and lignin AL and HL were examined by scanning electron microscopy (SEM) using a Regulus 8100 SEM (Japan Electronics Co., Ltd., Tokyo, Japan) after hydrothermal and alkaline pre-extractions and L-DES treatment to observe the difference between the crude cellulose and lignin obtained by hydrothermal and alkaline extractions. The differences between crude cellulose and lignin were observed.

A TG209F1 thermogravimetric analyzer (Bruker Corporation, Ettlingen, Germany) was used to test the residues (AB and HB) obtained after the hydrothermal and alkaline pre-extractions, respectively, and the thermogravimetric behavior of crude cellulose AC and HC treated by L-DES was analyzed. A sample of approximately 10 mg was weighed, the temperature was increased from 25 °C to 700 °C at a rate of 10 °C/min under N_2 protection, and the ventilation rate was 25 mL/min.

We accurately weighed 5 mg of cellulase lignin and the lignin samples AL and HL obtained by dissolving and separating with L-DES after alkaline and hydrothermal extractions, respectively, and set them in 5-mL centrifuge tubes. The pipette drew 5 mL of the chromatographically pure tetrahydrofuran (THF) to dissolve the lignin, and the centrifuge tube was placed in an ultrasonic wave for 2 h to dissolve the lignin and prepare the solution to be tested. A standard sample of polystyrene with a molecular weight of 6.9×10^4 g/mol was used to calibrate the existing molecular weight standard curve, and the supernatant was aspirated into the sample with an injection needle for analysis. Gel permeation chromatography (GPC, Agilent, Santa Clara, CA, USA) was performed on a small molecular weight column tandem with a chromatographically pure THF eluent and a differential refractive index detector. The molecular weights of the lignin samples were calculated from the standard curve of the standard samples.

An elemental analyzer (Vario EL III, Langenselbold, Hesse, Germany) was used to determine the C, H, O, and N in bagasse and crude lignin AL and HL, and the results are all in mass fraction.

3. Results and Discussion

3.1. Comparison of Lignin Removal Rates by Hydrothermal Extraction Method and Alkali Extraction Method

After the hydrothermal and alkali extractions, the lignin was removed by L-DES treatment, and the effects of the two methods on the lignin removal rate and the cellulose content and yield are shown in Figure 3.

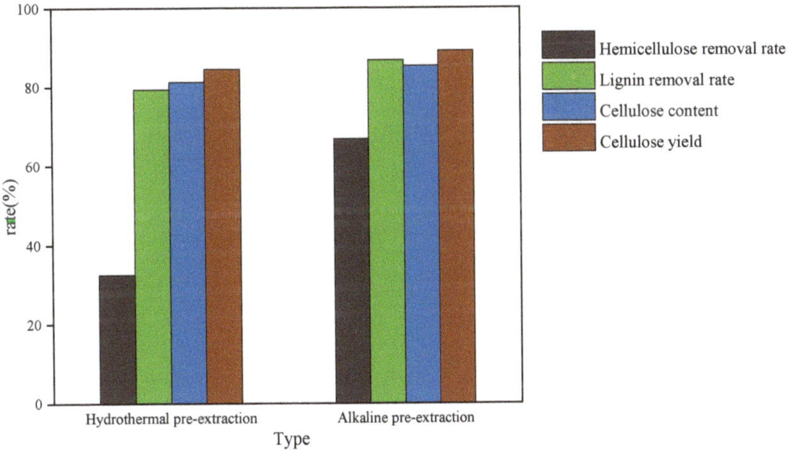

Figure 3. Effect of hydrothermal or alkaline pretreatment on the removal rate of lignin and hemicellulose, cellulose content and yield.

Figure 3 shows a comparison of the test indicators, such as the lignin removal rate, under the two pretreatment methods. Under the best experimental conditions: solvent L-DES, bagasse: DES = 1:25, temperature of 110 °C, 12 h, alkali pretreated hemicellulose removal rate of 66.84%, lignin removal rate of 86.7%, and respective cellulose content and yield of 85.3% and 89.1%. The removal rate of hemicellulose by hydrothermal pretreatment was 32.57%, the removal rate of lignin was 79.6%, and the respective cellulose content and yield rate were 81.2% and 84.5%. The comparison shows that the removal rate of lignin after alkaline pretreatment was higher than that after hydrothermal pretreatment because under the action of alkaline, new phenolic hydroxyl (phenoxy anion) could be derived from lignin, which is conducive to the dissolution of lignin due to the hydrophilic property of phenolic hydroxyl. In addition, according to the literature on the treatment of bagasse with DESs, L-DES is used to dissolve lignin in the residue after alkaline pretreatment. This work could obtain the highest removal rate of lignin according to the existing studies on lignin removal from bagasse with DESs treatment [6].

3.2. Structure Analysis of Crude Lignin and Cellulose by Infrared Spectroscopy

The infrared spectra of bagasse, bagasse after hydrothermal and alkali pre-extraction treatments, and bagasse after L-DES treatment are shown in Figures 4 and 5.

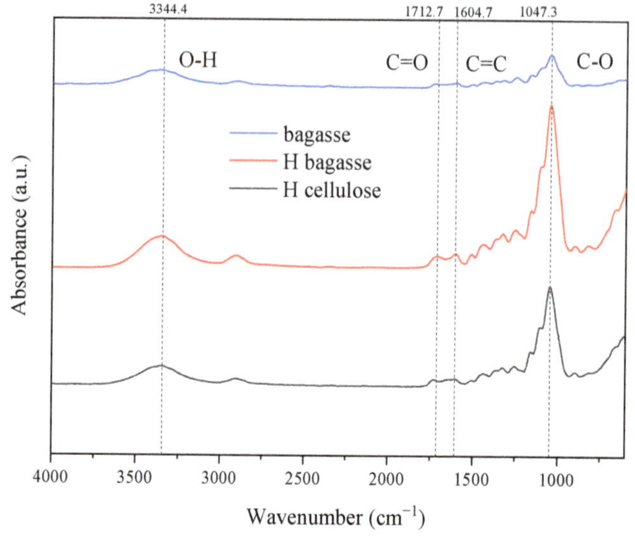

Figure 4. Infrared spectra of bagasse, HB and HC.

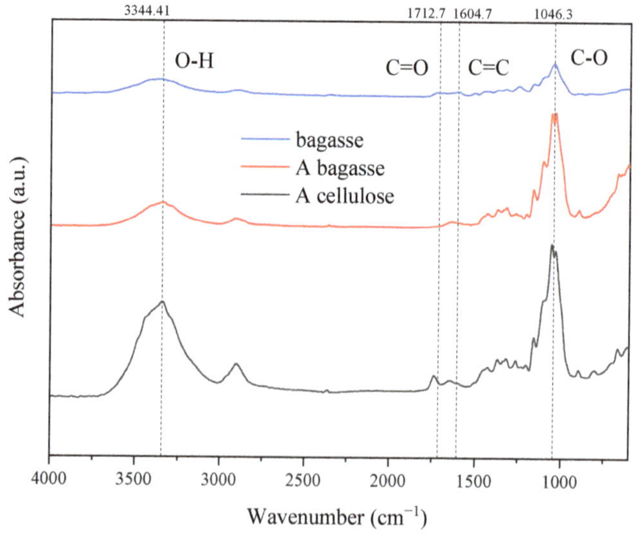

Figure 5. Infrared spectra of bagasse, bagasse after alkaline treatment (AB) and crude cellulose after L-DES treatment (AC).

The infrared spectra of bagasse, HB, and HC are shown in Figure 4. The broad peak at 3344 cm^{-1} is the -OH stretching vibration peak, and the -OH content is highest in cellulose HC. The -OH content increased relatively due to the removal of hemicellulose. The stretching vibration peak of C-H is at 2904 cm^{-1}, including the -CH and -CH_2 of the saturated hydrocarbon group and the -CH_3 of the methoxy group on the benzene ring. The bending vibration peak of C-H at 1330 cm^{-1}, 825 cm^{-1} is the C-H vibration peak linked with the benzene ring [17], 1604 and 1510 cm^{-1} are the skeleton vibration absorption peaks of benzene ring, 1242 cm^{-1} is the C-O vibration absorption peak of the phenolic hydroxyl group, and 1045 and 1101 cm^{-1} are the respective bending and stretching vibration peaks of the ether bond on the benzene ring, which represent the characteristic absorption peaks of lignin. The absorbance of these characteristic peaks in cellulose HC decreased significantly, indicating that lignin was removed in large quantities. However, the absorbance of these characteristic peaks in residue HB increased slightly because the lignin content in residue HB increased relatively with the removal of hemicellulose, so the characteristic peaks were enhanced slightly.

The strong absorption peak at 1101 cm^{-1} is also the stretching vibration peak of the alicyclic ether C-O-C, which is the alicyclic ether bond in cellulose and hemicellulose five- and six-carbon sugars. The intensity of the characteristic peak decreased obviously with the removal of lignin and hemicellulose. The C=O vibration absorption peak of hemicellulose is at 1712 cm^{-1}, which is the characteristic absorption peak of C=O on the acetyl group [18]. The absorbance of C=O in residue HB did not decrease obviously because the hemicellulose removal rate through the hydrothermal extraction method in this research was not high. The characteristic peak intensity in cellulose HC is weak because hemicellulose was dissolved in DESs. The β-1, 4-glycosidic bond of cellulose is located at 819 cm^{-1}. After L-DES treatment, a part of the glycosidic bond of cellulose was broken, so its strength decreased slightly.

Figure 5 shows the IR spectra of sugarcane bagasse, bagasse AB after alkali treatment, and the crude cellulose AC obtained after L-DES treatment. The characteristic peak of lignin weakened from the bagasse to residue AB to cellulose AC, indicating that the alkali extraction removed part of the lignin from the bagasse, while the L-DES solvent removed most of the lignin from the bagasse. The intensity of the C=O vibrational absorption peak of hemicellulose xylose at 1735 cm^{-1} showed a similar general trend, the hemicellulose in sugarcane bagasse was gradually removed by alkali extraction and the L-DES.

The peak at 3344 cm^{-1} is assigned to the -OH stretching of cellulose, and the peak at 2904 cm^{-1} is assigned to the C-H vibration of cellulose. The strength of the -OH and C-H stretching vibration peaks were all weakened significantly after the alkali extraction and the L-DES treatment, suggesting that the hydroxyl content in cellulose gradually decreased. The β-1, 4-glycosidic bond of cellulose is located at 819 cm^{-1}, the glycosidic bond in the residue AB was broken and its strength was significantly decreased, indicating that the alkali caused some degree of damage to the cellulose structure. The -OH stretching vibration bands blue-shifted from the bagasse to residue AB, and the wavelength from 3344 cm^{-1} to 3371 cm^{-1}; from residue AB to cellulose AC, the wavelength shifted from 3371 cm^{-1} to 3404 cm^{-1}; the bending vibration peaks of C-H red-shifted from the bagasse to residue AB, and the wavelength from 1321 cm^{-1} to 1316 cm^{-1}; from residue AB to cellulose AC, the wavelength shifted from 1316 cm^{-1} to 1313 cm^{-1}. These data suggest that the intermolecular hydrogen bonds of cellulose suffered from the disruption of alkali and L-DES, which also inevitably led to structural changes in the crystalline region of cellulose [19].

Figure 6a shows the spectra of AL and HL, and Figure 6b shows the spectra of cellulase lignin.

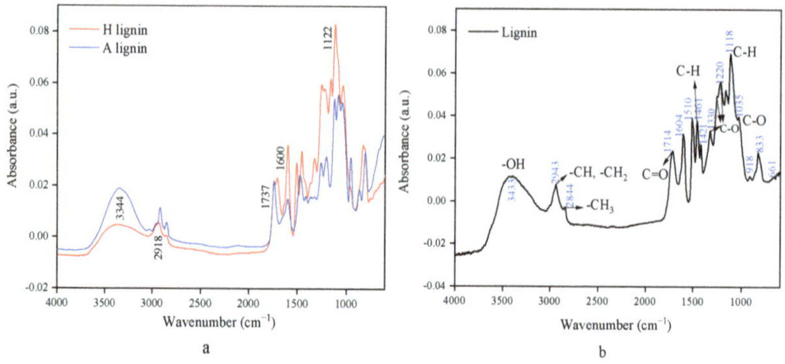

Figure 6. Infrared spectra of alkaline lignin AL and hydrothermal lignin HL separated by L-DES after alkaline pre-extraction and hydrothermal pre-extraction (**a**) and cellulase lignin (**b**).

Cellulase lignin is extracted by solubilizing cellulose and hemicellulose using cellulase enzymes, and its structure is the most similar to the original natural lignin structure. Figure 6b shows the peaks of the telescopic vibrations of the phenolic and aliphatic hydroxyl groups of lignin at 3433 cm^{-1}, the C-H telescopic vibrations of -CH and $-CH_2$ at 2943 cm^{-1}, the C-H telescopic vibrations of $-CH_3$ at 2844 cm^{-1}, and the C-H asymmetric bending vibrations of methoxy at 1461 cm^{-1} and 1421 cm^{-1}, and the fingerprint region at 833 cm^{-1} is the C-H vibration peak of the benzene ring linkage. The conjugated C=O stretching vibration peak in lignin is located at 1714 cm^{-1} [20], which is the C=O bond in the ketone or the aldehyde group of lignin. The basic structural units in bagasse lignin are the lilac- and guaiac-based phenylpropane units and a small amount of the p-hydroxyphenyl phenylpropane unit. Among the phenylpropane units, the lilac-based phenylpropane unit is the most abundant, and the C-O deformation vibration of this phenylpropane unit is located at 1330 cm^{-1}. The C-H in-plane bending vibration of the phenylpropane unit is located at 1118 cm^{-1}; 1263, 1220, 1220, and 1035 cm^{-1} are the peaks of the C-O stretching vibrations in lignin; 1220 cm^{-1} is the C-O stretching vibration in the guaiacyl phenylpropane unit [21]; 1035 cm^{-1} is the C-O stretching vibration of the secondary alcohol or fatty ether structure [22]; and 1604 and 1510 cm^{-1} are the absorption peaks of the benzene ring skeleton vibrations [23].

The comparison of Figure 6a, b indicates that the hydroxyl content of HL decreased slightly, which may be attributed to the reaction of the lignin hydroxyl group with the chloride ion of L-DES, and the reaction reached the high softening temperature of lignin with partial methoxy removal. The intensity of the bending vibrational peak within the C-H plane of the AL decreased significantly, showing that the phenyl propane unit was damaged. Moreover, the -OH stretching vibration peak in lignin blue-shifted and the C-H asymmetric bending vibration red-shifted, suggesting that the hydrogen bonding structure of alkali lignin was changed by the influence of alkali and the strength of hydrogen bonding interaction slightly decreased. The red-shift of the C-O stretching vibration peak in alkali lignin demonstrates that the electron cloud density between its C-O bonds is lower than that of the original lignin, the bond strength constant decreased, the bond strength weakened, and the vibration frequency decreased. This phenomenon indicates that alkali caused some C-O bonds to break.

3.3. Morphology Analysis and Comparison of Crude Cellulose AC and HC

After detecting the crude cellulose AC and HC by SEM, the difference between the crude cellulose HC and AC obtained by hydrothermal and alkaline extractions was observed. The scanning electron micrographs of crude cellulose HC and AC are shown in Figure 7.

Figure 7. Scanning electron micrographs of crude cellulose AC (**a**) and crude cellulose HC (**b**).

Figure 7a shows crude cellulose AC amplified by 1000 and 500 times; Figure 7b shows crude cellulose HC amplified by 500 and 200 times. Unlike that of crude cellulose AC, the surface of crude cellulose HC is relatively smooth. In addition, the difference between the structures of crude cellulose HC and AC is not obvious, and their diameters are also similar. The fiber bundles were subjected to L-DES. L-DES dissolved the hemicellulose as adhesive and degraded it to small molecule oligosaccharides. In addition, the ether bonds of lignin were broken and degraded into small molecule guaiac-based compounds, leading to the dispersion of the fiber bundles.

3.4. Analysis of Thermogravimetric Behavior

The thermogravimetric behaviors of residues AB and HB obtained after bagasse, hydrothermal and alkaline pre-extractions and those of crude cellulose AC and HC after L-DES treatment are shown in Figure 8.

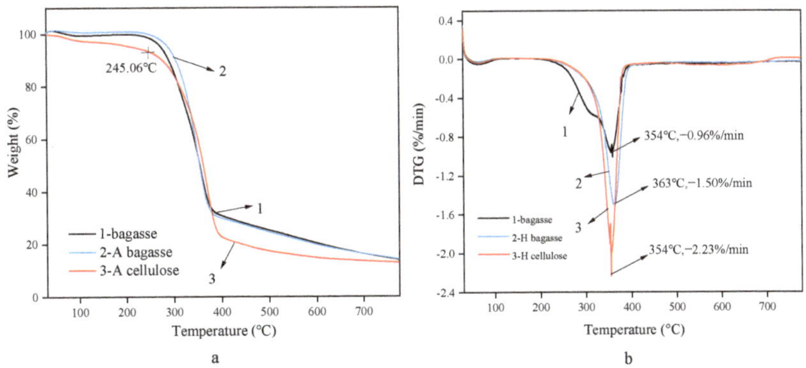

Figure 8. Thermogravimetric curves TG (**a**) and DTG (**b**) of two stages of hydrothermal extraction.

(Among them, 1-bagasse; 2-bagasse HB after hydrothermal extraction; 3-cellulose HC after hydrothermal extraction).

The comparison of the thermogravimetric curves of bagasse, residue HB, and crude cellulose HC is shown in Figure 8. The three curves between 260 °C and 380 °C show a significant decline in mass. The mass loss rate is very high, reaching the maximum rate of weightlessness between 350 °C and 365 °C. The maximum rates of weightlessness on the three curves are 0.96%/min, 1.50%/min, and 2.23%/min, respectively.

The thermogravimetric processes of bagasse and cellulose can be divided into three periods; they are turned into solid residue and gases, such as CH_4 and CO, through thermal decomposition. The first period is mostly the process of losing free water. During this process, the weight of residue HB and cellulose HC was basically unchanged, while the weight loss of bagasse was 2.38%. Bagasse had the highest free water content. The second period is the main period for the thermal decomposition of bagasse and cellulose. Bagasse had the lowest initial decomposition temperature, which was 260 °C, because the unstable hemicellulose and cellulose were decomposed under the influence of the high temperature and pressure. Therefore, the HB structure of the remaining cellulose is relatively stable and has a higher decomposition temperature. The maximum weight loss rate was also achieved in this period, and the maximum weight loss rate temperature is in the range of 350 °C to 365 °C. The weight loss of bagasse was approximately 70%, the weight loss of residual HB was 73%, and the weight loss of cellulose HC was 78%, suggesting that cellulose HC has the highest carbon content because it has the highest cellulose content. Bagasse, residue HB, and cellulose HC entered the third weightless period from approximately 377 °C to approximately 775 °C. Cellulose HC had the lowest quality of raw material residue, which was 7.83%, while those of residue HB and bagasse were 14.96% and 13.86%, respectively. It was observed that the HC particle size of the cellulose treated by DES decreased in the experiment and the specific surface and heated areas increased, facilitating the decomposition. In addition, the ash content of cellulose is less than that of lignin. After DESs treatment, the lignin content significantly decreased, so the residual content of cellulose HC was much lower than residue HB. The leftover content of residual HB is higher than that of bagasse because at the same weight, the lignin content in residual HB and the non-decomposed part is higher, so the residual quality is also higher.

The thermogravimetric curves of bagasse, residue AB, and crude cellulose AC are compared in Figure 9.

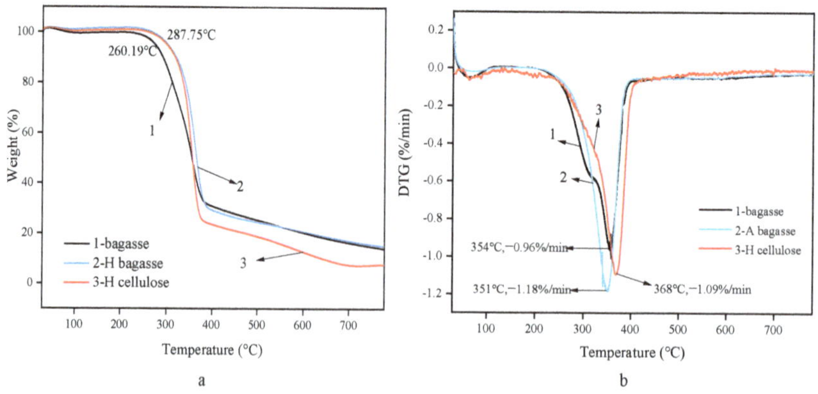

Figure 9. Thermogravimetric curves TG (**a**) and DTG (**b**) of two stages of alkaline extraction.

(Thereinto,1-bagasse; 2-bagasse extracted by alkali AB; 3-alkali cellulose Ac).

In the 240 °C to 390 °C range, the three curves exhibit a distinct descent and a very high mass loss rate. The respective maximum weight loss rates were 0.96%/min, 1.18%/min, and 1.09%/min in the 350 °C to 370 °C range.

The first period of thermogravimetry is mainly the loss of free water. In this process, the weight loss of bagasse and residual AB was very small, while the weight loss of cellulose AC was relatively high (6.31%), probably because cellulose absorbs water and expands under alkaline conditions, thereby increasing the free water content. The initial decomposition temperature of the three was approximately 245 °C, and they entered the second interval from 245 °C. The pyrolysis rates of bagasse and cellulose were basically the same, and the maximum weight loss rate was in the 350 °C to 370 °C range. The weight loss of bagasse was approximately 70%; that of residue AB was 71%; and that of cellulose AC was 73%, suggesting that cellulose AC had the highest carbon content. The third weightlessness period was from approximately 390 °C and terminated at about 775 °C. The weights of the raw material residues of the three are similar; the residue weights of bagasse and residual AC are the same weight, that is, 14.05%, and that of cellulose AC was slightly lower, that is, 12.93%. The residue quality of cellulose AC was higher than that of cellulose HC because the metal salt ions promoted coke formation.

3.5. Analysis of Lignin Molecular Weight

The results of the AL and HL samples obtained by CEL and alkaline and hydrothermal extractions followed by dissolution and separation with L-DES were measured by GPC, as shown in the table below. The number-average molecular weight (Mn), heavy-average molecular weight (Mw), Z-average molecular weight (Mz), Z+1-average molecular weight (Mz1), and polymerization distribution width index D of lignin were calculated from the calibration curve measured by the GPC system. Then, the heavy-average number-average ratio (Mw/Mn) and the Z-average heavy-average ratio Mz/Mw were calculated.

As shown in the Table 1, both lignin average molecular masses Mn obtained using L-DES treatment are slightly lower than that of CEL, ranging from 1900 to 3500. The molecular weight of lignin AL is lower than that of lignin HL because alkali disrupted the structure of lignin and the various ether bonds in lignin broke under the action of alkali, resulting in the decrease in polymerization and lignin molecular weight. The lower the numbers of aromatic ether bonds and condensed structures in lignin are, the lower the molecular weight is [24].

Table 1. Molecular weight of alkaline and hydrothermal lignin.

Sample	Mn	Mw	Mz	Mz1	Mw/Mn	Mz/Mw
CEL	3793	5319	6819	7942	1.40232	1.28200
AL	1904	2347	2782	3279	1.23267	1.18534
HL	3448	3973	4620	5346	1.15207	1.16294

High-temperature water extraction had little effect on the lignin structure, and the molecular weight of HL was not as large as the average molecular weight of natural lignin mainly because the reaction conditions were not mild enough. The solvent L-DES separation of lignin has a relatively strong effect on the lignin structure because the strong hydrogen bonding of lactic acid leads to a strong lignin degradation reaction [25], so that the molecular weight of the lignin obtained is usually small. However, the effect of L-DES on the structure of lignin is significantly smaller than that of lye.

Among the lignin separation and extraction methods, the difference in molecular weight between ground wood lignin (MWL) and CEL and the original lignin [26] is the smallest, although MWL and CEL have much lower molecular weights than the original lignin. For the Mw fraction, the molecular weight can be as high as 20,000 in order of magnitude, which is actually difficult to reach for the extracted lignin molecular weight. The polymerization distribution width index (D) can reflect the polydispersity of lignin and indicates the inhomogeneity of the relative molecular mass of lignin. The polymerization distribution width index of lignin was significantly lower after L-DES treatment, indicating that lignin formed subunits with a relatively small molecular weight and a more uniform structure through the degradation process [20]. The polymerization distribution width index (D) of lignin AL was 1.23, which is slightly higher than that of lignin HL of 1.15, indicating a highly uneven distribution of its lignin relative molecular mass. This phenomenon is due to the uneven destruction of the lignin structure by the alkali solution; the lignin structure on the surface was more damaged than that inside, which also led to the uneven distribution of its relative molecular masses. However, the polymerization distribution width index (D) values of 1.15–1.23 are still generally low, and the lignin AL and HL relative molecular masses are still distributed relatively uniformly.

3.6. Elemental Analysis

The results of the determination of the C, H, N, and O contents in crude cellulose and lignin using an elemental analyzer are shown in the Table 2.

Table 2. Elemental composition of alkaline and hydrothermal lignin.

Sample	C/%	H/%	O/%	N/%
Bagasse	52.32	8.51	33.71	1.26
CEL	62.10	6.32	27.94	1.15
AL	65.47	6.45	25.83	1.17
HL	63.53	6.09	27.41	1.05

As shown in Table 2, the carbon content of lignin AL was significantly higher than that of lignin HL, indicating that more lignin was extracted by alkaline pre-extraction. The result suggests that alkaline pre-extraction can dissolve lignin better than hydrothermal pre-extraction to obtain a higher rate of lignin extraction. This is also consistent with the TG analysis results that the molecular weight of lignin AL is low, and the residue mass of lignin AL is high. Both lignin AL and HL have a higher C content but a slightly lower O content than CEL because lignin is rich in methoxy, carbonyl, and hydroxyl groups and ether, carbon-oxygen, carbon-carbon, and other connecting bonds. The lower O content may indicate that reactions occurred, such as methoxy shedding and ether bond breaking. The H content of lignin AL was higher than that of lignin HL, probably because more phenolic hydroxyl groups were generated by the alkali method and the shedding of some of the carbonyl groups and breaking of some of the C=C bonds.

Given that lignin is an aromatic polymer with high unsaturation, the C content of all three lignin was higher and the H content was slightly lower than those of the sugarcane bagasse feedstock. The carbohydrates in the sugarcane bagasse feedstock contain many hydroxyl and alicyclic ether structures and are less unsaturated than lignin, so the O and H contents are higher than those of lignin.

4. Conclusions

Structural analysis revealed that both alkaline and hydrothermal pretreatments had good lignin removal efficiency. The hydroxyl content of hydrothermal lignin HL decreased slightly, the hydrogen bond strength of alkaline lignin AL weakened, and the phenyl propane unit was damaged, which indicated that both alkaline and hydrothermal pretreatments had some effects on the structure of lignin, while the alkaline pretreatment caused more effects on the structure of lignin. Alkali and L-DES caused some degree of damage to the β-1, 4 glycosidic bonds of cellulose, with alkali causing more damage to cellulose.

The L-DES-treated cellulose had an increased specific surface area, which is more easily decomposed with less residue mass and has a decreased lignin content. Compared with CEL, the polymerization distribution width index of lignin was significantly lower after L-DES treatment, indicating that lignin formed subunits with a relatively small molecular weight and a homogeneous structure through the degradation reaction.

Author Contributions: Writing—original draft preparation, N.W.; validation, B.X.; data curation, X.W.; investigation, J.L.; writing—review and editing, H.Z. All authors have read and agreed to the published version of the manuscript.

Funding: This research was funded by the Shandong Provincial Key Research and Development Program (SPKR&DP) (2019GGX102029) and the Research Fund Program of Guangdong Provincial Key Lab of Green Chemical Product Technology (Grant No. GC202112).

Institutional Review Board Statement: Not applicable.

Informed Consent Statement: Not applicable.

Data Availability Statement: Data sharing not applicable.

Acknowledgments: The author would like to graciously thank Yun Feng, Tong Yang, Linxi Chen, Jing Shi and Jie Bao for their insights, knowledge, and support.

Conflicts of Interest: The authors declare no conflict of interest.

References

1. Xie, G.H.; Wang, X.Y.; Ren, L.T. China's crop residues resources evaluation. *Chin. J. Biotechnol.* **2010**, *26*, 855–863.
2. Rajeshkumar, U.; Hariahran, V.; Sivakumar, G. Synthesis and characterization of activated carbon from unburned carbon bagasse fly ash. *Int. J. Phys.* **2016**, *2*, 25–28.
3. Isikgor, F.H.; Becer, C.R. Lignocellulosic biomass: A sustainable platform for the production of bio-based chemicals and polymers. *Polym. Chem.* **2015**, *6*, 4497–4559. [CrossRef]
4. Chandel, A.K.; Garlapati, V.K.; Singh, A.K.; Antunes, F.A.F.; Da Silva, S.S. The path forward for lignocellulose biorefineries: Bottlenecks, solutions, and perspective on commercialization. *Bioresour. Technol.* **2018**, *264*, 370–381. [CrossRef]
5. Hou, X.-D.; Smith, T.J.; Li, N.; Zong, M.-H. Novel Renewable Ionic Liquids as Highly Effective Solvents for Pretreatment of Rice Straw Biomass by Selective Removal of Lignin. *Biotechnol. Bioeng.* **2012**, *109*, 2484–2493. [CrossRef] [PubMed]
6. Satlewal, A.; Agrawal, R.; Das, P.; Bhagia, S.; Pu, Y.; Puri, S.K.; Ramakumar, S.S.V.; Ragauskas, A.J. Assessing the Facile Pretreatments of Bagasse for Efficient Enzymatic Conversion and Their Impacts on Structural and Chemical Properties. *ACS Sustain. Chem. Eng.* **2018**, *7*, 1095–1104. [CrossRef]
7. Huang, Z.J.; Yang, L.M.; Lin, K.P.; Feng, G.J.; Zhang, Y.H.; Pu, F.L.; Gu, R.R.; Hou, X.D. The difference for enzymatic hydrolysis of sugarcane bagasse pre-treated by different deep eutectic solvents. *Mod. Food Sci. Technol.* **2020**, *36*, 60–67.
8. Kumar, N.; Gautam, R.; Stallings, J.D.; Coty, G.G.; Lynam, J.G. Secondary Agriculture Residues Pretreatment Using Deep Eutectic Solvents. *Waste Biomass- Valorization* **2020**, *12*, 2259–2269. [CrossRef]
9. Liu, C., Li, M.-C.; Chen, W.; Huang, R.; Hong, S.; Wu, Q.; Mei, C. Production of lignin-containing cellulose nanofibers using deep eutectic solvents for UV-absorbing polymer reinforcement. *Carbohydr. Polym.* **2020**, *246*, 116548. [CrossRef]
10. Li, C.; Huang, C.; Zhao, Y.; Zheng, C.; Su, H.; Zhang, L.; Luo, W.; Zhao, H.; Wang, S.; Huang, L.-J. Effect of Choline-Based Deep Eutectic Solvent Pretreatment on the Structure of Cellulose and Lignin in Bagasse. *Processes* **2021**, *9*, 384. [CrossRef]
11. Li, Y.H. Study on Alkali Pre-Extraction of Bagasse Hemicellulose and Its Effect on Subsequent Pulping Performance. Master's Thesis, South China University of Technology, Guangzhou, China, 2019.
12. Sun, S.-L.; Wen, J.-L.; Ma, M.-G.; Song, X.-L.; Sun, R.-C. Integrated biorefinery based on hydrothermal and alkaline treatments: Investigation of sorghum hemicelluloses. *Carbohydr. Polym.* **2014**, *111*, 663–669. [CrossRef] [PubMed]
13. Sun, S.-N.; Cao, X.-F.; Li, H.-Y.; Xu, F.; Sun, R.-C. Structural characterization of residual hemicelluloses from hydrothermal pretreated Eucalyptus fiber. *Int. J. Biol. Macromol.* **2014**, *69*, 158–164. [CrossRef] [PubMed]
14. Jiang, B.; Cao, T.; Gu, F.; Wu, W.; Jin, Y. Comparison of the Structural Characteristics of Cellulolytic Enzyme Lignin Preparations Isolated from Wheat Straw Stem and Leaf. *ACS Sustain. Chem. Eng.* **2016**, *5*, 342–349. [CrossRef]
15. Nakagame, S.; Chandra, R.P.; Saddler, J.N. The effect of isolated lignins, obtained from a range of pretreated lignocellulosic substrates, on enzymatic hydrolysis. *Biotechnol. Bioeng.* **2010**, *105*, 871–879. [CrossRef] [PubMed]
16. Hu, Z.; Yeh, T.-F.; Chang, H.-M.; Matsumoto, Y.; Kadla, J.F. Elucidation of the structure of cellulolytic enzyme lignin. *Holzforschung* **2006**, *60*, 389–397. [CrossRef]
17. Schwanninger, M.; Rodrigues, J.; Pereira, H.; Hinterstoisser, B. Effects of short-time vibratory ball milling on the shape of FT-IR spectra of wood and cellulose. *Vib. Spectrosc.* **2004**, *36*, 23–40. [CrossRef]
18. Derkacheva, O.; Sukhov, D. Investigation of Lignins by FTIR Spectroscopy. *Macromol. Symp.* **2008**, *265*, 61–68. [CrossRef]
19. Xiong, L.; Yu, W.D. Analysis of the cellulose macromolecule structure after acid treatment by FITIR microspectroscopy. *J. Cellul. Sci. Technol.* **2013**, *21*, 59–62.
20. Li, L.; Wu, Z.; Xi, X.; Liu, B.; Cao, Y.; Xu, H.; Hu, Y. A Bifunctional Brønsted Acidic Deep Eutectic Solvent to Dissolve and Catalyze the Depolymerization of Alkali Lignin. *J. Renew. Mater.* **2021**, *9*, 219–235. [CrossRef]
21. Sun, R.; Fang, J.M.; Rowlands, P.; Bolton, J. Physicochemical and Thermal Characterization of Wheat Straw Hemicelluloses and Cellulose. *J. Agric. Food Chem.* **1998**, *46*, 2804–2809. [CrossRef]

22. Cheng, Y.; Zhai, S.-C.; Zhang, Y.-M.; Zhang, Y.-L. Simple evaluation of the degradation state of archaeological wood based on the infrared spectroscopy combined with thermogravimatery. *Spectrosc. Spectr. Anal.* **2020**, *40*, 2943–2950.
23. Van De Vyver, S.; Geboers, J.; Jacobs, P.A.; Sels, B.F. Recent Advances in the Catalytic Conversion of Cellulose. *ChemCatChem* **2010**, *3*, 82–94. [CrossRef]
24. Francisco, M.; Bruinhorst, A.V.D.; Kroon, M.C. New natural and renewable low transition temperature mixtures (LTTMs): Screening as solvents for lignocellulosic biomass processing. *Green Chem.* **2012**, *14*, 2153–2157. [CrossRef]
25. Loow, Y.-L.; New, E.K.; Yang, G.H.; Ang, L.Y.; Foo, L.Y.W.; Wu, T.Y. Potential use of deep eutectic solvents to facilitate lignocellulosic biomass utilization and conversion. *Cellulose* **2017**, *24*, 3591–3618. [CrossRef]
26. Meng, X.; Parikh, A.; Seemala, B.; Kumar, R.; Pu, Y.; Wyman, C.E.; Cai, C.; Ragauskas, A.J. Characterization of fractional cuts of co-solvent enhanced lignocellulosic fractionation lignin isolated by sequential precipitation. *Bioresour. Technol.* **2019**, *272*, 202–208. [CrossRef] [PubMed]

polymers

Article

Adsorption Mechanism of Chloropropanol by Crystalline Nanocellulose

Jinwei Zhao, Zhiqiang Gong, Can Chen, Chen Liang *, Lin Huang, Meijiao Huang, Chengrong Qin and Shuangfei Wang

Guangxi Key Laboratory of Clean Pulp & Papermaking and Pollution Control, School of Light Industrial and Food Engineering, Guangxi University, Nanning 530004, China; 1916391038@st.gxu.edu.cn (J.Z.); 15038502915@163.com (Z.G.); cc15125007558@163.com (C.C.); huanglin2841@163.com (L.H.); hmj0026@163.com (M.H.); qin_chengrong@163.com (C.Q.); wangsf@gxu.edu.cn (S.W.)

* Correspondence: liangchen@gxu.edu.cn

Abstract: Paper packaging materials are widely used as sustainable green materials in food packaging. The production or processing of paper materials is conducted in an environment that contains organic chlorides; therefore, potential food safety issues exist. In this study, the adsorption behavior of organic chlorides on paper materials was investigated. Chloropropanol, which has been extensively studied in the field of food safety, was employed as the research object. We studied the adsorption mechanism of chloropropanol on a crystalline nanocellulose (CNC) model. The results demonstrated that physical adsorption was the prevailing process, and the intermolecular hydrogen bonds acted as the driving force for adsorption. The adsorption effect assumed greatest significance under neutral and weakly alkaline conditions. A good linear relationship between the amount of chloropropanol adsorbed and the amount of CNC used was discovered. Thus, the findings of this study are crucial in monitoring the safety of products in systems containing chloropropanol and other chlorinated organic substances. This is particularly critical in the production of food-grade paper packaging materials.

Keywords: paper packaging; chloropropanol; crystalline nanocellulose; adsorption kinetics; quartz crystal microbalance

Citation: Zhao, J.; Gong, Z.; Chen, C.; Liang, C.; Huang, L.; Huang, M.; Qin, C.; Wang, S. Adsorption Mechanism of Chloropropanol by Crystalline Nanocellulose. *Polymers* **2022**, *14*, 1746. https://doi.org/10.3390/polym14091746

Academic Editors: Nathanael Guigo and Diego Antonioli

Received: 28 March 2022
Accepted: 20 April 2022
Published: 25 April 2022

1. Introduction

Plastic packaging offers protection and has a low packaging weight. Both factors have a positive impact on transportation and a longer service life. Therefore, plastic materials are widely used in food packaging [1]. Between 2018 and 2019, global plastic production reached 359 million tons [2], of which approximately 40% was used for packaging [3]. However, the applicability of plastic materials has been increasingly restricted owing to the non-degradability of plastics, release of hazardous chemicals during the process of manufacturing and use, as well as environmental pollution caused by landfills, incineration, or improper treatment after disposal [4–6]. Paper packaging materials possess the advantages of renewable resources [7] such as easy waste recycling ability and degradability [8]. They are considered the most promising green packaging materials. This can potentially play an important role both in the present and in the future [9]. The replacement of plastic products with paper packaging follows a recent development trend. Alternative products, such as paper bags, paper tableware, and paper straws, have been developed.

Food packaging materials provide physical protection to food. They exhibit excellent barrier properties as well as mechanical and optical properties. This helps in ensuring sufficient shelf life while maintaining the safety and quality of the packaged food [10]. However, chemical additives are added in the production of paper food packaging materials. Additionally, hazardous substances are present in the production environment [11,12]. These substances are at a risk of migrating into the packaged food. Therefore, the migration of hazardous materials in paper packaging is a major issue in the area of food safety. Among these substances, chloropropanols are the most common. Chloropropanols are toxic to various organisms [13]. They can attack various organs, cause diseases of various tissues, and affect the normal functions of living organisms. Therefore, several countries worldwide have conducted extensive research on the content and migration of chloropropanols in food and have imposed strict restrictions on them. Both the U.S. Food and Drug Administration and European Commission have stipulated limits on the content of chloropropanol in foods and food additives.

With the widespread applications of paper food packaging, the use of chloropropanol in paper materials has also received considerable attention [14]. Existing research has revealed that there exist two primary sources of organic chlorides in paper materials. One source is organic chlorides, introduced by various additives added in the production of paper food packaging materials. For example, polyamideamine epichlorohydrin (PAE) is often added to paper food packaging materials to improve their moisture resistance. This type of polymer with epichlorohydrin as a raw material also produces 3-monochloropropane diol and 1,3-dichloro-2-propanol during the process [15]. The other source is chlorine-containing organic matter in the production system, which enters the material via adsorption and other methods. Recent studies have discovered that paper materials produce chloropropanols during the chlorine-containing bleaching process [16]. White cardboard plays an important role in paper food packaging materials. White cardboard is a paper product made from bleached chemical pulp and a high-yield pulp. It possesses a high stiffness and surface strength and is widely used in food packaging materials [17,18]. Bleached chemical pulp is rich in sulfonic acid groups, whereas the high-yield pulp contains more carboxyl groups. The former is rich in sulfonic acid groups, whereas the latter contains more carboxyl groups. These two groups exhibit strong adsorptivity. Nanocellulose with sulfonic acid groups is often used for the adsorption of heavy metal ions in sewage and treatment of polluted water containing amines [19]. It exhibits a good adsorption effect. Carboxylated nanocellulose possesses the characteristics of a large specific surface area and high functional group density [20]. It is often used in environmental remediation and strongly adsorbs dye molecules and heavy-metal ions [21]. Therefore, it is useful in the production of paper-packaging materials. It can potentially adsorb the by-products of chemical additives and chlorinated organics in bleaching wastewater, which may cause potential food safety risks.

The adsorption process and properties of the adsorption layer are usually interpreted at the molecular level using methods such as quartz crystal microbalance (QCM) or ellipsometry [22]. QCM has ng-level quality inspection capabilities. In recent years, it has been widely used in the study of interfacial adsorption behavior [23]. The adsorption behavior of a composite material with crystalline nanocellulose (CNC) as the substrate for different molecules and ions was studied using QCM-D [24,25]. QCM-D was successfully applied to the adsorption of PAE on the surface of cellulose to improve the wet and dry strength of paper and effectively explain the adsorption behavior of amphoteric electrolytes. The adsorption performance of sulfonated nanocellulose with regard to metal ions was considerably enhanced by increasing the degree of substitution and by the dynamic process of adsorption and analysis of the same cellulase on cellulose.

Two types of CNC were selected for this study. Chloropropanol was used as a typical simulant of chloride organics. The possible adsorption mechanism and behavior of cellulose fibers on chloropropanol were investigated. We analyzed the samples using QCM-D, an atomic force microscope (AFM), a scanning electron microscope (SEM), and an electrophoretic light scattering analyzer (DELSA). The solution pH and CNC volume were considered in this study. Factors such as the influence of chloropropanol concentration on the adsorption capacity of CNC were investigated. Therefore, this provides a reference for the application of paper packaging materials in food packaging materials. To explore the adsorption capacity and adsorption mechanism of chlorine-containing organic compounds in paper materials, and to provide a theoretical basis for reducing the adsorption capacity of chlorine-containing organic compounds in paper materials.

2. Materials and Methods

2.1. Materials

Sulfonated nanocellulose (diameter 4–10 nm, length 100–500 nm) and carboxylated nanocellulose (diameter 4–10 nm, length 100–500 nm) were purchased from Macleans Reagent Co., Ltd. (Shanghai, China). Polyethylenimine (PEI, M_w = 7.5 × 10^5, 50 wt% in H_2O) was purchased from Aladdin Reagent Company (Shanghai, China). Chloropropanol (98%) was purchased from Sigma reagent (Shanghai, China).

2.2. Sample Configuration

PEI was dissolved in Milli-Q and stirred for 30 min to prepare a 0.2% PEI solution. Three uniformly dispersed CNC suspensions with concentrations of 0.01%, 0.05%, and 0.1%, respectively, were prepared with deionized water, which was dispersed at 3000 rpm for 5 min, using a T25 digital package high-speed disperser (T25, IKA, Guangzhou, China). Chloropropanol solutions of 0.01%, 0.02%, 0.05%, and 0.1% were prepared. All liquid solutions were bubbled using N_2 gas. Subsequently, they were degassed for 30 min and stored in a refrigerator at 4 °C.

2.3. QCM Detection

SiO_2-coated chips were cleaned with 0.2% sodium dodecyl sulfate (>60 min), rinsed with Milli-Q water, dried in the presence of N2, and treated with ultraviolet ozone (15 min). A crystal sensor (Biolin Scientific, Gothenburg, Sweden) was placed in the QCM (QHM401,Biolin Scientific, Gothenburg, Sweden) instrument, and deionized water was introduced into the QCM channel at a rate of 100 μL·min^{-1} to run the baseline. After the baseline was stable, the PEI solution was injected into the QCM channel. When the signal was stable, the CNC solution was injected into the channel. Subsequently, chloropropanol was injected into the QCM channel until the signal was stable, and the adsorption process was monitored. The frequency and dissipation of the third, fifth, and seventh channels were recorded to further analyze the adsorption kinetics of chloropropanol on CNC. The adsorbed quantity was usually very low, resulting in small differences between the channels. The calculations in this study were analyzed using the third channel frequency and Sauerbrey's equation. The QCM adsorption experiment was significantly impacted at a temperature of 25 °C owing to the density of the fluid in the QCM instrument [26]. Therefore, the temperature was maintained at 25 ± 0.02 °C in this study. The flow rate was maintained at 0.1 mL·min^{-1}. All tests were conducted in parallel for 3 times. The error range of CNC adsorption amount is between 1.4% and 4.5%. The error range of the adsorption capacity of chloropropanol was between 2.8% and 12.5%.

2.4. SEM Characterization

After the experiment, the SiO_2 sensor ((Biolin Scientific, Gothenburg, Sweden)was vacuum dried for 24 h. It was sprayed with gold using sputtering coating machine (Vapor Technologies, Longmont, CO, USA). The microstructure and size of CNC adsorbed onto the SiO_2 sensor were analyzed using SEM (TESCAN MIRA4, TESCAN, Brno, Czech Republic). The SEM was equipped with an energy spectrometer (EDS) (TESCAN, Brno, Czech Republic) to map the elements of the sample and analyze the relevant components.

2.5. AFM Characterization

CNC was analyzed before and after adsorption on the SiO_2 sensor using an AFM (Dimension Edge, Bruker Co, Ltd., Berlin, Germany). Commercial probes were used with a spring constant of 20–80 N·m^{-1} and a resonant frequency of 300–340 kHz. The scanned dimensions were 1 × 1 μm^2 and 1 × 1 nm^2.

2.6. Zeta Potential Measurement

The zeta potentials of the CNC suspensions and chloropropanol solutions at different pH values were studied using an electrophoretic light scattering analyzer (DelsaMax,

Beckman Coulter, Indianapolis, IN, USA). The stability of the suspension and the functional groups on the surface of the nanocellulose were analyzed. A homogeneous suspension (25 mL) was tested at 20 °C. The accuracy of this method was 5% [27].

3. Results and Discussion

3.1. Surface Analysis before and after CNC Adsorption

The adsorption of CNC on the SiO_2 sensor was analyzed using an SEM. The results are demonstrated in Figure 1a. CNC forms a uniformly distributed thin film on the SiO_2 sensor. The cellulose nanocrystals are tightly interwoven. A large number of micropores are distributed on the nanocellulose coating, with a diameter of approximately 0.05 μm, as depicted in Figure 1b,c. This can be attributed to the large number of hydroxyl groups in CNC. The high density of hydroxyl groups allows CNC to be combined through hydrogen bonds. A strong porous structure is formed. Following the adsorption of CNC, the SiO_2 sensor attains a certain capacity of physical adsorption. As pointed out by Shuai Li in 2011, porous films have physical adsorption capacity [28].

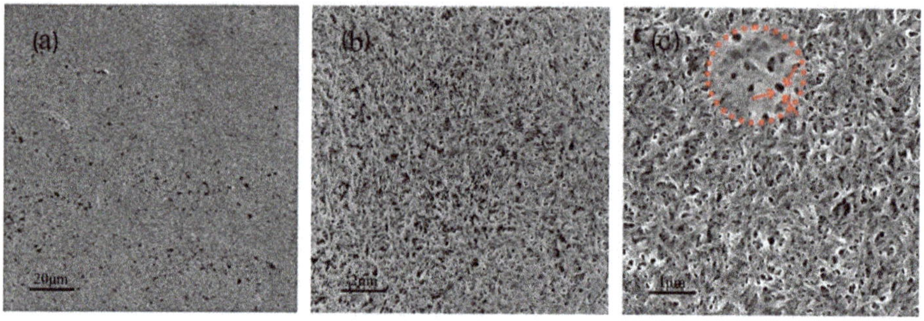

Figure 1. SEM of SiO_2 sensor before and after CNC adsorption.

The dimensional information and surface morphology were observed using an AFM. From the AFM image of the stable adsorption of CNC on the SiO_2 sensor (Figure 2a), it can be observed that the surface of the CNC film is not uniform and smooth. There exist voids of approximately 50 nm; this is consistent with the SEM data. Figure 2b illustrates the accumulation of CNC on the surface of the SiO_2 sensor. We can observe that the surface of the CNC film is irregular. Combined with the longitudinal section data analysis, it can be observed that there exists a certain height difference on the surface of the cellulose film (Figure 2d). Additionally, there exists a height difference between the fiber bundles of the clusters (Figure 2e) [29].

The CNC film on the SiO_2 sensor was analyzed before and after the adsorption of chloropropanol. The results of the EDS energy spectrum analysis are illustrated in Figure 3. Figure 3b reveals that the energy spectrum of the CNC film, after the addition of the chloropropanol solution, is densely packed with chlorine. (The red dots represent chlorine). However, the energy spectrum of the CNC film illustrated in Figure 3e, before the addition of the chloropropanol solution, depicts only scattered chlorine elements. The mapping chlorine analysis image (Figure 3b) and the original SEM image (Figure 3a) were overlapped to obtain Figure 3c. Figure 3d,e were combined to obtain Figure 3f. Next, Figure 3c,f were compared and analyzed. As can be observed, the chloropropanol solution is adsorbed after it is passed through the CNC film. No chlorine distribution can be observed in the CNC accumulation holes. Chlorine is intensively adsorbed onto the fiber. Therefore, chloropropanol was stably adsorbed on the surface of the CNC. Moreover, the adsorption capacity was high.

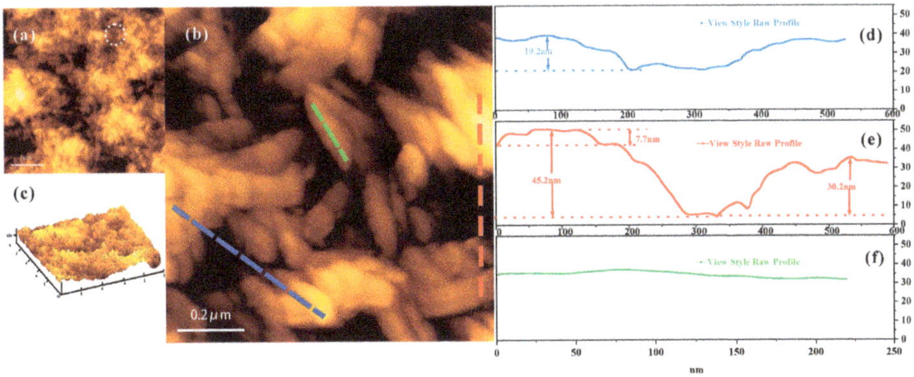

Figure 2. AFM morphology of CNC and the cross section of the fiber. ((**c**) is 3D image of CNC film. (**d**–**f**) are the length and section height data of the corresponding color line segment in C respectively. The three line segments of red, yellow and blue are randomly selected, and the height difference of the surface of nano cellulose film is measured).

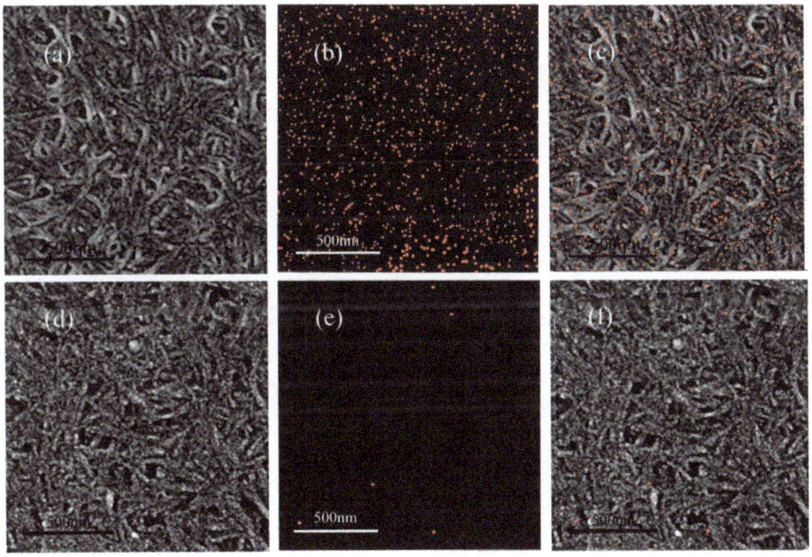

Figure 3. Mapping analysis of CNC before and after the adsorption of chloropropanol.

3.2. *Effect of Chloropropanol Concentration on the Adsorption Capacity of CNC*

We investigated the extent to which the adsorption effect on CNC was influenced by the concentration of chloropropanol. The adsorption capacity of CNC was maintained at a constant value by controlling the concentration of the CNC solution in the QCM. The results are depicted in Figure 4. At CNC concentrations of 0.01%, 0.02%, 0.05%, and 0.1%, the SiO_2 sensor adsorbs 740 ± 20 ng, 1240 ± 40 ng, 1780 ± 80 ng, and 2130 ± 30 ng of CNC, respectively. Different concentrations of chloropropanol were passed into CNC films of the same quality. The amount of chloropropanol adsorbed on the CNC film was maintained at 120 ± 15 ng, 330 ± 30 ng, 350 ± 10 ng, and 680 ± 30 ng, respectively. The change in the concentration of chloropropanol had a minimal effect on the chloropropanol adsorption capacity of CNC when the amount of CNC was constant. The same amount of CNC adsorbed the same amount of chloropropene at different concentrations of the chloropropanol solution. Therefore, the concentration of chloropropanol did not significantly affect its adsorption on the nanofibers. The concentration of chloropropanol is, therefore, not a major influencing factor in the process of adsorption on CNC.

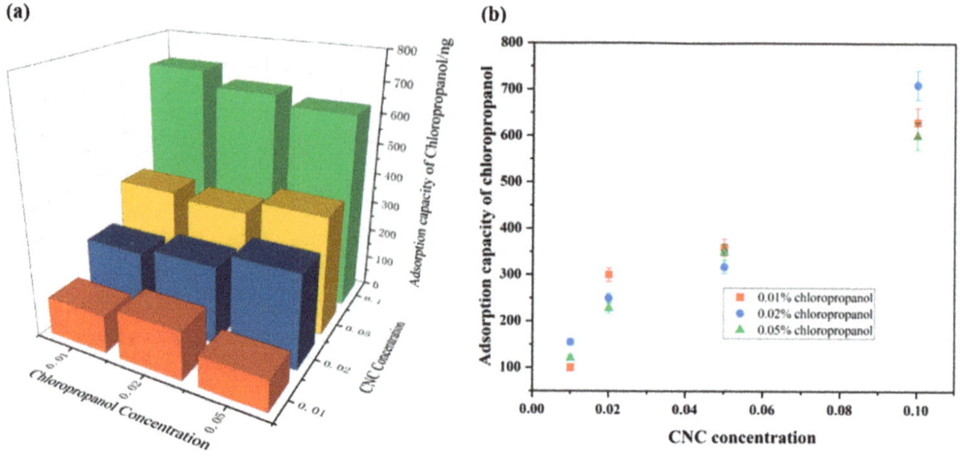

Figure 4. The effect of chloropropanol concentration on its adsorption on CNC. (**a**) The effects of CNC concentration and chloropropanol concentration on the adsorption capacity of chloropropanol; (**b**) The effect of CNC Concentration on Adsorption Capacity of Chloropropanol.

3.3. Effect of CNC Dosage pH on the Adsorption Capacity

The dosage of CNC on the SiO_2 sensor was controlled by adjusting the concentration of the CNC solution in the QCM. Figure 4b indicates that as the adsorption capacity of CNC increases, the adsorption capacity of CNC for chloropropanol simultaneously increases. The relationship between the adsorption amount of CNC and the amount of chloropropanol adsorbed was studied. The results in Figure 5b indicate that the adsorption of chloropropanol on sulfonated CNC exhibits a good linear relationship under different pH conditions. Thus, it can be concluded that there exists a good linear relationship between the adsorption capacities of CNC and chloropropanol. To eliminate the influence of sulfonic acid groups, carboxylated nanocellulose was selected. The obtained results under the same experimental conditions are presented in Figure 5. From Figure 5a,b, we can observe that the quantity of chloropropanol does not start from zero. This can be attributed to the presence of hydrophilic groups (–OH) on the CNC surface. The water molecules in the solution are associated with the hydroxyl groups in the form of hydrogen bonds. A hydration layer with a thickness equivalent to that of water molecules is formed on the membrane surface. The water molecules in the hydration layer are highly structured. They are in a dynamic equilibrium with free water molecules. Therefore, chloropropanol is not effectively adsorbed at the beginning of the CNC adsorption process. The initial mass is the mass of the hydration layer. In summary, the adsorption capacity of CNC is an important factor in determining the adsorption capacity of chloropropanol. There exists a good linear relationship between the adsorption capacities of CNC and chloropropanol. The adsorption capacity of chloropropanol increases with an increase in the CNC adsorption capacity.

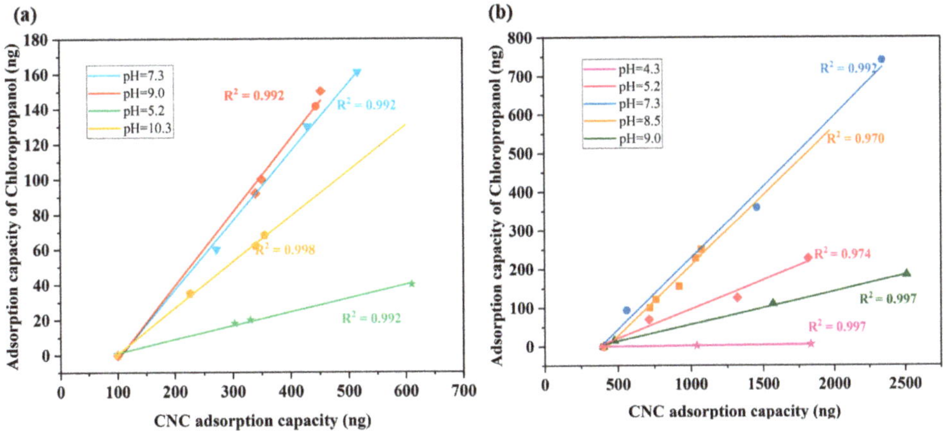

Figure 5. Adsorption capacity of chloropropanol on sulfonated and carboxylated nanocellulose samples.

3.4. Effect of pH on the Adsorption Capacity

As presented in Figure 5a,b, the adsorbed amount of chloropropanol, with the same CNC adsorption capacity, can be changed by adjusting the pH. Under varying pH conditions, the linear relationship between the adsorption amount of CNC and the adsorption amount of chloropropanol was found to be different. Therefore, we studied the influence of the pH conditions on the adsorption of chloropropanol on CNC. Figure 6 depicts the amount of chloropropanol adsorbed per unit volume of CNC under different pH conditions for the two types of CNC. The pH of the solution significantly affects the adsorption of chloropropanol on CNC; the adsorption is optimized in neutral and weakly alkaline environments. An acidic environment inhibits the adsorption of chloropropanol on CNC. When the pH of the adsorption system is approximately four, the chloropropanol adsorption capacity of CNC is zero. With an increase in the pH value, the adsorption capacity of CNC is gradually increased. However, when the alkalinity is strong, the adsorption of chloropropanol on CNC is inhibited. This is because the adsorption process of CNC on chloropropanol involves physical adsorption. No charge was observed on the surface of the chloropropanol. Therefore, the driving force between the CNC and chloropropanol molecules can be attributed to the intermolecular hydrogen bond and van der Waals forces. Particularly, the number of donors and acceptors of the hydrogen bond in chloropropanol are both one. The adsorption of chloropropanol on CNC is inhibited under acidic conditions owing to the existence of sulfonic acid groups (R–SO_3^-) and carboxyl groups (R–COO^-), which alter the ionization balance equation. The number of free hydrogen ions is increased, negative atoms on both sides are covered, and formation of hydrogen bonds is inhibited. Under strongly alkaline conditions, there exist excess hydroxide radicals in the solution, and the hydrolysis equilibrium shifts to the right owing to the active free hydrogen ions being covered. Hydrogen bond formation is inhibited by the absence of sufficient free hydrogen ions. This results in the inhibition of chloropropanol adsorption and a reduction in the adsorption capacity of CNC [30].

3.5. Adsorption Kinetics

The adsorption behavior of chloropropanol on the CNC model was investigated. Figure 7a illustrates the frequency of chloropropanol adsorption on CNC as measured by QCM. The adsorption of chloropropanol proceeds rapidly, and the entire adsorption process occurs in 200 s. This is because there are several adsorption sites on the CNC surface; the adsorption driving force is large, and there is no competition between the adsorption molecules. Therefore, chloropropanol can be adsorbed quickly on the CNC model. The adsorption gradually approaches an equilibrium after 200 s. This is because the adsorption driving force between the two phases was weakened, the adsorption active sites were occupied by chloropropanol molecules, and the adsorption process gradually reached the equilibrium state. Therefore, the adsorption capacity for chloropropanol remains unchanged after 200 s.

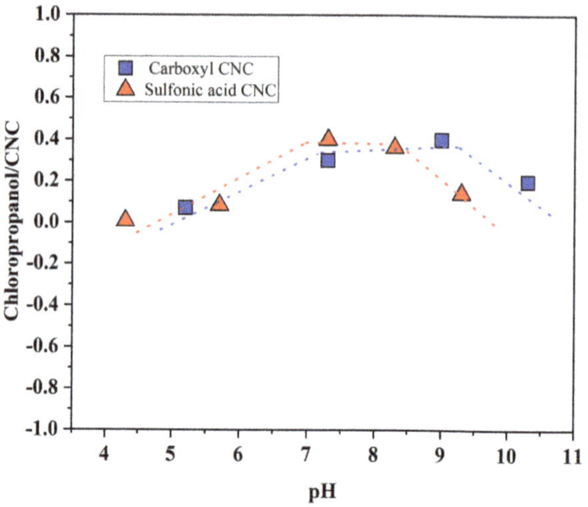

Figure 6. The effect of pH on the adsorption of chloropropanol on CNC.

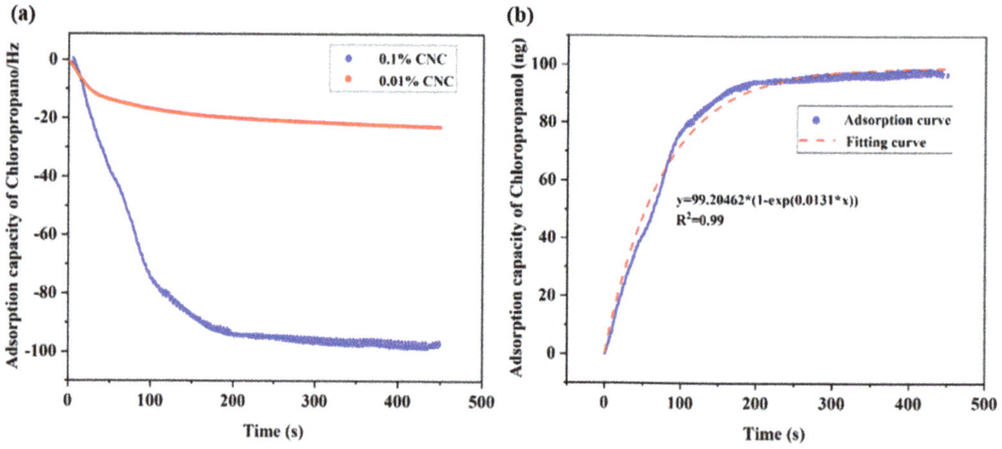

Figure 7. (**a**) Adsorption of chloropropanol at different CNC dosages. (**b**) Fitting curve of chloropropanol adsorption kinetics.

Usually, the study of the adsorption kinetic process simulation of a solid-liquid two-phase system is composed of the intra-particle diffusion model equation, quasi-first-order kinetic equation, and quasi-second-order kinetic equation. The quasi-first-order kinetic model can be expressed as follows [31]:

$$\ln(q_e - q_t) = \ln q_e - k_{1t} \quad (1)$$

where k_1 is the rate constant of quasi-first-order adsorption. q_t and q_e are the amounts of chloropropanol adsorbed on the CNC model at time *t* and at equilibrium, respectively.

The quasi-second-order kinetic model can be represented by the following equation:

$$t/q_t = 1/k_2 q_e^2 + t/q_e \tag{2}$$

where k_2 is the rate constant of quasi-second order adsorption.

The intraparticle diffusion model can be defined by the following equation

$$q_t = k_i t_1/2 + C \tag{3}$$

where k_i is the rate constant of the intraparticle diffusion model, and C is a constant related to the thickness of the boundary layer.

These three equations were applied to simulate the CNC adsorption process of chloropropanol molecules. When the particle diffusion simulation equation was used, linear fitting of the equation was not ideal. Therefore, the quasi-first-order kinetic model was used for the simulation; consequently, it demonstrated a better fit. The fitted image is depicted in Figure 7b. The fitting was found to be poor when the quasi-second-order kinetic equation was adopted to analyze the adsorption data. Therefore, it can be concluded that the adsorption of chloropropanol on CNC ideally follows a quasi-first-order adsorption curve.

The driving force for the adsorption of chloropropanol was studied, and zeta potentials of CNC solutions under different pH conditions were determined. For CNC solutions containing carboxyl groups, the zeta potential moved in a positive direction with an increase in pH. For CNC solutions containing sulfonic acid groups, the zeta potentials did not significantly vary at different pH conditions. This is inconsistent with the changing trend observed in the ratio of the adsorption capacity of chloropropanol to that of CNC. This indicates that the zeta potential of the CNC solution has no significant effect on the adsorption of chloropropanol. Therefore, it can be inferred that the driving force of adsorption between CNC and chloropropanol is not electrostatic but an intermolecular force. In summary, the adsorption of chloropropanol on CNC relies primarily on intermolecular forces. The entire adsorption process is primarily physical, and the adsorption process can be described by a quasi-first-order kinetic equation.

3.6. Saturated Adsorption Capacity of Chloropropanol by CNC

The maximum adsorbed amount of chloropropanol per unit mass of CNC was studied. Table 1 lists the amount of chloropropanol adsorbed per unit mass of the two types of CNC under different pH conditions, which were studied using quartz microbalance technology. As presented in Table 1, the adsorption effect of chloropropanol per unit mass of CNC is superior. This is considerably higher than the limit of adsorbable organic halogen content of paper in China's national standard GB/T36420–2018 (15 $mg \cdot kg^{-1}$ or 1.5×10^{-5}), especially under weakly alkaline and neutral conditions; it is 20,000 times that of the allowable limit. Therefore, it is important to control the concentration of organic chlorides in paper production systems. Modern papermaking processes are typically performed under weakly alkaline conditions [18]. Chlorinated organics are extremely likely to be encountered during paper production. The fiber adsorption speed can achieve adsorption saturation in a short time. This is extremely likely to cause excessive levels of chloropropanol in paper and food packaging. Therefore, the potential risks are worthy of attention.

Table 1. Saturated adsorption capacity of chloropropanol on CNC.

Carboxylated CNC		Sulfonated CNC	
pH	**Chloropropanol ng/CNC ng**	**pH**	**Chloropropanol ng/CNC ng**
5.2	7×10^{-2}	5.7	8.0×10^{-2}
7.3	3×10^{-1}	7.3	4.0×10^{-1}
9.0	4×10^{-1}	8.3	3.6×10^{-1}
10.3	2×10^{-1}	9.3	1.4×10^{-1}

4. Conclusions

The adsorption behavior of chloropropanol on CNC was studied, and the adsorption kinetics was established. The morphology of CNC and distribution of chloropropanol were analyzed after the introduction of chloropropanol. The results indicate that chloropropanol is physically adsorbed on CNC. Moreover, intermolecular hydrogen bonds are the primary driving force of adsorption. The adsorption process conforms to the quasi-first-order kinetic equation. The adsorption capacity of CNC and the pH of the solution are critical factors that affect the adsorption of chloropropanol. There exists a good linear relationship between the adsorption capacity of CNC and adsorption capacity of chloropropanol. The amount of chloropropanol adsorbed correspondingly increases with the amount of CNC used. The adsorption of chloropropanol on CNC is optimal under weak alkali and neutral conditions. The maximum adsorption capacity is 20,000-times that of the national standard; therefore, the potential risk caused is noteworthy.

Author Contributions: Conceptualization, C.L.; methodology, J.Z.; resources, C.C. and Z.G.; data curation, L.H. and M.H.; writing—original draft preparation, J.Z.; writing—review and editing, S.W.; project administration, C.Q. All authors have read and agreed to the published version of the manuscript.

Funding: This project was sponsored by the National Natural Science Foundation of China (22168007). This project was supported by the Guangxi Natural Science Foundation (2020GXNSFAA159027).

Data Availability Statement: Not applicable.

Conflicts of Interest: The authors declare no conflict of interest. The funders had no role in the design of the study; in the collection, analyses, or interpretation of data; in the writing of the manuscript, or in the decision to publish the results.

References

1. Bezerra, M.A.; Santelli, R.E.; Oliveira, E.P.; Villar, L.S.; Escaleira, L.A. Response surface methodology (RSM) as a tool for optimization in analytical chemistry. *Talanta* **2008**, *76*, 965–977. [CrossRef] [PubMed]
2. Taherimehr, M.; YousefniaPasha, H.; Tabatabaeekoloor, R.; Pesaranhajiabbas, E. Trends and challenges of biopolymer-based nanocomposites in food packaging. *Compr. Rev. Food Sci. Food Saf.* **2021**, *20*, 5321–5344. [CrossRef] [PubMed]
3. Humbert, S.; Rossi, V.; Margni, M.; Jolliet, O.; Loerincik, Y. Life cycle assessment of two baby food packaging alternatives: Glass jars vs. plastic pots. *Int. J. Life Cycle Assess.* **2009**, *14*, 95–106. [CrossRef]
4. Jambeck, J.R.; Geyer, R.; Wilcox, C.; Siegler, T.R.; Perryman, M.; Andrady, A.; Narayan, R.; Law, K.L. Plastic waste inputs from land into the ocean. *Science* **2015**, *347*, 768–771. [CrossRef] [PubMed]
5. Groh, K.J.; Backhaus, T.; Carney-Almroth, B.; Geueke, B.; Inostroza, P.A.; Lennquist, A.; Leslie, H.A.; Maffini, M.; Slunge, D.; Trasande, L.; et al. Overview of known plastic packaging-associated chemicals and their hazards. *Sci. Total Environ.* **2019**, *651*, 3253–3268. [CrossRef] [PubMed]
6. Ge, J.; Wu, Y.; Han, Y.; Qin, C.; Nie, S.; Liu, S.; Wang, S.; Yao, S. Effect of hydrothermal pretreatment on the demineralization and thermal degradation behavior of eucalyptus. *Bioresour. Technol.* **2020**, *307*, 123246. [CrossRef] [PubMed]
7. Siqueira, G.; Bras, J.; Dufresne, A. Cellulosic Bionanocomposites: A Review of Preparation, Properties and Applications. *Polymers* **2010**, *2*, 728. [CrossRef]
8. Weber, C.J.; Haugaard, V.; Festersen, R.; Bertelsen, G. Production and applications of biobased packaging materials for the food industry. *Food Addit. Contam.* **2002**, *19*, 172–177. [CrossRef]
9. Rai, P.; Mehrotra, S.; Priya, S.; Gnansounou, E.; Sharma, S.K. Recent advances in the sustainable design and applications of biodegradable polymers. *Bioresour. Technol.* **2021**, *325*, 124739. [CrossRef]
10. Marsh, K.; Bugusu, B. Food packaging-Roles, materials, and environmental issues. *J. Food Sci.* **2007**, *72*, R39–R55. [CrossRef]
11. Yao, S.; Nie, S.; Zhu, H.; Wang, S.; Song, X.; Qin, C. Extraction of hemicellulose by hot water to reduce adsorbable organic halogen formation in chlorine dioxide bleaching of bagasse pulp. *Ind. Crops Prod.* **2017**, *96*, 178–185. [CrossRef]
12. Luo, Y.; Li, Y.; Cao, L.; Zhu, J.; Deng, B.; Hou, Y.; Liang, C.; Huang, C.; Qin, C.; Yao, S. High efficiency and clean separation of eucalyptus components by glycolic acid pretreatment. *Bioresour. Technol.* **2021**, *341*, 125757. [CrossRef] [PubMed]
13. Schilter, B.; Scholz, G.; Seefelder, W. Fatty acid esters of chloropropanols and related compounds in food: Toxicological aspects. *Eur. J. Lipid Sci. Technol.* **2011**, *113*, 309–313. [CrossRef]
14. Mezouari, S.; Liu, W.Y.; Pace, G.; Hartman, T.G. Development and validation of an improved method for the determination of chloropropanols in paperboard food packaging by GC-MS. *Food Addit. Contam. Part. A-Chem. Anal. Control. Expo. Risk Assess.* **2015**, *32*, 768–778. [CrossRef] [PubMed]

15. Pace, G.V.; Hartman, T.G. Migration studies of 3-chloro-1,2-propanediol (3-MCPD) in polyethylene extrusion-coated paperboard food packaging. *Food Addit. Contam. Part. A-Chem. Anal. Control. Expo. Risk Assess.* **2010**, *27*, 884–891. [CrossRef] [PubMed]
16. Korte, R.; Schulz, S.; Brauer, B. Chloropropanols (3-MCPD, 1,3-DCP) from food contact materials: GC-MS method improvement, market survey and investigations on the effect of hot water extraction. *Food Addit. Contam. Part. A-Chem. Anal. Control. Expo. Risk Assess.* **2021**, *38*, 904–913. [CrossRef]
17. Koljonen, K.; Mustranta, A.; Stenius, P. Surface characterisation of mechanical pulps by polyelectrolyte adsorption. *Nord. Pulp Pap. Res. J.* **2004**, *19*, 495–505. [CrossRef]
18. Avitsland, G.A.; Wagberg, L. Flow resistance of wet and dry sheets used for preparation of liquid packaging board. *Nord. Pulp Pap. Res. J.* **2005**, *20*, 345–353. [CrossRef]
19. Dong, C.; Wang, B.; Meng, Y.; Pang, Z. Preparation, structural changes and adsorption performance of heavy metal ions on sulfonated cellulose with varying degrees of substitution. *Holzforschung* **2019**, *73*, 501–507. [CrossRef]
20. De Nino, A.; Tallarida, M.A.; Algieri, V.; Olivito, F.; Costanzo, P.; De Filpo, G.; Maiuolo, L. Sulfonated Cellulose-Based Magnetic Composite as Useful Media for Water Remediation from Amine Pollutants. *Appl. Sci.-Basel* **2020**, *10*, 8155. [CrossRef]
21. Tang, F.; Yu, H.; Abdalkarim, S.Y.H.; Sun, J.; Fan, X.; Li, Y.; Zhou, Y.; Tam, K.C. Green acid-free hydrolysis of wasted pomelo peel to produce carboxylated cellulose nanofibers with super absorption/flocculation ability for environmental remediation materials. *Chem. Eng. J.* **2020**, *395*, 125070. [CrossRef]
22. Ahola, S.; Osterberg, M.; Laine, J. Cellulose nanofibrils-adsorption with poly(amideamine) epichlorohydrin studied by QCM-D and application as a paper strength additive. *Cellulose* **2008**, *15*, 303–314. [CrossRef]
23. Marx, K.A. Quartz crystal microbalance: A useful tool for studying thin polymer films and complex biomolecular systems at the solution-surface interface. *Biomacromolecules* **2003**, *4*, 1099–1120. [CrossRef]
24. Zhang, X.; Zhu, Y.; Wang, X.; Wang, P.; Tian, J.; Zhu, W.; Song, J.; Xiao, H. Revealing Adsorption Behaviors of Amphoteric Polyacrylamide on Cellulose Fibers and Impact on Dry Strength of Fiber Networks. *Polymers* **2019**, *11*, 1886. [CrossRef] [PubMed]
25. Hu, F.; Zhang, Y.; Wang, P.; Wu, S.; Jin, Y.; Song, J. Comparison of the interactions between fungal cellulases from different origins and cellulose nanocrystal substrates with different polymorphs. *Cellulose* **2018**, *25*, 1185–1195. [CrossRef]
26. Larsson, C.; Rodahl, M.; Hook, F. Characterization of DNA immobilization and subsequent hybridization on a 2D arrangement of streptavidin on a biotin-modified lipid bilayer supported on SiO_2. *Anal. Chem.* **2003**, *75*, 5080–5087. [CrossRef]
27. Julien, F.; Baudu, M.; Mazet, M. Relationship between chemical and physical surface properties of activated carbon. *Water Res.* **1998**, *32*, 3414–3424. [CrossRef]
28. Li, S.; Gao, Y.; Bai, L.; Tian, W.; Zhang, L. Research on the Porous Structures and Properties of Composite Membranes of Polysulfone and Nanocrystalline Cellulose. In Proceedings of the 7th International Forum on Advanced Material Science and Technology, Dalian, China, 26–28 June 2011; pp. 391–394.
29. Soni, B.; Hassan, E.B.; Mahmoud, B. Chemical isolation and characterization of different cellulose nanofibers from cotton stalks. *Carbohydr. Polym.* **2015**, *134*, 581–589. [CrossRef]
30. Dong, C.; Zhang, H.; Pang, Z.; Liu, Y.; Zhang, F. Sulfonated modification of cotton linter and its application as adsorbent for high-efficiency removal of lead(II) in effluent. *Bioresour. Technol.* **2013**, *146*, 512–518. [CrossRef]
31. Barnea, O.; Gillon, G. Cavernosometry: A theoretical analysis. *Int. J. Impot. Res.* **2004**, *16*, 154–159. [CrossRef]

polymers

MDPI

Article

Optimization of Demineralization and Pyrolysis Performance of Eucalyptus Hydrothermal Pretreatment

Jiatian Zhu, Yuqi Bao, Luxiong Lv, Fanyan Zeng, Dasong Du, Chen Liang, Jiayan Ge, Shuangfei Wang and Shuangquan Yao *

Guangxi Key Laboratory of Clean Pulp & Papermaking and Pollution Control, School of Light Industrial and Food Engineering, Guangxi University, Nanning 530004, China; 2016301053@st.gxu.edu.cn (J.Z.); lb04100126@163.com (Y.B.); luxionglv@163.com (L.L.); 2116301005@st.gxu.edu.cn (F.Z.); ggbond19114788464@163.com (D.D.); liangchen@gxu.edu.cn (C.L.); hebo1104@outlook.com (J.G.); liubaojie@st.gxu.edu.cn (S.W.)

* Correspondence: yaoshuangquan@gxu.edu.cn

Abstract: The preparation of bio-oil through biomass pyrolysis is promoted by different demineralization processes to remove alkali and alkaline earth metal elements (AAEMs). In this study, the hydrothermal pretreatment demineralization was optimized by the response surface method. The pretreatment temperature, time and pH were the response elements, and the total dissolution rates of potassium, calcium and magnesium were the response values. The interactions of response factors for AAEMs removal were analyzed. The interaction between temperature and time was significant. The optimal AAEMs removal process was obtained with a reaction temperature of 172.98 °C, time of 59.77 min, and pH of 3.01. The optimal dissolution rate of AAEMs was 47.59%. The thermal stability of eucalyptus with and without pretreatment was analyzed by TGA. The hydrothermal pretreatment samples exhibit higher thermostability. The composition and distribution of pyrolysis products of different samples were analyzed by Py-GC/MS. The results showed that the content of sugars and high-quality bio-oil (C6, C7, C8 and C9) were 60.74% and 80.99%, respectively, by hydrothermal pretreatment. These results show that the removal of AAEMs through hydrothermal pretreatment not only improves the yield of bio-oil, but also improves the quality of bio-oil and promotes an upgrade in the quality of bio-oil.

Keywords: eucalyptus; demineralization; hydrothermal pretreatment; thermostability; pyrolysis products

Citation: Zhu, J.; Bao, Y.; Lv, L.; Zeng, F.; Du, D.; Liang, C.; Ge, J.; Wang, S.; Yao, S. Optimization of Demineralization and Pyrolysis Performance of Eucalyptus Hydrothermal Pretreatment. *Polymers* **2022**, *14*, 1333. https://doi.org/10.3390/polym14071333

Academic Editor: Susana C. M. Fernandes

Received: 11 March 2022
Accepted: 22 March 2022
Published: 25 March 2022

1. Introduction

The global demand for fossil-fuel-derived energy has increased significantly with rapid economic and global population growth [1,2]. Therefore, the search for alternative energy sources presents a global challenge [3,4]. Biomass can be converted into liquid fuels or chemicals, thus effectively relieving the energy crisis and environmental pressure [5–7]. The pyrolysis of biomass to produce bio-oil represents one of the main methods to utilize biomass resources [8]. However, there are many technical difficulties in the fast pyrolysis process, which limits its popularization and application. Typical limitations include poor volatility, poor thermal stability, high viscosity, and low calorific value of bio-oil [9,10]. Bio-oil quality is mainly affected by the physicochemical structure of the biomass and the pyrolysis conditions [11]. In particular, alkali and alkaline earth metals (AAEMs) comprise important components of biomass, and they have a significant effect on the pyrolysis reaction [12,13]. The formation of carbonyl compounds, water, and acids are improved by the presence of AAEMs, which act as catalysts. However, this leads to a reduction in bio-oil production [14].

AAEMs in biomass can be removed by demineralization [15,16]. There are two main demineralization methods: water washing and acid leaching [17]. Although water washing can effectively remove water-soluble AAEMs, the removal of acid-soluble AAEMs is inefficient. Chen et al. [14] reported that the removal rates of water-soluble AAEMs, such as potassium in cotton stalks reached more than 80% by water washing. However, the removal rates of acid-soluble AAEMs, such as calcium and magnesium, measured 29% and 48%, respectively. On the other hand, acid leaching is very effective in removing alkaline earth metals. Ma et al. [18] studied the effect of acid leaching on the rapid pyrolysis of rice husk to produce bio-oil. The results showed that acid leaching could effectively enhance the removal of alkaline earth metals. Meanwhile, acid leaching can effectively improve the yield of bio-oil produced by rapid pyrolysis. However, some functional groups in biomass can also be damaged. Dong et al. [19] found that the hydrogen bonds in the chemical structure of the biomass were broken. In addition, the hemicellulose in the biomass is removed in large quantities. In addition, it has been reported that the structure of lignin and cellulose is also damaged, and the crystallinity is reduced. In contrast, hydrothermal pretreatment can effectively remove AAEMs while giving the biomass higher cellulose crystallinity and thermal stability [20]. Chang et al. [21] studied the effect of hydrothermal pretreatment on the rapid pyrolysis of biomass to produce bio-oil. The results showed that temperature had a significant effect on the removal of AAEMs. However, the dissolution of cellulose, hemicellulose and lignin increased with the increase of temperature. In our previous study, it was discovered that the main influencing factors of hydrothermal pretreatment were temperature, pH and time [22]. However, there are few reports on the interaction between different factors during hydrothermal pretreatment. In addition, previous studies have found that when the AAEMs removal rate of eucalyptus is equal between hydrothermal pretreatment and hydrochloric acid leaching, the yield and composition distribution of bio-oil produced by rapid pyrolysis of the biomass treated by the two pretreatment methods are different. Moreover, the forms of AAEMs in biomass may affect the yield and composition distribution of bio-oil [23]. Therefore, it is of great significance to study the dissolution rule of AAEMs in different forms during pretreatment to control the yield and component distribution of bio-oil.

In this study, the demineralization of the eucalyptus wood by hydrothermal pretreatment was optimized using the response surface method. A quadratic polynomial mathematical model of the AAEMs dissolution rate was established using temperature, pH and time as response factors and the AAMEs dissolution rate as response values. The changes in AAEMs (potassium, calcium and magnesium) content were analyzed by inductively coupled plasma atomic emission spectrometry (ICP-OES). The thermal stability of samples was analyzed by a thermogravimetric analyzer (TGA) and pyrolysis–gas chromatography combined instrument (Py-GC/MS). The chemical composition of the eucalyptus before and after pretreatment was determined by sulfuric acid hydrolysis according to the NREL method. The contents of AAEMs in different forms were analyzed by chemical fractionation analysis (CFA). The removal ability of different AAEMs by hydrothermal pretreatment and acid pickling was compared. This work aims to evaluate the demineralization of the two pretreatments on eucalyptus. The composition and distribution of pyrolysis products of eucalyptus via two pretreatments was analyzed.

2. Materials and Methods

2.1. Materials

Eucalyptus chips (20 mm × 5 mm) were provided by a local company (Guangxi, China). The chemical composition of the eucalyptus were determined by the sulfuric acid hydrolysis according to the NREL method [22]. The contents of cellulose, hemicellulose and lignin and ash from eucalyptus were 49.55%, 12.93%, 34.53% and 0.35%, respectively. Potassium, calcium, and magnesium standard solutions were purchased from Agilent Technologies, Inc. (Santa Clara, CA, USA). These analytical chemicals were purchased from Aladdin (Shanghai, China).

2.2. Demineralization

Hydrothermal pretreatment demineralization was carried out in a rotary digester with six stainless steel cylindrical reactors (Green Wood, Brooklyn, NY, USA). The solid–liquid ratio was 1:6. Hydrothermal pretreatment was performed at different temperatures, times, and pH values. The hydrolysate was collected and centrifuged at 10,000 rpm for 10 min after reaction. The supernatant was filtered using a 0.45 um filtration membrane to obtain the filtrate liquid. The hydrolytic solution after membrane separation was cryopreserved [24]. The demineralization of acid leaching was analyzed. The method was described in the previous study [22].

2.3. Sample Preparation by Microwave-Assisted Digestion

A 1 mL volume of hydrolysate was pipetted into the polytetrafluoroethylene microwave digestion tube, and 5 mL of concentrated nitric acid and 2 mL of hydrogen peroxide were added. After the mixture was completely mixed for 15 min, the samples were digested in a microwave digestion system for 3 min at 220 °C, and cooled for 2 h. The samples were diluted for analysis [25].

2.4. ICP-OES Measurements

The standard solutions of potassium, calcium and magnesium were each accurately absorbed into and gradually diluted with 5% dilute nitric acid solution to configure a series of standard solutions at concentrations: 0 $mg{\cdot}L^{-1}$, 0.5 $mg{\cdot}L^{-1}$, 5 $mg{\cdot}L^{-1}$, 10 $mg{\cdot}L^{-1}$, and 20 $mg{\cdot}L^{-1}$. The standard curves for different metal ions were plotted by ICP-OES (Optima 5300 DV, Agilent Technologies Inc., Santa Clara, CA, USA) [26]. The concentration of AAEMs in the sample was detected.

2.5. Process Optimization

Based on previous studies [22], the dissolution rate of AAEMs was most affected by the reaction temperature, time and pH. Therefore, the above three reaction conditions and the total dissolution rate of the three AAEMs (potassium, calcium, and magnesium) were taken as the response factors and response values, respectively. Three-factor, three-level response surface optimization experiments were designed (Table 1). The interaction between different factors was studied.

Table 1. Response factors and design levels.

Code	Response Factor		
	Temperature (x_1) °C	Time (x_2) min	pH (x_3)
−1	160	50	3
0	170	60	4
1	180	70	5

2.6. Component Analysis

The main components of the samples with and without pretreatment were analyzed. First, 40–60 mesh wood powder was obtained by screening. Then, 20 g of the powder was reacted with benzyl alcohol for 8 h, and later subjected to a two-step acidolysis process. The specific methods and processes used were demonstrated by Ge and co-authors [22]. The relative contents of cellulose, hemicellulose and lignin were analyzed [27].

2.7. Chemical Fractionation

The chemical fractionation was carried out according to Pettersson and co-authors [28]. It is a step-by-step leaching method resulting in selective extraction of inorganic elements, based on the solubility of their association forms in the samples (Figure 1). The experimental procedure consists of three successive extractions. First, the water-soluble compounds such as alkaline salts are removed using pure water. Then, addition of ammonium acetate dissolves ion exchangeable elements, such as sodium, calcium and magnesium. The third extraction step with hydrochloric acid removes acid soluble compounds. The solid residue fraction consists of silicates, oxides, sulphides and other minerals. After each step the solid sample was washed two times by deionized water. The washing water was added to the leachate prior to analysis.

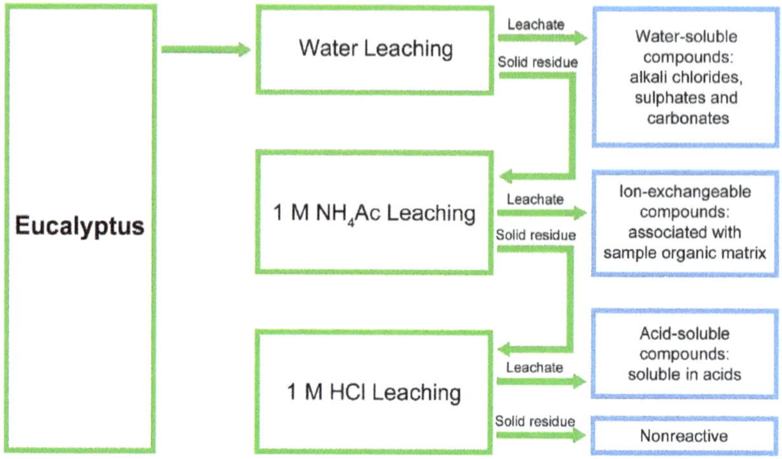

Figure 1. Chemical fractionation general procedure.

2.8. *Rapid Pyrolysis of Eucalyptus*

Eucalyptus samples with and without pretreatment were crushed by a grinder. The sawdust with a particle size range of 30–80 mesh was selected for the rapid pyrolysis reaction. The pyrolysis reaction was carried out by a small, fixed bed pyrolysis device built in the laboratory. High purity nitrogen (500 mL·min^{-1}) was continuously injected to provide an inert environment for pyrolysis, and 10 g samples were selected for each pyrolysis experiment. The pyrolysis temperature was 500 °C, and the pyrolysis time was 10 min. The pyrolysis gas product (non-condensable gas) was collected with a collector bag. The liquid product (bio-oil) was collected by a condenser directly connected to the pyrolysis reactor. The solid product (biochar) remained in the reactor [29].

2.9. *Pyrolysis Performance Characterization*

The thermo-gravimetric analysis (TG) and differential thermal gravity (DTG) of the samples were analyzed via thermal-gravimetric analyzer (STA 449 F5 Jupiter, Netzsch, Germany). A 10.0 mg mass of the sample was placed in an alumina crucible at nitrogen atmosphere. The temperature was increased from 30 °C to 800 °C at 10 °C·min^{-1} [30].

The yield of bio-oil and biochar was calculated by the differential method. The gas velocity was calculated from the gas pressure drop value inside the fixed bed. The difference between the gas velocity and carrier gas velocity in the fixed bed provided the velocity of pyrolysis gas. The amount and yield of non-condensable gas was calculated [31]. The water content of bio-oil was measured using Karl Fischer hydrometer (KF DL31, Mettler-Toledo, Zurich, Switzerland). The viscosity of bio-oil was measured by ChemTron Viscolead rotational viscometer (ChemTron, Celle, Germany) [32]. The chemical composition and distribution of bio-oil was determined by Py-GC/MS. Pyrolysis was performed at 550 °C. Analytical Py-GC/MS experiments were performed using a pyrolysis furnace (a VF-1701 MS column) connected to the Agilent 7890 A gas chromatograph. The basic method and process were described by Gu and co-authors [33].

3. Results and Discussion

3.1. Response Surface Design and Results

Box–Behnken was used for evaluating the effect of concentration of reaction temperature (x_1), time (x_2), pH (x_3) on the total dissolution rate of AAEMs. The experimental design and results are shown in Table 2.

Table 2. Response surface experiment design and results.

Run	Factor			Response
	x_1 (°C)	x_2 (min)	x_3	Total Removal Rate Y (%)
1	170	60	4	43.10
2	180	70	4	35.99
3	170	50	5	32.81
4	180	60	3	47.42
5	170	60	4	44.44
6	160	70	4	32.57
7	180	50	4	37.62
8	170	60	4	44.26
9	170	50	3	43.18
10	180	60	5	35.31
11	170	70	3	45.68
12	160	60	3	40.46
13	170	70	5	33.66
14	170	60	4	43.51
15	170	60	4	44.32
16	160	50	4	36.65
17	160	60	5	29.9

The experimental data were analyzed by regression. The responses and independent variables were correlated by the resulting second-order polynomial Equation (1).

$$Y(\%) = 41.73 + 2.10x_1 - 0.29x_2 - 5.63x_3 + 0.61x_1x_2 - 0.39x_1x_3 - 0.41x_2x_3 - 3.29x_1^2 - 2.73x_2^2 - 0.16x_3^2 \quad (1)$$

3.2. Interaction between Reaction Factors

A change in the color of the 3D response surface graph from blue to red indicates a change in the extraction quality from less to more. The faster the change, the greater the slope, and the more significant the impression of the test result. The optimum process parameters and the interaction between the parameters were studied. The results are shown in Figure 2.

Figure 2a shows the interaction between reaction temperature and holding time on the total removal rate of AAEMs at a fixed pH 4.0. The total removal rate of AAEMs varies with time with a similar change rule at different temperatures, increasing with time from 50 min to 60 min. However, it decreases with time from 60 min to 70 min. This is due to the fact that the dissolution of AAEMs was promoted with increased cell wall damage as the reaction progressed [34]. At the same time, carbohydrate degradation increased with reaction time [35,36]. The formation of organic acids was promoted. The complexation reaction between AAEMs and organic acids was intensified with the increase of organic acid concentration [37]. The dissolution of AAEMs was inhibited by intracellular accumulation of complexes. The results also show that the variation range of the total removal rate of AAEMs decreases within the same temperature range with the increase of temperature. The total AAEMs removal rate at 160 °C for 50 min and 70 min measured 36.65% and 32.57%, respectively. It was 37.62% and 35.99% at 180 °C for 50 min and 70 min. The degree of cell wall damage increases with increasing temperature. The removal rate of AAEMs increased at the same time. The formation of organic acids was facilitated by the increased acidity of hydrolysates at high temperatures. However, the residual AAEMs in the cell were reduced due to increased damage to the cell wall. The complexation reaction between AAEMs and organic acids was reduced. Compared with low temperature, the total removal rate of AAEMs increased. The results showed that temperature and time exhibited a significant interaction effect on the removal of AAEMs.

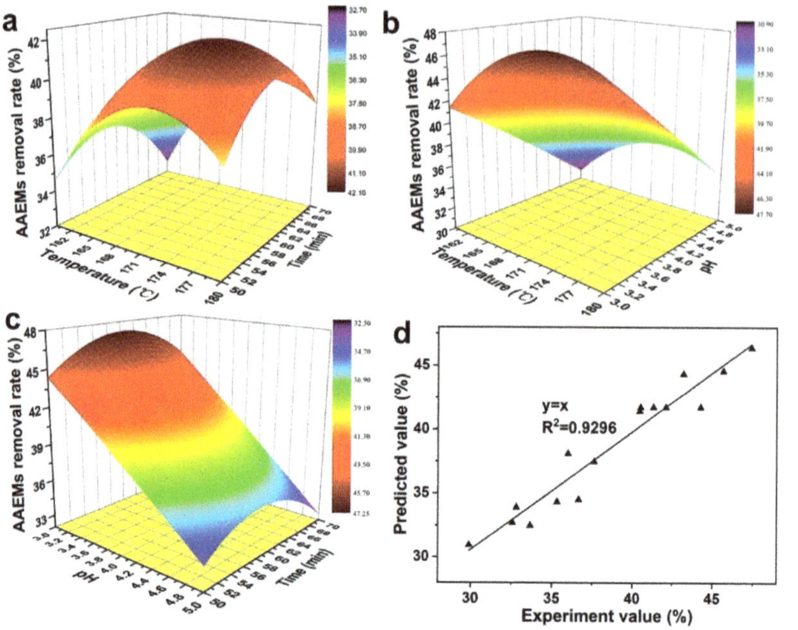

Figure 2. Interaction of reaction temperature, time and pH on the total removal rate of AAEMs. ((**a**), interaction between temperature and time on the total removal rate of AAEMs. (**b**), interaction between temperature and pH on the total removal rate of AAEMs. (**c**), interaction between pH and time on the total removal rate of AAEMs. (**d**), fitting relationship between the predicted value and the experiment value).

The interactive influence of reaction temperature and pH value on the total removal rate of AAEMs at fixed reaction time (60 min) is shown in Figure 2b. The effect of pH on AAEMs removal was similar at different temperatures. The total removal rate of AAEMs increased with an increase of pH at low pH values (3.0–4.0). However, the total removal rate of AAEMs decreased with pH values between 4 and 5. This is due to the extent of cell wall damage being exacerbated by strong acidity [38]. Specifically, the AAEMs removal rates of pH 3 and 5 were 40.46% and 29.90%, respectively, at 160 °C. However, the removal effect of AAEMs was influenced by the complexation of AAEMs and organic acids inside the cell. Previous studies have shown that the optimal extraction of hemicellulose was obtained during hydrothermal pretreatment at pH 4 [39]. This means that the maximum cellulose extraction yield was obtained while the degradation was inhibited. This results in a decrease in organic acid content. The degree of complexation reaction between AAEMs and organic acid was reduced. Under the interaction of cell wall rupture and organic acid complexation reaction, AAEMs removal rate was higher at pH 4. Figure 2b also shows that the total removal rate of AAEMs increases with an increase of temperature at the same pH. The AAEMs removal rates of pH 3 and 5 were 47.42% and 35.31%, respectively, at 180 °C. This is because the acidity of the hydrolysate and the steam pressure increase with increasing temperature [40]. The damage to the cell wall was exacerbated and the removal of AAEMs was facilitated. The results showed that pH was dominant in the interaction between pH and temperature on AAEMs removal.

Figure 2c shows the interaction of pH and time with AAEMs removal at a fixed temperature (170 °C). The effect of time on AAEMs removal was similar to the effect of time on AAEMs removal in Figure 2a. In addition, the removal rate of AAEMs decreases with an increase of pH at the same time. The reasons for this have been explained above. Different from Figure 2b, AAEMs removal rate was not abnormal at pH 4. This is because the consumption of AAEMs in the complexation reaction was much lower than the total amount of AAEMs released by the cells at 170 °C. At this point, the removal of AAEMs was mainly affected by the amount of AAEMs dissolution after cell wall rupture. The above results indicate that pH dominates the interaction between time and pH on AAEMs removal.

The optimal process of AAEMs removal in hydrothermal pretreatment was obtained by response surface design, namely at a temperature of 172.98 °C, time of 59.77 min, and pH of 3.01. The optimal AAEMs total removal rate was 47.59%. In addition, the accuracy of the response surface model was analyzed. The results are shown in Figure 2b. There is a good linear correlation between the experimental value and the predicted value. The correlation coefficient R^2 of the linear regression equation between them was 0.9296. This means that the model exhibits a high degree of accuracy. The prediction data of the model are real and effective.

3.3. Thermal Stability Analysis

The thermal stability of woody biomass improved with the removal of AAEMs. The rapid pyrolysis of biomass was promoted. High quality and high yield of bio-oil was obtained [41]. Thermo-gravimetric analysis (TGA) of eucalyptus during hydrothermal pretreatment and acid leaching was studied and compared at the same removal rate of AAEMs. The results are shown in Figure 3.

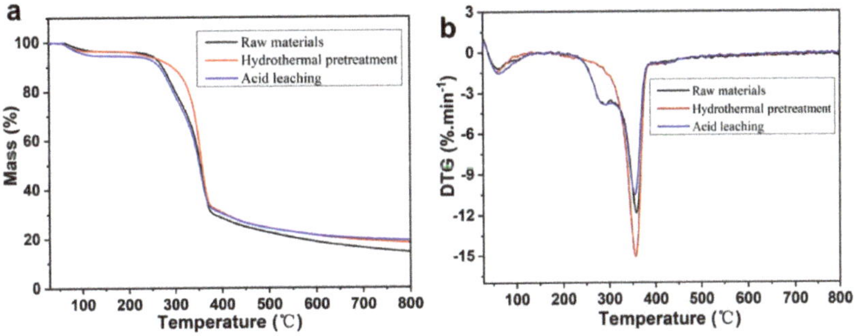

Figure 3. TG (**a**) and DTG (**b**) of eucalyptus with or without acid leaching and hydrothermal pretreatment.

There are several significant changes in Figure 3a. The first is that the initial decomposition temperature of the sample is different. The initial decomposition temperatures of raw materials and acid leaching samples were 254 °C and 255 °C, respectively. Remarkably, the initial decomposition temperature of the hydrothermal pretreatment sample significantly increased to 296 °C. More importantly, the maximum weight loss of the samples differed although their final decomposition temperature was similar at 377 °C. The maximum weight loss of raw material, acid leaching and hydrothermal pretreatment samples was 69.49%, 67.47% and 66.90%, respectively. As is well known, the effect of acid leaching on the physicochemical structure of woody biomass was significant. The dissolution and degradation of cellulose, hemicellulose and lignin was improved at the same removal effect of AAEMs. This resulted in a reduction of the number of decomposable components. The maximum weight loss of acid leaching sample was reduced. However, AAEMs were effectively removed while hemicellulose was selectively removed during hydrothermal pretreatment. The initial decomposition temperature was increased due to the decrease of hemicellulose content after pretreatment. Correspondingly, the relative contents of cellulose and lignin in the pretreated sample was increased. Thus, it also bears a higher residual mass than the raw material. The results show that the hydrothermal pretreated sample has higher initial decomposition temperature. However, this is insufficient to indicate an improvement in thermal stability, and is also related to the maximum rate of weight loss.

The maximum weight loss rates of different samples are shown in Figure 3b. The maximum weight loss rate of raw materials was 11.85 $\%\cdot min^{-1}$. It decreased to 10.53 $\%\cdot min^{-1}$ with acid leaching. This indicates that the removal of AAEMs is accompanied by the loss of more effective pyrolysis components (cellulose). This is inconducive to the thermal and chemical utilization of woody biomass. Contrary to acid leaching, hydrothermal pretreated samples exhibit a higher maximum weight loss rate (15.06 $\%\cdot min^{-1}$). This means that the sample has higher thermal stability. In addition, the DTG curves of different samples provide another important piece of information. The "shoulder peak" exists in the DTG curve of raw materials and acid leaching samples. It is a significant marker of the presence of hemicellulose in samples. However, there was no such peak in the sample after hydrothermal pretreatment. This also verifies our previous inference that hydrothermal pretreatment can efficiently remove AAEMs while selectively removing hemicellulose. The results showed that the hydrothermal pretreated sample exhibits higher thermal stability than the acid leaching sample at the same removal effect of AAEMs.

3.4. Pyrolysis Performance Analysis

It is well known that the pyrolysis properties and products of cellulose, hemicellulose and lignin differ. Therefore, the effect of different pretreatments on the fast pyrolysis performance of eucalyptus was studied. The results are shown in Table 3.

Table 3. Pyrolysis characteristics of eucalyptus with and without acid leaching and hydrothermal pretreatment.

Samples	Bio-Oil Yield (%)	Biochar Yield (%)	Non-Condensable Gas Yield (%)	Bio-Oil Moisture (%)	Bio-Oil Viscosity (mPa·s)
Raw material	49.56	18.24	26.43	27.08	53.17
Acid leaching	59.59	14.31	20.46	21.14	95.25
Hydrothermal	65.87	12.89	15.65	19.54	89.51

Table 3 shows that the yield of bio-oil was increased by acid leaching. The yield of biochar and non-condensable gas was decreased. This is consistent with previous research [42,43]. Significantly, a higher yield of bio-oil was obtained by hydrothermal pretreatment demineralization, while the generation of biochar and non-condensable gas was inhibited. A large amount of hemicellulose was selectively removed during hydrothermal pretreatment [36,39]. The pyrolysis reaction was facilitated by the higher cellulose and lignin content in the sample. In addition, the biomass pyrolysis reaction was affected by the synergies between the three components (cellulose, hemicellulose and lignin) [44]. The synergistic effect of hemicellulose and cellulose was not evident. The synergetic effect of cellulose and lignin pyrolysis was conspicuous [45]. The formation of laevoglucose was inhibited during cellulose pyrolysis due to the presence of lignin. Therefore, the formation of low molecular weight products was promoted, and the yield of biochar was reduced. The formation of secondary carbon products during lignin pyrolysis was inhibited, and the formation of lignin pyrolysis products, such as o-methoxyphenol and 4-methyl guaiacol, was promoted due to the presence of cellulose. Meanwhile, the gas products (CO, H_2, CH_4 and C_2H_4) were clearly inhibited by the synergistic effect between cellulose and lignin. Based on the above research conclusions, the synergistic effect of cellulose and lignin in the samples after hydrothermal pretreatment further promoted the yield of bio-oil and decreased the yield of biochar and non-condensable gas.

The effects of acid leaching and hydrothermal pretreatment on moisture in bio-oil are shown in Table 3. The moisture in bio-oil with acid leaching decreases to 5.94%. However, the moisture reduction effect with hydrothermal pretreatment was more significant, measured at 7.54%. In fact, the pyrolysis water mainly originates from the dehydration reaction of the structural units in cellulose and hemicellulose [46]. For example, ketones and ethers were produced by the dehydration reactions between adjacent cellulose chains. In addition, the pyrolysis water was generated from the dehydration of lignin molecules. The content of pyrolysis water formed by dehydration reaction of hemicellulose was greatly reduced, which was due to the selective removal of hemicellulose by hydrothermal pretreatment.

Table 3 shows that a higher concentration of bio-oil was obtained by acid leaching and hydrothermal pretreatment. The viscosity of bio-oil with acid leaching was higher under the same removal capacity of AAEMs. This was due to an increase in its bio-oil "superheavy components" (solids that do not decompose by heat). As shown in Figure 3, the thermal stability of solid residues in the sample with acid leaching was higher than that of the corresponding components in the sample with untreated or hydrothermal pretreatment.

3.5. Chemical Composition and Distribution of Bio-Oil

The chemical composition and distribution of bio-oils are altered by demineralization [47]. Figure 4a shows that the bio-oils from raw material samples were mainly composed of sugars, phenols, ketones and hydrocarbons. The contents were 42.50%, 16.57%, 14.73% and 12.61%, respectively. The sugars of bio-oil decreased to 30.59% after acid leaching (Figure 4b). This was attributed to changes in the composition of eucalyptus during acid leaching for AAEMs removal. Table 4 shows the contents of cellulose, hemicellulose and lignin in eucalyptus after acid leaching decreased to 37.55%, 9.68% and 16.54%, respectively. In fact, high yield of laevoglucose is obtained in the pyrolysis of cellulose [48]. The pyrolysis products of hemicellulose mainly include hydrocarbons, acids and ketones [49]. Phenols are obtained from lignin pyrolysis [50]. Cellulose, hemicellulose and lignin were significantly degraded by acid leaching. Therefore, the content of main components of bio-oil was reduced. In addition, hydrothermal pretreatment has little effect on the composition of eucalyptus. Hemicellulose was selectively removed (Table 4). Therefore, the sugars increased to 60.74% after hydrothermal pretreatment (Figure 4c). The bio-oils from raw material samples are mainly divided into high-quality bio-oils (C6, C7, C8 and C9), light bio-oils (C4 and C5) and heavy bio-oils (C10, C11 and C12+), as shown in Figure 4d. The contents were 66.80%, 13.06% and 20.14%, respectively. The content of high-quality bio-oil decreased to 65.40% after acid leaching (Figure 4e). Its content increased to 80.99% in the hydrothermal pretreatment sample (Figure 4f). This was due to the protection of cellulose. The content of acids and ketones in hydrothermal pretreatment sample was reduced to 0.83% and 9.43%, respectively. This reduces the light bio-oil content of the hydrothermal pretreatment sample to 7.17%. Therefore, the quality of bio-oil was improved by hydrothermal pretreatment.

Table 4 shows that the lignin content of hydrothermal pretreatment sample was higher than that of acid leaching sample. However, the concentration of phenols in the hydrothermal pretreatment sample were low, measured at 11.38% (Figure 4c). This means that there are other factors affecting the composition and distribution of bio-oil. The content of AAEMs in different forms was changed by demineralization [51]. In fact, hydrothermal pretreatment and acid leaching exert different effects on the dissolution of AAEMs in different forms. Therefore, different forms of AAEMs in eucalyptus after hydrothermal pretreatment and acid leaching were analyzed. Figure 5a shows that the content of potassium, calcium, and magnesium were the same in eucalyptus after different treatments. This verifies that acid leaching and hydrothermal pretreatment exhibit the same AAEMs removal rate. However, more water-soluble AAEMs were obtained in hydrothermal pretreatment sample (Figure 5b–d). The deoxidation of lignin during pyrolysis was promoted to utilize water-soluble AAEMs [52]. Therefore, the yield of phenols in hydrothermal pretreatment samples were reduced. In addition, Figure 5c shows that acid-soluble Ca^{2+}

content was higher in acid leaching sample. Calcium carboxylate was formed as a result of acid-soluble Ca^{2+} binding to esters during pyrolysis. The yield of ester products was reduced [53]. In addition, the yield of ketones was increased by further decomposition of calcium carboxylate into linear ketones [54]. Decreased lipid content and increased ketones in hydrothermal pretreatment samples were explained. This indicates that biomass was effectively protected during hydrothermal pretreatment, and more water-soluble AAEMs were retained. The pyrolysis efficiency of biomass showed improvement and higher quality bio-oils were obtained.

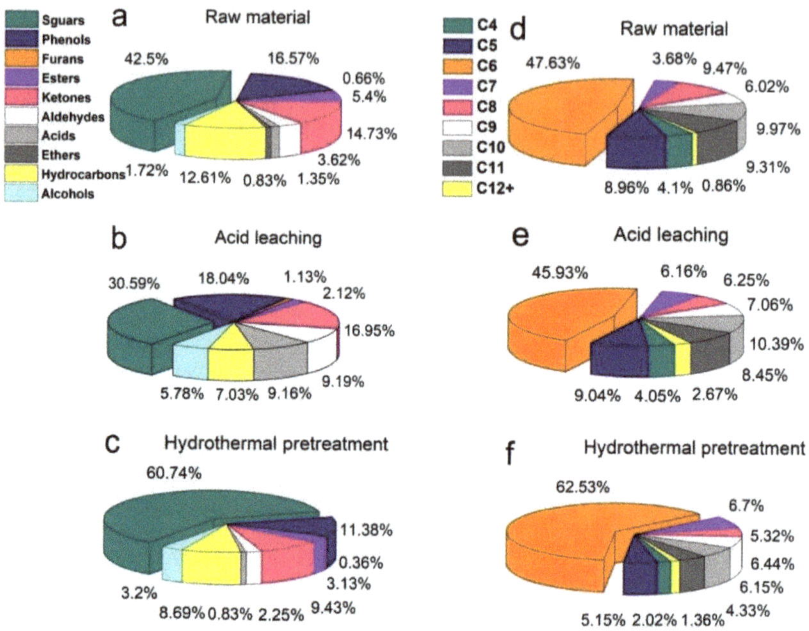

Figure 4. Composition and distribution of pyrolytic bio-oil from different eucalyptus samples (**a–c**), components of pyrolytic bio-oil from raw material, acid leaching eucalyptus and hydrothermal pretreatment eucalyptus. (**d–f**), distribution of pyrolytic bio-oil from raw material, acid leaching eucalyptus and hydrothermal pretreatment eucalyptus.

Table 4. Chemical composition of eucalyptus before and after hydrothermal pretreatment and acid leaching.

Samples	Cellulose (%)	Hemicellulose (%)	Lignin (%)
Raw material	49.55	14.93	32.53
Acid leaching	37.55	9.68	16.54
Hydrothermal pretreatment	47.02	6.24	33.17

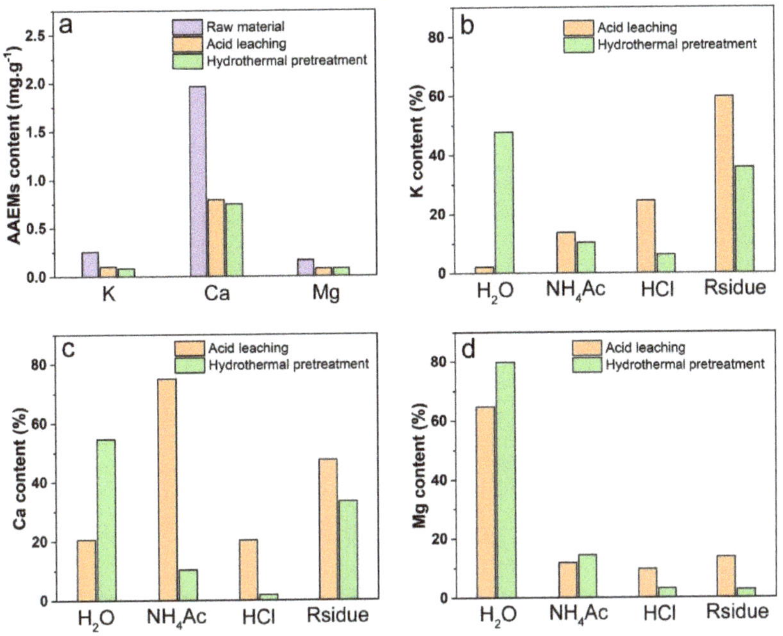

Figure 5. Forms of main metal elements in eucalyptus before and after hydrothermal pretreatment and acid leaching ((**a**), contents of K, Ca and Mg in different eucalyptus samples. (**b**), contents of K in different forms in eucalyptus after acid pickling and hydrothermal pretreatment. (**c**), contents of Ca in different forms in eucalyptus after acid pickling and hydrothermal pretreatment. (**d**), contents of Mg in different forms in eucalyptus after acid pickling and hydrothermal pretreatment).

4. Conclusions

The interaction of temperature, time, and pH value on AAEMs removal during hydrothermal pretreatment was studied by the response surface method. The optimal AAEMs removal process and the best removal rate were obtained. Compared with acid leaching, the hydrothermal pretreated samples exhibited higher thermal stability. The content of sugars and the yield of high-quality bio-oil in the pyrolysis products were significantly increased. The results show that the hydrothermal pretreatment bears high effective demineralization and practical application value.

Author Contributions: Conceptualization, S.Y.; methodology, J.Z. and Y.B.; resources, L.L. and F.Z.; data curation, C.L.; writing—original draft preparation, J.G.; writing—review and editing, D.D. and S.W.; project administration, S.Y. All authors have read and agreed to the published version of the manuscript.

Funding: This research was funded by the National Natural Science Foundation of China, grant number 22078075 and 21968004.

Data Availability Statement: The data presented in this study are available in the manuscript's figure.

Conflicts of Interest: The authors declare no conflict of interest. The funders had no role in the design of the study; in the collection, analyses, or interpretation of data; in the writing of the manuscript, or in the decision to publish the results.

References

1. Johnsson, F.; Kjärstad, J.; Rootzén, J. The threat to climate change mitigation posed by the abundance of fossil fuels. *Clim. Policy* **2019**, *19*, 258–274. [CrossRef]
2. Zeng, H.; Liu, B.; Li, J.; Li, M.; Peng, M.; Qin, C.; Liang, C.; Huang, C.; Li, X.; Yao, S. Efficient separation of bagasse lignin by freeze–thaw-assisted p-toluenesulfonic acid pretreatment. *Bioresour. Technol.* **2022**, *351*, 126951. [CrossRef] [PubMed]
3. Feng, C.; Zhu, J.; Hou, Y.; Qin, C.; Chen, W.; Nong, Y.; Liao, Z.; Liang, C.; Bian, H.; Yao, S. Effect of temperature on simultaneous separation and extraction of hemicellulose using p-toluenesulfonic acid treatment at atmospheric pressure. *Bioresour. Technol.* **2022**, *348*, 126793. [CrossRef] [PubMed]
4. Luo, Y.; Li, Y.; Cao, L.; Zhu, J.; Deng, B.; Hou, Y.; Liang, C.; Huang, C.; Qin, C.; Yao, S. High efficiency and clean separation of eucalyptus components by glycolic acid pretreatment. *Bioresour. Technol.* **2021**, *341*, 125757. [CrossRef] [PubMed]
5. Bian, H.; Chen, L.; Dong, M.; Fu, Y.; Wang, R.; Zhou, X.; Wang, X.; Xu, J.; Dai, H. Cleaner production of lignocellulosic nanofibrils: Potential of mixed enzymatic treatment. *J. Clean. Prod.* **2020**, *270*, 122506. [CrossRef]
6. Nguyen, T.-V.; Clausen, L.R. Techno-economic analysis of polygeneration systems based on catalytic hydropyrolysis for the production of bio-oil and fuels. *Energy Convers. Manag.* **2019**, *184*, 539–558. [CrossRef]
7. Patel, M.; Kumar, A. Production of renewable diesel through the hydroprocessing of lignocellulosic biomass-derived bio-oil: A review. *Renew. Sustain. Energy Rev.* **2016**, *58*, 1293–1307. [CrossRef]
8. Xiong, Z.; Wang, Y.; Syed-Hassan, S.S.A.; Hu, X.; Han, H.; Su, S.; Xu, K.; Jiang, L.; Guo, J.; Berthold, E.E.S.; et al. Effects of heating rate on the evolution of bio-oil during its pyrolysis. *Energy Convers. Manag.* **2018**, *163*, 420–427. [CrossRef]
9. Dai, L.; Wang, Y.; Liu, Y.; Ruan, R.; He, C.; Yu, Z.; Jiang, L.; Zeng, Z.; Tian, X. Integrated process of lignocellulosic biomass torrefaction and pyrolysis for upgrading bio-oil production: A state-of-the-art review. *Renew. Sustain. Energy Rev.* **2019**, *107*, 20–36. [CrossRef]
10. Liang, J.; Shan, G.; Sun, Y. Catalytic fast pyrolysis of lignocellulosic biomass: Critical role of zeolite catalysts. *Renew. Sustain. Energy Rev.* **2021**, *139*, 110707. [CrossRef]
11. Zhang, L.; Li, S.; Li, K.; Zhu, X. Two-step pyrolysis of corncob for value-added chemicals and high quality bio-oil: Effects of pyrolysis temperature and residence time. *Energy Convers. Manag.* **2018**, *166*, 260–267. [CrossRef]
12. Deng, L.; Ye, J.; Jin, X.; Che, D. Transformation and release of potassium during fixed-bed pyrolysis of biomass. *J. Energy Inst.* **2018**, *91*, 630–637. [CrossRef]
13. Persson, H.; Kantarelis, E.; Evangelopoulos, P.; Yang, W. Wood-derived acid leaching of biomass for enhanced production of sugars and sugar derivatives during pyrolysis: Influence of acidity and treatment time. *J. Anal. Appl. Pyrolysis* **2017**, *127*, 329–334. [CrossRef]
14. Chen, D.; Mei, J.; Li, H.; Li, Y.; Lu, M.; Ma, T.; Ma, Z. Combined pretreatment with torrefaction and washing using torrefaction liquid products to yield upgraded biomass and pyrolysis products. *Bioresour. Technol.* **2017**, *228*, 62–68. [CrossRef] [PubMed]
15. Asadieraghi, M.; Wan Daud, W.M.A. Characterization of lignocellulosic biomass thermal degradation and physiochemical structure: Effects of demineralization by diverse acid solutions. *Energy Convers. Manag.* **2014**, *82*, 71–82. [CrossRef]
16. Dai, L.; Wang, Y.; Liu, Y.; Ruan, R.; He, C.; Duan, D.; Zhao, Y.; Yu, Z.; Jiang, L.; Wu, Q. Bridging the relationship between hydrothermal pretreatment and co-pyrolysis: Effect of hydrothermal pretreatment on aromatic production. *Energy Convers. Manag.* **2019**, *180*, 36–43. [CrossRef]
17. Yao, Z.; Ma, X.; Xiao, Z. The effect of two pretreatment levels on the pyrolysis characteristics of water hyacinth. *Renew. Energy* **2020**, *151*, 514–527. [CrossRef]
18. Ma, Y.; Zhang, H.; Yang, H.; Zhang, Y. The effect of acid washing pretreatment on bio-oil production in fast pyrolysis of rice husk. *Cellulose* **2019**, *26*, 8465–8474. [CrossRef]
19. Dong, Q.; Zhang, S.; Zhang, L.; Ding, K.; Xiong, Y. Effects of four types of dilute acid washing on moso bamboo pyrolysis using Py–GC/MS. *Bioresour. Technol.* **2015**, *185*, 62–69. [CrossRef]
20. Mosqueda, A.; Wei, J.; Medrano, K.; Gonzales, H.; Ding, L.; Yu, G.; Yoshikawa, K. Co-gasification reactivity and synergy of banana residue hydrochar and anthracite coal blends. *Appl. Energy* **2019**, *250*, 92–97. [CrossRef]
21. Chang, S.; Zhao, Z.; Zheng, A.; Li, X.; Wang, X.; Huang, Z.; He, F.; Li, H. Effect of hydrothermal pretreatment on properties of bio-oil produced from fast pyrolysis of eucalyptus wood in a fluidized bed reactor. *Bioresour. Technol.* **2013**, *138*, 321–328. [CrossRef] [PubMed]
22. Ge, J.; Wu, Y.; Han, Y.; Qin, C.; Nie, S.; Liu, S.; Wang, S.; Yao, S. Effect of hydrothermal pretreatment on the demineralization and thermal degradation behavior of eucalyptus. *Bioresour. Technol.* **2020**, *307*, 123246. [CrossRef] [PubMed]
23. Li, X.; Bai, Z.-Q.; Bai, J.; Zhao, B.-B.; Li, P.; Han, Y.-N.; Kong, L.-X.; Li, W. Effect of Ca2+ species with different modes of occurrence on direct liquefaction of a calcium-rich lignite. *Fuel Processing Technol.* **2015**, *133*, 161–166. [CrossRef]
24. Yao, S.; Nie, S.; Zhu, H.; Wang, S.; Song, X.; Qin, C. Extraction of hemicellulose by hot water to reduce adsorbable organic halogen formation in chlorine dioxide bleaching of bagasse pulp. *Ind. Crops Prod.* **2017**, *96*, 178–185. [CrossRef]
25. Santos, H.M.; Coutinho, J.P.; Amorim, F.A.C.; Lôbo, I.P.; Moreira, L.S.; Nascimento, M.M.; de Jesus, R.M. Microwave-assisted digestion using diluted HNO_3 and H_2O_2 for macro and microelements determination in guarana samples by ICP OES. *Food Chemistry* **2019**, *273*, 159–165. [CrossRef]
26. Chauhan, G.; de Klerk, A. Dissolution Methods for the Quantification of Metals in Oil Sands Bitumen. *Energy Fuels* **2020**, *34*, 2870–2879. [CrossRef]

27. Yang, Q.; Huo, D.; Si, C.; Fang, G.; Liu, Q.; Hou, Q.; Chen, X.; Zhang, F. Improving enzymatic saccharification of eucalyptus with a pretreatment process using $MgCl_2$. *Ind. Crops Prod.* **2018**, *123*, 401–406. [CrossRef]
28. Pettersson, A.; Åmand, L.-E.; Steenari, B.-M. Chemical fractionation for the characterisation of fly ashes from co-combustion of biofuels using different methods for alkali reduction. *Fuel* **2009**, *88*, 1758–1772. [CrossRef]
29. Chan, Y.H.; Loh, S.K.; Chin, B.L.F.; Yiin, C.L.; How, B.S.; Cheah, K.W.; Wong, M.K.; Loy, A.C.M.; Gwee, Y.L.; Lo, S.L.Y.; et al. Fractionation and extraction of bio-oil for production of greener fuel and value-added chemicals: Recent advances and future prospects. *Chem. Eng. J.* **2020**, *397*, 125406. [CrossRef]
30. An, L.; Si, C.; Bae, J.H.; Jeong, H.; Kim, Y.S. One-step silanization and amination of lignin and its adsorption of Congo red and Cu(II) ions in aqueous solution. *Int. J. Biol. Macromol.* **2020**, *159*, 222–230. [CrossRef]
31. Kang, K.; Zhu, M.; Sun, G.; Qiu, L.; Guo, X.; Meda, V.; Sun, R. Codensification of Eucommia ulmoides Oliver stem with pyrolysis oil and char for solid biofuel: An optimization and characterization study. *Appl. Energy* **2018**, *223*, 347–357. [CrossRef]
32. Cai, W.; Liu, R.; He, Y.; Chai, M.; Cai, J. Bio-oil production from fast pyrolysis of rice husk in a commercial-scale plant with a downdraft circulating fluidized bed reactor. *Fuel Processing Technol.* **2018**, *171*, 308–317. [CrossRef]
33. Gu, J.; Fan, H.; Wang, Y.; Zhang, Y.; Yuan, H.; Chen, Y. Co-pyrolysis of xylan and high-density polyethylene: Product distribution and synergistic effects. *Fuel* **2020**, *267*, 116896. [CrossRef]
34. Zhuang, X.; Zhan, H.; Song, Y.; He, C.; Huang, Y.; Yin, X.; Wu, C. Insights into the evolution of chemical structures in lignocellulose and non-lignocellulose biowastes during hydrothermal carbonization (HTC). *Fuel* **2019**, *236*, 960–974. [CrossRef]
35. Bian, H.; Luo, J.; Wang, R.; Zhou, X.; Ni, S.; Shi, R.; Fang, G.; Dai, H. Recyclable and Reusable Maleic Acid for Efficient Production of Cellulose Nanofibrils with Stable Performance. *ACS Sustain. Chem. Eng.* **2019**, *7*, 20022–20031. [CrossRef]
36. Yao, S.; Nie, S.; Yuan, Y.; Wang, S.; Qin, C. Efficient extraction of bagasse hemicelluloses and characterization of solid remainder. *Bioresour. Technol.* **2015**, *185*, 21–27. [CrossRef]
37. Karley, A.J.; White, P.J. Moving cationic minerals to edible tissues: Potassium, magnesium, calcium. *Curr. Opin. Plant Biol.* **2009**, *12*, 291–298. [CrossRef]
38. Phyo, P.; Gu, Y.; Hong, M. Impact of acidic pH on plant cell wall polysaccharide structure and dynamics: Insights into the mechanism of acid growth in plants from solid-state NMR. *Cellulose* **2019**, *26*, 291–304. [CrossRef]
39. Huang, L.; Yang, Z.; Li, M.; Liu, Z.; Qin, C.; Nie, S.; Yao, S. Effect of Pre-Corrected pH on the Carbohydrate Hydrolysis of Bamboo during Hydrothermal Pretreatment. *Polymers* **2020**, *12*, 612. [CrossRef]
40. Kratky, L.; Jirout, T. The effect of process parameters during the thermal-expansionary pretreatment of wheat straw on hydrolysate quality and on biogas yield. *Renew. Energy* **2015**, *77*, 250–258. [CrossRef]
41. Karnowo; Zahara, Z.F.; Kudo, S.; Norinaga, K.; Hayashi, J.-I. Leaching of Alkali and Alkaline Earth Metallic Species from Rice Husk with Bio-oil from Its Pyrolysis. *Energy Fuels* **2014**, *28*, 6459–6466. [CrossRef]
42. Hwang, H.; Lee, J.-H.; Moon, J.; Kim, U.-J.; Choi, I.-G.; Choi, J.W. Influence of K and Mg Concentration on the Storage Stability of Bio-Oil. *ACS Sustain. Chem. Eng.* **2016**, *4*, 4346–4353. [CrossRef]
43. Wu, L.; Yang, Y.; Yan, T.; Wang, Y.; Zheng, L.; Qian, K.; Hong, F. Sustainable design and optimization of co-processing of bio-oil and vacuum gas oil in an existing refinery. *Renew. Sustain. Energy Rev.* **2020**, *130*, 109952. [CrossRef]
44. Hosoya, T.; Kawamoto, H.; Saka, S. Cellulose–hemicellulose and cellulose–lignin interactions in wood pyrolysis at gasification temperature. *J. Anal. Appl. Pyrolysis* **2007**, *80*, 118–125. [CrossRef]
45. Giudicianni, P.; Cardone, G.; Ragucci, R. Cellulose, hemicellulose and lignin slow steam pyrolysis: Thermal decomposition of biomass components mixtures. *J. Anal. Appl. Pyrolysis* **2013**, *100*, 213–222. [CrossRef]
46. Zhou, Y.; Chen, Z.; Gong, H.; Chen, L.; Yu, H.; Wu, W. Characteristics of dehydration during rice husk pyrolysis and catalytic mechanism of dehydration reaction with $NiO/\gamma\text{-}Al_2O_3$ as catalyst. *Fuel* **2019**, *245*, 131–138. [CrossRef]
47. Jiang, L.; Zheng, A.; Zhao, Z.; He, F.; Li, H. Comprehensive utilization of glycerol from sugarcane bagasse pretreatment to fermentation. *Bioresour. Technol.* **2015**, *196*, 194–199. [CrossRef]
48. Collard, F.-X.; Blin, J. A review on pyrolysis of biomass constituents: Mechanisms and composition of the products obtained from the conversion of cellulose, hemicelluloses and lignin. *Renew. Sustain. Energy Rev.* **2014**, *38*, 594–608. [CrossRef]
49. Wang, S.; Ru, B.; Lin, H.; Sun, W. Pyrolysis behaviors of four O-acetyl-preserved hemicelluloses isolated from hardwoods and softwoods. *Fuel* **2015**, *150*, 243–251. [CrossRef]
50. Wang, S.; Li, Z.; Bai, X.; Yi, W.; Fu, P. Influence of inherent hierarchical porous char with alkali and alkaline earth metallic species on lignin pyrolysis. *Bioresour. Technol.* **2018**, *268*, 323–331. [CrossRef]
51. He, Y.; Gao, X.; Qiao, Y.; Xu, M. Occurrence forms of key ash-forming elements in defatted microalgal biomass. *Fuel* **2017**, *200*, 182–185. [CrossRef]
52. Wang, W.-L.; Ren, X.-Y.; Chang, J.-M.; Cai, L.-P.; Shi, S.Q. Characterization of bio-oils and bio-chars obtained from the catalytic pyrolysis of alkali lignin with metal chlorides. *Fuel Processing Technol.* **2015**, *138*, 605–611. [CrossRef]
53. Wang, D.; Xiao, R.; Zhang, H.; He, G. Comparison of catalytic pyrolysis of biomass with MCM-41 and CaO catalysts by using TGA–FTIR analysis. *J. Anal. Appl. Pyrolysis* **2010**, *89*, 171–177. [CrossRef]
54. Chen, X.; Chen, Y.; Yang, H.; Chen, W.; Wang, X.; Chen, H. Fast pyrolysis of cotton stalk biomass using calcium oxide. *Bioresour. Technol.* **2017**, *233*, 15–20. [CrossRef]

polymers

MDPI

Article

Hemicellulose and Nano/Microfibrils Improving the Pliability and Hydrophobic Properties of Cellulose Film by Interstitial Filling and Forming Micro/Nanostructure

Yan Li [1], Mingzhu Yao [1], Chen Liang [1,2], Hui Zhao [1,2,3], Yang Liu [1,2,4,*] and Yifeng Zong [1]

1 College of Light Industry and Food Engineering, Guangxi University, Nanning 530004, China; 1916391015@st.gxu.edu.cn (Y.L.); yaomingzhu@st.gxu.edu.cn (M.Y.); liangchen@gxu.edu.cn (C.L.); zhh@gxu.edu.cn (H.Z.); 2016391055@st.gxu.edu.cn (Y.Z.)
2 Guangxi Key Laboratory of Clean Pulp & Papermaking and Pollution Control, Guangxi University, Nanning 530004, China
3 State Key Laboratory of Biocatalysis and Enzyme Engineering, School of Life Sciences, Hubei University, Wuhan 430062, China
4 Guangxi Bossco Environmental Protection Technology Co., Ltd., Nanning 530000, China
* Correspondence: xiaobai@gxu.edu.cn; Tel.: +86-1557-832-3385

Abstract: In this paper, nano/microfibrils were applied to enhance the mechanical and hydrophobic properties of the sugarcane bagasse fiber films. The successful preparation of nano/microfibrils was confirmed by scanning electron microscope (SEM), X-ray diffraction (XRD), fiber length analyzer (FLA), and ion chromatography (IC). The transparency, morphology, mechanical and hydrophobic properties of the cellulose films were evaluated. The results show that the nanoparticle was formed by the hemicellulose diffusing on the surface of the cellulose and agglomerating in the film-forming process at 40 °C. The elastic modulus of the cellulose film was as high as 4140.60 MPa, and the water contact angle was increased to 113°. The micro/nanostructures were formed due to hemicellulose adsorption on nano/microfilament surfaces. The hydrophobicity of the films was improved. The directional crystallization of nano/microfibrous molecules was found. Cellulose films with a high elastic modulus and high elasticity were obtained. It provides theoretical support for the preparation of high-performance cellulose film.

Keywords: bagasse; nano/microfibrils; hydrophobicity; high pliability; high-consistency refiner

Citation: Li, Y.; Yao, M.; Liang, C.; Zhao, H.; Liu, Y.; Zong, Y. Hemicellulose and Nano/Microfibrils Improving the Pliability and Hydrophobic Properties of Cellulose Film by Interstitial Filling and Forming Micro/Nanostructure. *Polymers* **2022**, *14*, 1297. https://doi.org/10.3390/polym14071297

Academic Editor: Tomasz Tański

Received: 12 February 2022
Accepted: 10 March 2022
Published: 23 March 2022

1. Introduction

Due to the extensive use of petroleum-based materials, a great deal of white pollution brings a critical menace to the ecotope [1]. In recent years, people have paid more and more attention to the development and application of green, sustainable, and renewable environment-friendly materials [2,3]. Cellulose films (i.e., nanofibrils film [4], microfibrils film [5], and nano/microfibrils film) had attracted intensive interest during the past decade due to its green sustainable nature and constant advances in micro and nanoscale patterning [6] for many applications of functional materials [7,8], including the filler [9], composite material manufacturing [10–12], packaging coating [13,14], medicine [15,16], high-performance green flexible electronics [17], and biotechnology [18].

Cellulose can be pretreated and machined to produce nanocrystalline cellulose (CNC), nanofibrils cellulose (NFC) [19], and microfibrils cellulose (MFC) [20]. It has unique attributes, such as enhanced capabilities [21], high mechanical strength, and adjustable self-assembly in an aqueous solution, due to its unique surface chemistry, size, shape, and high crystallinity [22]. At present, cellulose film research mainly focuses on the film properties of cellulose uniform system (i.e., uniform nanocellulose system, uniform microcellulose system), such that combining nanofibrils with renewable polymers could prepare partially degradable materials to improve material performance deficiencies and to address white pollution [23,24]. Shu et al. [25] prepared cellulose-based bioplastics, which were reorganized by the aggregation structure of cellulose. However, the hydrophobicity of regenerated cellulose membrane was not studied.

However, the application of cellulose films developed to date has been impeded by their poor mechanical performance and poor water resistance [26]. In the literature, a number of approaches have been reported to improve the increased hydrophobicity from cellulose films, such as low surface energy organic compounds [27], e.g., esterification [28], silanization [29], amidation [30], carboxymethylation [31], and fluoropolymer [32]. Unfortunately, this approach had a poor modification effect and resulted in problems related to organic pollution. The performance of nano/microfibril films dictates their hydrophobicity and pliability in relevant applications; however, utilizing the inherent structure of nano/microfibrils within the films for materials applications was an approach still in its infancy.

The method of the nano/microfibrils film for improving hydrophobicity and pliability, which aims to enhance the hydrophobic and pliability of the nano/microfibrils film without surface modification (i.e., unitary nano/microfibrils component), has not been reported. This study provides a simple pathway to improve hydrophobic and pliability properties of sugarcane bagasse nano/microfibrils film isolated by a high-consistency refiner and exhibiting a great potential for the further utilization of cellulose. We used enzymatic pretreatments to investigate the grinding of sugarcane bagasse fiber using a high-consistency refiner grinding to reproduce nano/microfibrils films to determine the mechanical and hydrophobicity properties. The main advantage compared to previous methods is its simplicity in preparation with less environmental pollution and a better economic benefit. The morphology, particle size distribution, structure, and transparency of the obtained nano/microfibrils were analyzed with a scanning electron microscope (SEM), an atomic force microscopy (AFM), a fiber length analyzer (FLA), a zetasizer nanoanalyzer, an X-ray diffraction (XRD), an ion-exchange chromatography (IC), and an ultraviolet and visible spectrophotometer (UV–Vis). The water resistance and mechanical properties of the nano/microfibrils films were analyzed by contact angle determination (CA), water vapor transmission rate (WVTR), and tensile tests.

2. Materials and Methods

2.1. Materials and Chemicals

Bleached sugarcane bagasse pulp was procured from Guitang Co., Ltd. (Guangxi, China). Celluclast (1.5 L, from *T. reesei*) was purchased from Novoxin Biotechnology Co., Ltd. (Beijing, China).

2.2. Methods

2.2.1. Enzyme Pretreatment

Enzymatic hydrolysis pretreatment was carried out with untreated sugarcane bagasse samples. The fibers were treated with cellulase. The fibers and the appropriate amount of cellulase were mixed at a ratio of 1:500. A 1 mL enzyme solution was added to every 500 mL 30 wt% bagasse samples. The enzyme activity was 16 FPU/mL. The most suitable pH was 6.0–8.0. After constant temperature stirring at 50 °C for 30 min, the enzyme activity was eliminated with a constant temperature water bath at 100 °C for 10 min and washed with distilled water multiple times to remove residual enzyme liquid.

2.2.2. Preparation of Nano/Microfibrils by Mechanical Grinding

The enzyme pretreated sugarcane bagasse pulp was diluted to 30% (*w*/*w*) with distilled water and ground with a high-consistency refiner at a disc gap of 0.1 mm, and with 10, 20, 25, 30 rounds of grinding. After grinding, a mixture of nanocellulose and micro-cellulose was stored in a refrigerator at 4 °C for subsequent analysis.

2.2.3. Preparation of Nano/Microfibrils Film

A nano/microfibrils film was formed by pouring nano/microfibrils into a polystyrene template by the flow-edge method, which could be prepared by drying nano/microfibrils in oven at 40 °C for 24 h, as shown in Figure 1.

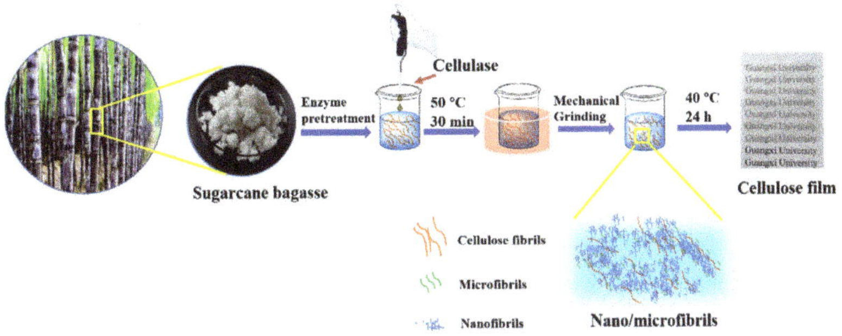

Figure 1. Schematic representation of process for preparing nano/microfibril film.

2.3. Characterization

To evaluate the properties of samples and understand the changes caused by mechanical processing, a raw material analysis of fiber components was performed on the specimens. Then, the hydrophobic and mechanical properties of cellulose film samples were analyzed.

2.3.1. Characterization of Morphological Features

AFM (SI-DFP2, Hitachi, Tokyo, Japan) and SEM (SU8220, Hitachi, Tokyo, Japan) are often combined with cellulose image analysis. In the SEM test, samples were observed at a voltage of 10 kV after spraying gold under vacuum for 90 s. The magnification was 500× and 2000×, respectively. AFM could evaluate their surfaces and morphological changes at the test pressure of 299 kHz, an accelerating voltage of 8.1 V, and a cantilever elastic coefficient (C) of 32 N/m.

2.3.2. XRD Analysis

The nano/microfibril films were cut into small pieces of 1.5 cm × 1.5 cm and used XRD (Miniflex600, Rigaku, Tokyo, Japan) under $Cu_{K\alpha}$ ray radiation (λ = 0.15418 nm) for inspection. The scanning range was 2θ = 5°–35°, and the speed was 5°/min. The tube pressure level and the tube flow rate were 40 kV and 30 mA, respectively. Segal's empirical formula (*CrI*) was used to calculate the crystallization index of the samples [33].

$$CrI\ (\%) = (I_{002} - I_{am})/I_{002} \tag{1}$$

where, *CrI* is the crystal index, I_{002} is the diffraction intensity level obtained when $2\theta = 22.6°$, that is, the diffraction intensity of the crystal region. I_{am} is the diffraction intensity level obtained when $2\theta = 16.0°$, that is, the diffraction intensity of the amorphous region.

2.3.3. Particles Size Analysis

The particle size and its distribution of the nano/microfibril suspension were measured by Malvern Zetasizer nano (ZS90X, Malvern Panalytical, London, UK). A 0.5 wt % nano/microfibril suspension was tested after magnetic stirring for 12 h.

The raw cellulose properties including length-weighted distribution (Ln), weight-weighted distribution (Lw), and fines were measured with a fiber length analyzer (FLA, Kajaani FS-300, Metso Automation, Helsinki, Finland).

2.3.4. IC Analysis

The content of hemicellulose sugar in sugarcane bagasse pulp was determined by ion chromatography (ICS-5000+SP, Thermo Science, Sunnyvale, CA, USA). The chromatographic column was a PA20, 3 × 150 mm, and the protection column was a PA10, 4 × 50 mm. A mix of 80% Milli-Q water and 20% NaOH was used for the mobile phase. The flow rate was 0.3 mL/min. Specific steps conformed to the National Renewable Energy Laboratory standard method TP-510-42618.

2.3.5. UV–Vis Analysis

UV–Vis (SPECORD-PLUS-50, Analytik Jena, Berlin, Germany) was used to measure the transmittance of the samples. The film sample was carefully cut into a rectangle of 40 mm × 9 mm and placed in a quartz cuvette 25 cm from the entrance of the integrating sphere. A quartz cuvette was placed as a blank reference. The wavelength range of the sample was 190–1100 nm, and the transmittance of the visible band (400–800 nm) was analyzed.

2.3.6. Water Resistance Analysis

The water contact angle of the sample was measured using a Drop Shape Analyzer (DSA100, KRUSS, Berlin, Germany) to evaluate the surface hydrophobicity of the film samples. The film samples were fixed on the glass slide with double-sided tape. An automatic pipette was used to carefully apply a drop (4 μL) of distilled water onto the film surface. Parallel tests were performed six times and the results were averaged.

The WVTR (W3/031, Labthink, Jinan, China) measurements were performed according to the standard ASTM Standard E96/E96M-05 for cup method water vapor permeability [34] testing at 25 °C with 90% RH. Weights were monitored every 30 min until constant.

2.3.7. Mechanical Properties

The mechanical properties consisting of folding endurance, elasticity modulus, and elongation at break of the films were measured with a universal testing system (3367, INSTRON, Sunnyvale, CA, USA). The test was conducted at a crosshead speed of 2 mm/min at 23 °C and 55% relative humidity. The sample length was 50 mm and the width was 10 mm.

3. Results

3.1. SEM Analysis of Fibers after Different Stages of Treatments

The SEM analysis of the fiber suspension obtained after different grinding stages are shown in Figure 2a–e. The obtained images exhibit substantial differences in surface and morphological changes. In C-0 without grinding treatment, the fiber bundle structure was compact, and the fibers were relatively compact, such as in Figure 2a. The images obviously show that the sugarcane-bagasse-bleached pulp fibers were nearly smashed into nano/microfibrils after the high-consistency refiner, although a few microfibrils bundles with correspondingly larger fibers still exist. After the mechanical processes, the tightly bound fibers divided into smaller fiber bundles, as shown in Figure 2b–e. In the higher right corner of Figure 2b–e are 2 μm images, which show that a large number of nano/microfibers were partly stripped from the surface due to the effects of grinding. The fiber diameters decreased with each treatment grinding times, and high-speed grinding gradually declined the fiber length and layers' thickness. After grinding more than 25 times, we observed that cellulose fibers were almost completely decomposed into nano/microfibrils aggregates around the fiber (Figure 2d). It formed many fine cellulose aggregates in Figure 2c,d, possibly due to the soluble hemicellulose being exposed during the grinding process and because hemicellulose was loosely bound to the fiber surface. Echoing the results of Tenhunen [35], xylose, one of the principal components of hemicellulose from sugarcane bagasse, had an affinity for cellulose and readsorbs to the cellulose surface during the grinding process. The presence of xylan with a negative charge facilitates the liberation of fibrils from the pulp by generating repulsion between fibrils. Accordingly, in the case of higher levels of hemicellulose in sugarcane bagasse, the preparation of nano/microfibers by the high-consistency refiner became more efficient.

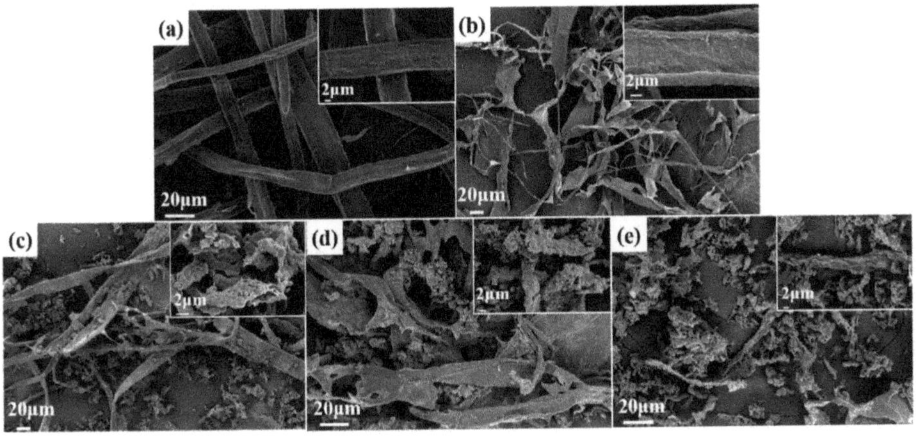

Figure 2. SEM micrographs of (**a**) the raw material C-0, (**b**) 10 times ground C-10, (**c**) 20 times ground C-20, (**d**) 25 times ground C-25, and (**e**) 30 times ground C-30. C-0 represents raw pulp fibers and C-10, C-20, C-25, and C-30 represent fibers that have been grounded 10, 20, 25, and 30 times by a high-consistency refiner, respectively.

3.2. *Particle Size and Components Analysis*

The particle size and particle size distribution of the nano/microfibrils and untreated fibers (C-0) suspension were measured by a zetasizer nano analyzer (Figure 3) and FLA (Table 1). Under the action of mechanical force, bagasse fiber was easy to crack along the axial direction and formed fibrillary cellulose. Fibrillated ultrafine fiber was an anon-uniform system composed of fibers of different sizes and shapes. Without homogenization, the fines fiber (0–0.2 mm) content was approximately 46.16% and when grinding 10 more times, the C-30 fines fiber content was approximately 53.52% (Table 1). L(n) and L(w) of untreated fibers were 0.34 mm and 1.20 mm, respectively; after 10 grindings, L(n) was 0.29 mm and L(w) was 0.86 mm. These data indicate that grinding had an important effect on the L(w) of fibers, especially in the longitudinal direction.

Fibrillated ultrafine fiber is a complex system, which contains a large number of submicron and nanofibers in the range of ultrafine fibers. The zetasizer nano analyzer was used to analyze the particle size range of nano/microfibrils in the fibril suspension system (Figure 3). Different sizes and size distributions of fibrillated ultrafine fiber had an important influence on the film-forming properties. After grinding 10 times, the particle size distribution was in the range of 539–1591 nm. The particle size distribution of C-20 and C-25 fibers had three distinct peaks, one at 100–500 nm, another at 500–1900 nm, and the last one at 3500–6500 nm (Figure 3). This indicates that the cellulose suspension was a multisize mixed system. After grinding 30 times, the particle size distribution was mainly concentrated in the range of 130–400 nm, and the fiber suspension system was relatively uniform. Therefore, compared with unground bagasse cellulose, the size distribution of ground bagasse cellulose was more extensive and richer. The average length of unmilled cellulose was about 650–2170 μm. After milling, the cellulose system had a lot of nanocellulose (1–10 μm) and microcellulose (10–200 μm), and some millimeter-grade cellulose (200–7000 μm). There were many sizes of cellulose, such as nanometer (13–36%), micron (17–32%), and millimeter (46–54%), in the cellulose suspensions. The fiber size distribution of the fibrillated ultrafine fiber was more abundant after 25 times of grinding. The fibrillated ultrafine fiber suspension system of C-30 was relatively uniform. It was feasible and highly efficient to prepare a hybrid system of fibrillated nano/microfibrils with the high-consistency refiner.

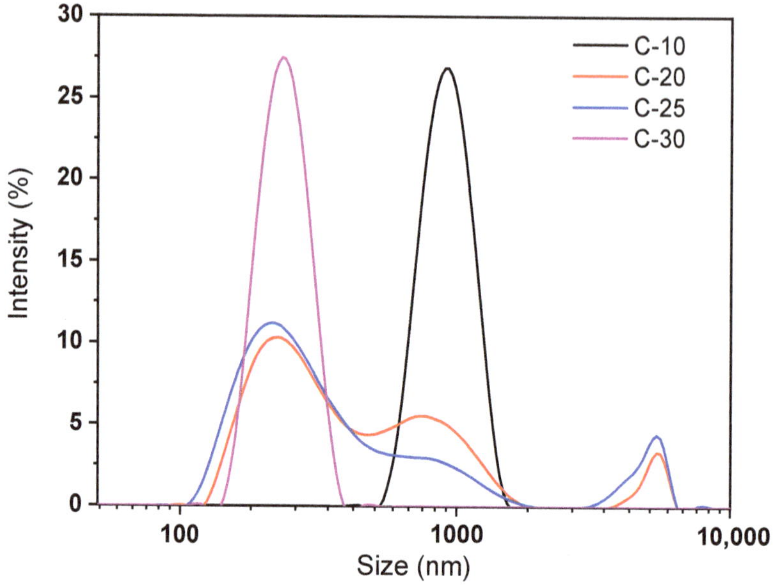

Figure 3. The particle size distribution of fibers: fibers ground 10, 20, 25, and 30 times fibers with a high-consistency refiner, respectively.

Table 1. Fiber size of different treatment stages.

	C-0	C-10	C-20	C-25	C-30
L(n), mm	0.34 ± 0.08	0.29 ± 0.02	0.28 ± 0.07	0.27 ± 0.02	0.27 ± 0.03
L(w), mm	1.20 ± 0.07	0.86 ± 0.04	0.80 ± 0.03	0.80 ± 0.05	0.78 ± 0.08
Fines, %	42.08 ± 1.46	46.16 ± 1.21	47.94 ± 0.98	50.96 ± 1.82	53.52 ± 1.63

The analysis of the sugar composition content of fibers was obtained by ion chromatography. Since the hemicellulose in the bagasse bleaching pulp was mainly xylose, there was also a small amount of arabinose and galactose. As shown in Table 2, all samples contained three carbohydrates: glucose, xylose, and arabinose. Compared with C-0 and C-E, the glucose and arabinose contents were reduced after grinding; instead, the xylose content was increased. The hemicellulose contents had a significant effect on the film formation, wettability, and mechanical properties of the film, through the interaction between hemicellulose and cellulose [36]. The hemicellulose content of C-0 and C-E was 21.21%, and 20.97%, respectively, which indicated that cellulase treatment did not affect the dissolution of hemicellulose. The content of xylose and arabinose in the cellulose suspension after grinding decreased slightly, which proved that the hemicellulose between the secondary wall fiber bundles was separated in the nano/microfibrils suspension after grinding, and a part of the hemicellulose polysaccharide was soluble in water, so the measured xylose and arabinose content decreased. The results prove that exposing sugarcane bagasse to suitable grinding times when grinding with high-consistency refiner plays a key role that can increase the hemicellulose content in microfibers.

Table 2. Sugar composition of fibers after different treatment stages.

Sample	Neutral Sugars and Acidic Oligomers (%)		
	Glucose	Xylose	Arabinose
C-0	74.95 ± 4.43	21.21 ± 0.90	1.16 ± 0.0006
C-E	75.48 ± 5.48	20.97 ± 0.56	1.14 ± 0.0006
C-10	72.64 ± 5.32	19.52 ± 1.52	1.05 ± 0.0006
C-20	72.07 ± 8.89	17.46 ± 1.45	1.01 ± 0.0011
C-25	73.02 ± 7.65	17.16 ± 0.96	0.86 ± 0.0018
C-30	73.39 ± 4.09	18.66 ± 0.98	0.96 ± 0.0004

3.3. Crystallinity of Fibers after Different Treatment Stages

Figure 4 shows crystallinity of nano/microfibrils obtained after various treatments. Two evident diffraction peaks were obtained at 16.0°and 22.5°, corresponding to the (110) and (002) crystal planes of a prototypical cellulose I structure [37]. The highest crystallinity of the cellulose (C-0) was 53.4%. After cellulase treatment of the raw materials (C-E), the crystallinity was 55.2%. There was no significant change in crystallinity after cellulase treatment, indicating that cellulase pretreatment had little effect on the crystallinity of cellulose. Crystallinity indices of C-10, C-20, C-25, and C-30 were 45.0%, 47.3%, 48.2%, and 48.3%, respectively, increasing a little with each grinding time. This may be related to the different content of hemicellulose in cellulose suspension. Hemicellulose was an amorphous polymer. Therefore, the XRD analysis uncovered that fibers had a striking decrease in crystallinity index after the high-consistency refiner pretreatment. The XRD analysis indicated that the crystallization peak of fibers did not change through the mechanical process stages. Part of the hemicellulose was exposed to the fiber suspension during the high-concentration refining process, thereby reducing the crystallinity. However, the number of grindings had little effect on crystallinity.

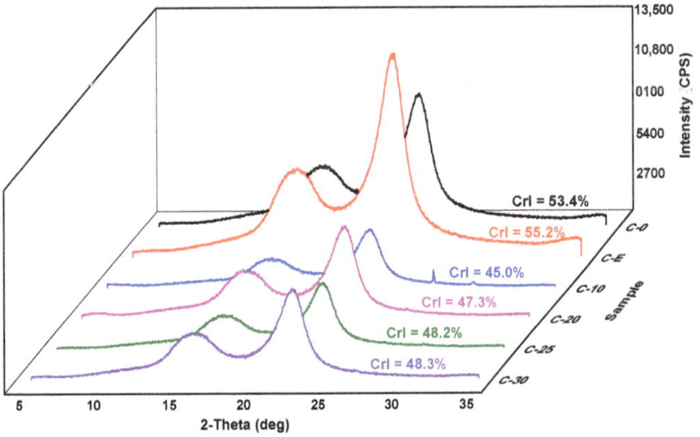

Figure 4. XRD patterns of nano/microfibrils films with different treatment stages.

3.4. Optical Transparency of Nano/Microfibrils Films

Transparency was an important advantage in the use of packaging materials. In Figure 5, the optical properties of the nano/microfibrils films were surveyed and compared with photos from a camera. Moreover, the impact of grinding times on the transmittance properties of nano/microfibrils were researched using a UV–Vis method in Figure 5a. At a 400–800 nm wavelength, the light transmittance of C-0 without mechanical treatment was about 0.97% at room temperature. Compared with C-0, all nano/microfibrils films exhibited high levels of transparency and it significantly increased the transmittance to a maximum of 18.58% (C-25) after mechanical grinding by increasing the nano/microfibrils content. This proved that the fine fibers play a decisive role in the transparency of the film. The higher the fine fiber content, the higher the transparency. However, the transparency of the C-30 film was reduced a little, by 2.36%, as the hemicellulose could enhance the reflected light, and decrease the transmittance of the film [38].

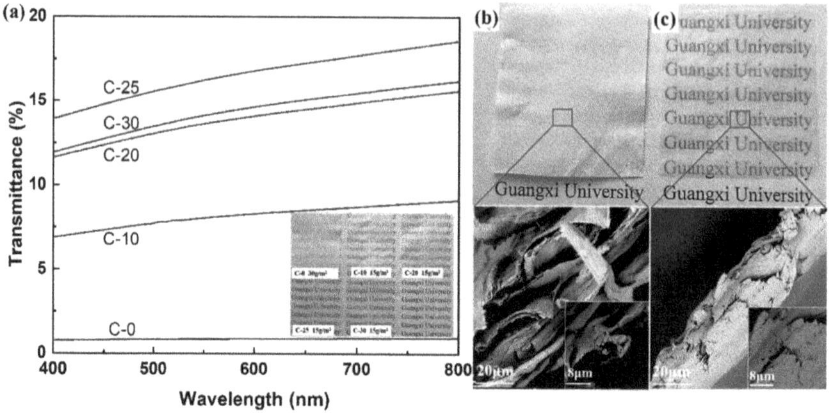

Figure 5. (**a**) Photo and optical transparency of films after different treatment stages, and (**b**) SEM micrographs and digital photographs of the raw material of C-0 film and (**c**) C-30 film after mechanical treatment.

SEM images of the film cross-section showed that the surface of the untreated fiber film was nonuniform with a large number of fiber fragments and a big porosity between fibers (Figure 5b). However, a more uniform surface morphology, smaller fibers and nano/microfibrils bundles, and a more compact structure of the cross-section were observed on the films after grinding (Figure 5c). This phenomenon indicated that as the grinding time increased, more nano/microfibrils were obtained, and denser and better structure films were formed. The tight overlap between the fibers eliminated the scattering and reflection of light inside the film, so that the transmittance of the film was significantly increased. Therefore, the densification and refinement of the nano/microfibrils structure directly led to the higher transmittance and strength of films [39].

3.5. Hydrophobic Properties and Principle of Nano/Microfibrils Film

The surface hydrophobicity of raw materials and nano/microfibrils films were evaluated through the measurements of the surface contact angles, and the hydrophobic mechanism of the cellulose membrane was analyzed (in Figure 6). The CA of the C-10 film was 98.7°, indicating that the nano/microfibrils film was hydrophobic (CA > 90°). With the increase of grinding times, the hydrophobicity first increased and then decreased (in Figure 6a). The CA of the nano/microfibrils film was the highest at 113° after grinding 25 times (Figure 6a). This may be due to the cross-linking of hemicellulose and nano/microfibrils during the film-forming process. Hemicellulose adsorbed on the pores of nano/microfibrils as fillers, and the surplus hemicellulose on the surface of the cellulose to form a large number of micro/nanostructures. Thus, it had high hydrophobicity. The surface wettability of the material depended on the surface chemistry and the surface micro/nanostructure [40]. These two parameters determined the level of adhesion between the droplet and the surface. The nano/microfibrils film was hydrophobic after grinding (in Figure 6b). The primary reason was that part of the hemicellulose existed in the form of single molecules or colloidal aggregates in the cellulose suspension after the grinding treatment. Then, hemicellulose diffused and adsorbed to the cellulose surface to form nanoparticles during the film-forming process at 40 °C. The film surface owned a certain roughness and was hydrophobic.

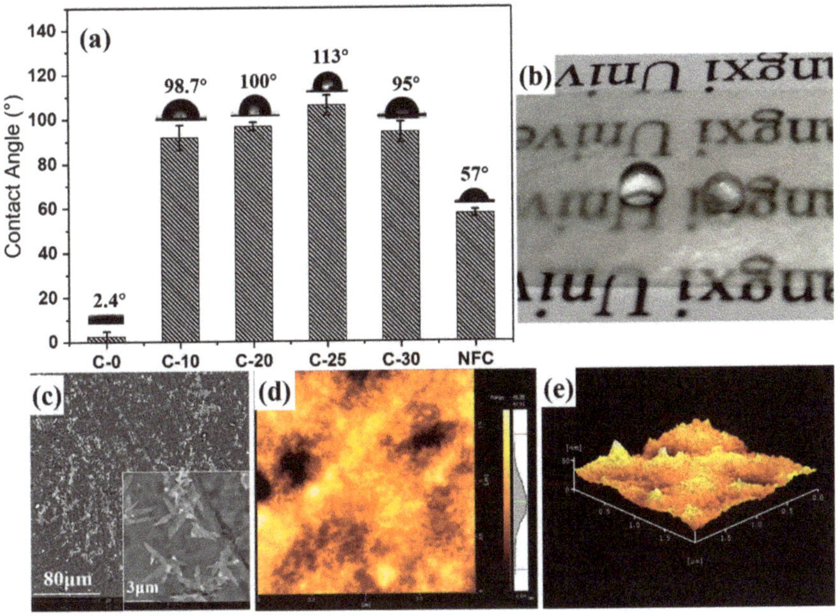

Figure 6. (**a**) Contact angle of nano/microfibrils films, (**b**) digital photographs of C-25 film hydrophobicity, (**c**) SEM micrographs of the C-25 film, (**d**) AFM surface topography (2 × 2 μm^2), and (**e**) 3D image of the C-25 film.

In Figure 6c, many dense white micro/nanostructures on the surface of the film exist, which may be because the exposed hemicellulose was adsorbed on the surface of nanocellulose driven by entropy [41]. As previously [42] reported, the xylan is likely to assemble into nanoparticulate aggregates in solution first and it was deposited like this onto the cellulose fibrils, forming a large number of white micro/nanoprotrusions consistent with the C-25 film surface 3D image analysis in Figure 6d,e. The surface roughness (Sa) was 4.493 nm.

However, the CA measurements of C-30 film was reduced to 95°, which could be due to the increase of fine fiber, which broke the balance between hemicellulose and fine cellulose in the fiber suspension system. After hemicellulose adsorbed on the pores of nano/microfibrils, the surplus hemicellulose on the surface of cellulose decreased. Because the micro/nanostructure formed on the surface of the film decreased, the contact angle of C-30 slightly decreased. This phenomenon indicated that the particle size of the fiber reduced after grinding, and that the nano/microfibrils became the principal component of the fiber suspension. This showed that the size of fibrils determined the macrostructure properties of the film and the hemicellulose content in the suspension determined the microstructure properties of the film. The micro/nanostructure formed by the exposure of the hemicellulose played the leading role in affecting the cellulose film hydrophobicity properties.

3.6. Barrier Performance of Nano/Microfibrils Film

The water vapor barrier performance of the substrate side (contacting the polystyrene template) as a moisture barrier was studied by the water vapor transmission rate (Figure 7a). The water vapor transmission rate of the untreated cellulose film was about 2426.59 g/(m^2·24 h). It significantly improved the barrier properties of the film after grinding treatment. The fine fiber content of the C-25 film was 50.96%, its permeability pass rate was 1038.16 g/(m^2·24 h), and the barrier performance was about twice as high as before. The water vapor transmittance was not a linear function of the fine fiber content, although, with the increase of the fine fiber content, the water vapor transmittance decreased (Figure 7b). This highlighted that the content of fine fibers increased the barrier property of the film. The water vapor transmission rate of the C-25 film was as high as 1038.16 g/(m^2·24 h) under these conditions. According to previous studies, biodegradable films have been studied to use nanocellulose as a coating. Compared with the moisture resistance of nanocellulose film, the WVTR of CNF-coated paper was about 300 g/(m^2·24 h), and the moisture resistance was improved by 55%. The water vapor permeability of CNF was about 960–980 g/(m^2·24 h) [43]. It was the same as that of traditional CNF film. However, the preparation process of nano/microfibrils film in this study was simple. After grinding treatment, the barrier properties of the film were significantly improved. This phenomenon was primarily due to the fact that the fine fibers became the main component of the fiber slurry after the grinding treatment. During the film-forming process, microfibers and nanofibers were inserted between the large fiber bundles, the pores between the large fibers were filled with fine fibrils, which separated a part of the hemicellulose into the nano/microfiber suspension during the grinding process. There was adsorption between hemicellulose and cellulose fibers, and hemicellulose was used as the adhesive to fill the pores between the fine fibers again in the film-forming process, so the barrier property of the film was improved remarkably. However, when the content of fine fibers was 53.52%, the water vapor transmission rate of the film increased. This was because the increase of fine fibers was much greater than the increase in hemicellulose, and a large number of pores between the fine fibers could not be filled by the increased hemicellulose, resulting in a slight decrease in the barrier properties of the film. Therefore, fine fiber and hemicellulose were the key factors to improve the water vapor barrier performance of nano/microfiber films.

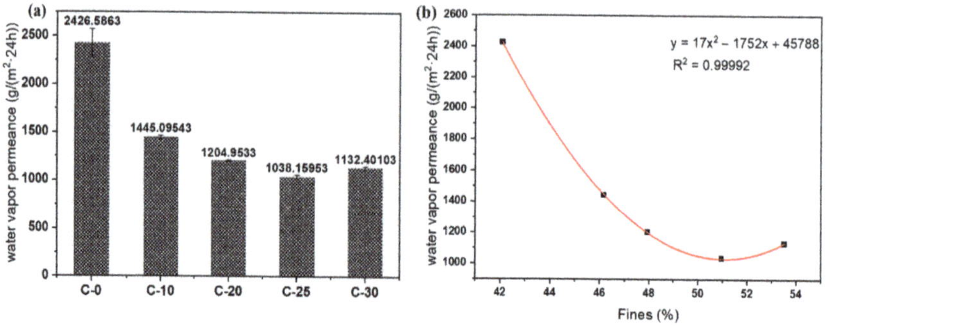

Figure 7. (**a**) Water vapor transmittance of nano/microfibrils films. (**b**) Fines (x) as a function of water vapor transmission rate (y). The solid line represents the nonlinear fit, with correlation coefficient R^2 = 0.99992.

3.7. Mechanical Properties of Films

The thickness, folding endurance, and elastic modulus of the nano/microfibril films are recorded in Figure 8. C-10 displayed a higher thickness, lower folding endurance, and lower elastic modulus, which were 31.1 μm, 1.59, and 2671.18 MPa, respectively. Compared to the films after different treatment stages, C-30 displayed a lower thickness, higher folding endurance, and higher elastic modulus, at 25.9 μm, 1.50, and 4140.60 MPa, respectively, which indicated that the film was more compact and flexible. In the lower right corner of Figure 8 is a paper airplane made of C-30 film, which showed high pliability. This was because the content of nanocellulose in the cellulose system was 36.34% after grinding 30 times, which was about triple the content of nanocellulose in the 10-time grinding system. The elastic modulus of the films increased by about 62% when the content of nanocellulose was 13–36%. The tensile strength tripled. Moreover, because the cellulose particle size distribution was relatively rich, nanocellulose and microcellulose [(0–9)] could provide a certain degree of strength and stiffness for the film. The folding resistance decreased from 1.7 to 1.5 and decreased with the decrease of large-size cellulose. This showed that a larger size of cellulose could make the film have a certain degree of flexibility. Therefore, the thin film prepared by this method had not only a certain strength and rigidity but also certain folding properties.

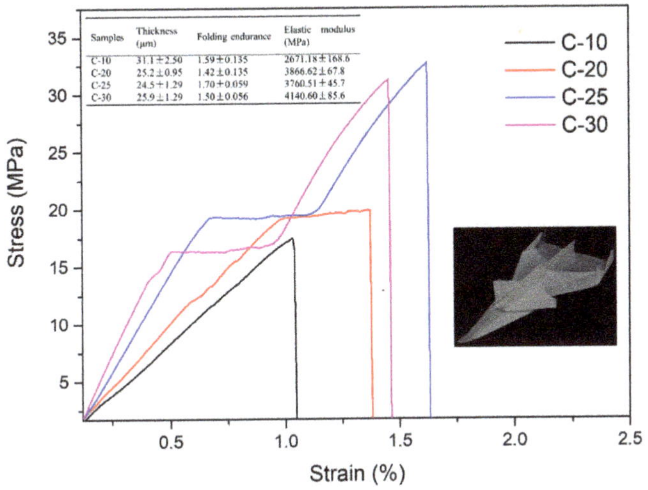

Figure 8. Stress–strain curves of the nano/microfibrils-based films.

Then, according to Figure 8, the C-20, C-25, and C-30 films had a stress platform at about 0.5–1.5% strain, with yielding remaining unchanged with the increase of strain during the stretching process, which was a nonuniform stretching stage. The elastic moduli were 3866.62 MPa, 3760.51 MPa, and 4140.60 MPa, respectively. This was because the film contained fibers of different sizes. During the stretching process, the crystal structure inside the film changed (Figure 9). Under the action of an external force, the crystal lattice molecules of micron-sized fibers and nano-sized fibers began to adjust and orient in the direction of the external force, destroying large-sized cellulose [44,45]. The reduced tensile strength of the lattice failure offset the increased tensile strength after the orientation of the nano/microfibers molecules, forming a curved platform area [46]. The findings indicated that the increase of a certain number of fine fibers could not only increase the hardness of the film, but also could give the film flexibility and plasticity. After passing through the plateau area, the C-25 and C-30 films appeared in a strain hardening stage, the stress rising sharply. This was because the molecules oriented under the stronger stress were highly oriented, and the materials formed a new higher-order crystal structure [46]. The tensile strength of C-25 and C-30 could be as high as 31.21 MPa and 32.70 MPa.

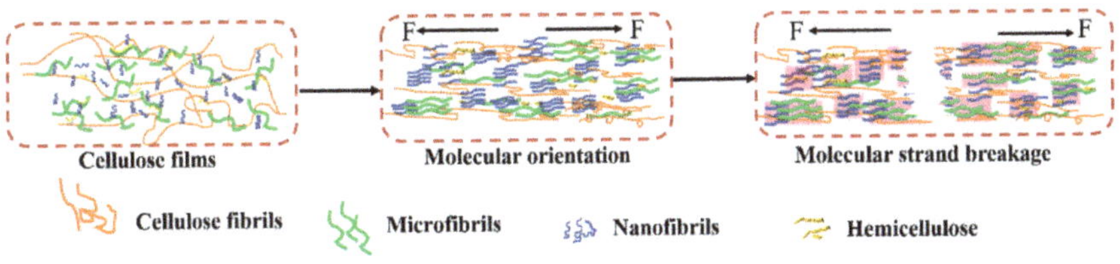

Figure 9. Schematic of tensile mechanism of films.

4. Conclusions

In this study, a novel process was proposed to obtain a hydrophobic sugarcane bagasse film with high pliability. The C-25 film had the best comprehensive performance, with better hydrophobic properties (CA = 113°) and a better elastic module of 4140.60 MPa. Fibrils with different hemicellulose and nano/microfibrils content were prepared by adjusting the number of times of the grinding. In addition, unique hydrophobic properties of films were obtained, due to the hemicellulose adsorption in the temperature-driven film-forming process. A better hydrophobic effect and high pliability were attained, and the adsorption effect of hemicellulose was effectively demonstrated. This research greatly improves the practical value of nonuniform cellulose in the field of packaging materials, which lays a foundation for the preparation of functional biomaterials without modification.

Author Contributions: Conceptualization, Y.L. (Yang Liu); methodology, C.L.; software, M.Y. and Y.Z.; validation, Y.L. (Yang Liu) and H.Z.; formal analysis, M.Y.; investigation, M.Y. and Y.L. (Yan Li); resources, Y.L. (Yang Liu); data curation, M.Y. and Y.L. (Yang Liu); writing—original draft preparation, M.Y.; writing—review and editing, Y.L. (Yang Liu); visualization, M.Y.; supervision, Y.L. (Yang Liu); project administration, H.Z.; funding acquisition, Y.L. (Yang Liu), C.L. and H.Z. All authors have read and agreed to the published version of the manuscript.

Funding: This study was supported by the National Natural Science Foundation of China (NSFC, 22068004) and the Natural Science Foundation of Guangxi, China (2020GXNSFAA159027, 2019GXNSFBA185006, 2020GXNSFBA159023), Foundation (No.2019ZR03) of Guangxi Key La-boratory of Clean Pulp & Papermaking and Pollution Control, College of Light Industry and Food Engineering, Guangxi University and Open Funding Project of the State Key Laboratory of Bio-catalysis and Enzyme Engineering (SKLBEE2020009).

Informed Consent Statement: Informed consent was obtained from all subjects involved in the study.

Data Availability Statement: The data presented in this study are available in the manuscript's figure.

Conflicts of Interest: The authors declare no conflict of interest.

References

1. Mohanty, A.K.; Vivekanandhan, S.; Pin, J.M.; Misra, M. Composites from renewable and sustainable resources: Challenges and innovations. *Science* **2018**, *362*, 536–542. [CrossRef]
2. Vanitjinda, G.; Nimchua, T.; Sukyai, P. Effect of xylanase-assisted pretreatment on the properties of cellulose and regenerated cellulose films from sugarcane bagasse. *Int. J. Biol. Macromol.* **2019**, *122*, 503–516. [CrossRef] [PubMed]
3. Xu, J.T.; Chen, X.Q. Preparation and characterization of spherical cellulose nanocrystals with high purity by the composite enzymolysis of pulp fibers. *Bioresour. Technol.* **2019**, *291*, 121842. [CrossRef] [PubMed]
4. Molina, N.F.; Brito, Y.F.; Benavides, J.M.P. Recycling of Residual Polymers Reinforced with Natural Fibers as a Sustainable Alternative: A Review. *Polymers* **2021**, *13*, 3612. [CrossRef] [PubMed]
5. Hassan, M.M.; Fowler, I.J. Thermal, mechanical, and rheological properties of micro-fibrillated cellulose-reinforced starch foams crosslinked with polysiloxane-based cross-linking agents. *Int. J. Biol. Macromol.* **2022**, *205*, 55–65. [CrossRef] [PubMed]
6. Kontturi, E.; Spirk, S. Ultrathin Films of Cellulose: A Materials Perspective. *Front. Chem.* **2019**, *7*, 488. [CrossRef] [PubMed]
7. Botta, L.; Titone, V.; Mistretta, M.C.; La Mantia, F.P.; Modica, A.; Bruno, M.; Sottile, F.; Lopresti, F. PBAT Based Composites Reinforced with Microcrystalline Cellulose Obtained from Softwood Almond Shells. *Polymers* **2021**, *13*, 2643. [CrossRef] [PubMed]
8. Garrido-Romero, M.; Aguado, R.; Moral, A.; Brindley, C.; Ballesteros, M. From traditional paper to nanocomposite films: Analysis of global research into cellulose for food packaging. *Food Packag. Shelf Life* **2022**, *31*, 100788. [CrossRef]
9. Haney, R.; Kollarigowda, R.H.; Wiegart, L.; Ramakrishnan, S. Surface-Functionalized Cellulose Nanocrystals as Nanofillers for Crosslinking Processes: Implications for Thermosetting Resins. *ACS Appl. Nano Mater.* **2022**, *5*, 1891–1901. [CrossRef]
10. Tekinalp, H.L.; Meng, X.; Lu, Y.; Kunc, V.; Love, L.J.; Peter, W.H.; Ozcan, S. High modulus biocomposites via additive manufacturing: Cellulose nanofibril networks as "microsponges". *Compos. Part B Eng.* **2019**, *173*, 106817. [CrossRef]
11. Zhao, K.; Wang, W.; Teng, A.; Zhang, K.; Ma, Y.; Duan, S.; Li, S.; Guo, Y. Using cellulose nanofibers to reinforce polysaccharide films: Blending vs layer-by-layer casting. *Carbohydr. Polym.* **2020**, *227*, 115264. [CrossRef]
12. Chowdhury, R.A.; Clarkson, C.; Youngblood, J. Continuous roll-to-roll fabrication of transparent cellulose nanocrystal (CNC) coatings with controlled anisotropy. *Cellulose* **2018**, *25*, 1769–1781. [CrossRef]
13. Li, K.; Jin, S.; Li, J.; Chen, H. Improvement in antibacterial and functional properties of mussel-inspired cellulose nanofibrils/gelatin nanocomposites incorporated with graphene oxide for active packaging. *Ind. Crop. Prod.* **2019**, *132*, 197–212. [CrossRef]

14. Chowdhury, R.A.; Nuruddin, M.; Clarkson, C.; Montes, F.; Howarter, J.; Youngblood, J.P. Cellulose Nanocrystal (CNC) Coatings with Controlled Anisotropy as High-Performance Gas Barrier Films. *ACS Appl. Mater. Interfaces* **2019**, *11*, 1376–1383. [CrossRef] [PubMed]
15. Azeredo, H.M.C.; Rosa, M.F.; Mattoso, L.H.C. Nanocellulose in bio-based food packaging applications. *Ind. Crop. Prod.* **2017**, *97*, 664–671. [CrossRef]
16. Zhang, S.; Li, J.; Chen, S.; Zhang, X.; Ma, J.; He, J. Oxidized cellulose-based hemostatic materials. *Carbohydr. Polym.* **2020**, *230*, 115585. [CrossRef]
17. Hao, P.; Zhao, Z.; Tian, J.; Li, H.; Sang, Y.; Yu, G.; Cai, H.; Liu, H.; Wong, C.P.; Umar, A. Hierarchical porous carbon aerogel derived from bagasse for high performance supercapacitor electrode. *Nanoscale* **2014**, *6*, 12120–12129. [CrossRef]
18. Thomas, B.; Raj, M.C.; Athira, K.B.; Rubiah, M.H.; Joy, J.; Moores, A.; Drisko, G.L.; Sanchez, C. Nanocellulose, a Versatile Green Platform: From Biosources to Materials and Their Applications. *Chem. Rev.* **2018**, *118*, 11575–11625. [CrossRef] [PubMed]
19. Huang, P.; Zhao, Y.; Kuga, S.; Wu, M.; Huang, Y. A versatile method for producing functionalized cellulose nanofibers and their application. *Nanoscale* **2016**, *8*, 3753–3759. [CrossRef]
20. Siró, I.; Plackett, D. Microfibrillated cellulose and new nanocomposite materials: A review. *Cellulose* **2010**, *17*, 459–494. [CrossRef]
21. Yang, Q.; Saito, T.; Berglund, L.A.; Isogai, A. Cellulose nanofibrils improve the properties of all-cellulose composites by the nano-reinforcement mechanism and nanofibril-induced crystallization. *Nanoscale* **2015**, *7*, 17957–17963. [CrossRef] [PubMed]
22. Klemm, D.; Kramer, F.; Moritz, S.; Lindstrom, T.; Ankerfors, M.; Gray, D.; Dorris, A. Nanocelluloses: A new family of nature-based materials. *Angew. Chem. Int. Ed. Engl.* **2011**, *50*, 5438–5466. [CrossRef] [PubMed]
23. Pinem, M.P.; Wardhono, E.Y.; Nadaud, F.; Clausse, D.; Saleh, K.; Guenin, E. Nanofluid to Nanocomposite Film: Chitosan and Cellulose-Based Edible Packaging. *Nanomaterials* **2020**, *10*, 660. [CrossRef] [PubMed]
24. Zhang, Y.; Liu, W.; Huang, W.; Ding, Y.; Song, L.; Zheng, S.; Wang, Z. The toughening of polymeric glasses using cellulose without sacrificing transparency. *Ind. Crop. Prod.* **2019**, *142*, 111842. [CrossRef]
25. Shu, L.; Zhang, X.F.; Wang, Z.G.; Yao, J.F. Structure reorganization of cellulose hydrogel by green solvent exchange for potential plastic replacement. *Carbohydr. Polym.* **2022**, *275*, 118695. [CrossRef] [PubMed]
26. Yang, W.; Qi, G.; Kenny, J.M.; Puglia, D.; Ma, P. Effect of Cellulose Nanocrystals and Lignin Nanoparticles on Mechanical, Antioxidant and Water Vapour Barrier Properties of Glutaraldehyde Crosslinked PVA Films. *Polymers* **2020**, *12*, 1364. [CrossRef] [PubMed]
27. Rol, F.; Belgacem, M.N.; Gandini, A.; Bras, J. Recent advances in surface-modified cellulose nanofibrils. *Prog. Polym. Sci.* **2019**, *88*, 241–264. [CrossRef]
28. Sehaqui, H.; Zimmermann, T.; Tingaut, P. Hydrophobic cellulose nanopaper through a mild esterification procedure. *Cellulose* **2013**, *21*, 367–382. [CrossRef]
29. Dhali, K.; Daver, F.; Cass, P.; Adhikari, B. Surface modification of the cellulose nanocrystals through vinyl silane grafting. *Int. J. Biol. Macromol.* **2022**, *200*, 397–408. [CrossRef] [PubMed]
30. Calderón-Vergara, L.A.; Ovalle-Serrano, S.A.; Blanco-Tirado, C.; Combariza, M.Y. Influence of post-oxidation reactions on the physicochemical properties of TEMPO-oxidized cellulose nanofibers before and after amidation. *Cellulose* **2019**, *27*, 1273–1288. [CrossRef]
31. Farhat, W.; Venditti, R.A.; Hubbe, M.; Taha, M.; Becquart, F.; Ayoub, A. A Review of Water-Resistant Hemicellulose-Based Materials: Processing and Applications. *ChemSusChem* **2017**, *10*, 305–323. [CrossRef] [PubMed]
32. Chen, Y.; Chen, D.; Ma, Y.; Yang, W. Multiple levels hydrophobic modification of polymeric substrates by UV-grafting polymerization with TFEMA as monomer. *J. Polym. Sci. Part A Polym. Chem.* **2014**, *52*, 1059–1067. [CrossRef]
33. Henriksson, M.; Henriksson, G.; Berglund, L.A.; Lindström, T. An environmentally friendly method for enzyme-assisted preparation of microfibrillated cellulose (MFC) nanofibers. *Eur. Polym. J.* **2007**, *43*, 3434–3441. [CrossRef]
34. Hult, E.-L.; Iotti, M.; Lenes, M. Efficient approach to high barrier packaging using microfibrillar cellulose and shellac. *Cellulose* **2010**, *17*, 575–586. [CrossRef]
35. Tenhunen, T.-M.; Peresin, M.S.; Penttilä, P.A.; Pere, J.; Serimaa, R.; Tammelin, T. Significance of xylan on the stability and water interactions of cellulosic nanofibrils. *React. Funct. Polym.* **2014**, *85*, 157–166. [CrossRef]
36. Claro, F.C.; Matos, M.; Jordao, C.; Avelino, F.; Lomonaco, D.; Magalhaes, W.L.E. Enhanced microfibrillated cellulose-based film by controlling the hemicellulose content and MFC rheology. *Carbohydr. Polym.* **2019**, *218*, 307–314. [CrossRef] [PubMed]
37. Besbes, I.; Alila, S.; Boufi, S. Nanofibrillated cellulose from TEMPO-oxidized eucalyptus fibres: Effect of the carboxyl content. *Carbohydr. Polym.* **2011**, *84*, 975–983. [CrossRef]
38. Meng, Q.; Fu, S.; Lucia, L.A. The role of heteropolysaccharides in developing oxidized cellulose nanofibrils. *Carbohydr. Polym.* **2016**, *144*, 187–195. [CrossRef]
39. Nogi, M.; Iwamoto, S.; Nakagaito, A.N.; Yano, H. Optically Transparent Nanofiber Paper. *Adv. Mater.* **2009**, *21*, 1595–1598. [CrossRef]
40. Bosmans, T.J.; Stepan, A.M.; Toriz, G.; Renneckar, S.; Karabulut, E.; Wagberg, L.; Gatenholm, P. Assembly of debranched xylan from solution and on nanocellulosic surfaces. *Biomacromolecules* **2014**, *15*, 924–930. [CrossRef] [PubMed]
41. Yao, M.; Liang, C.; Yao, S.; Liu, Y.; Zhao, H.; Qin, C. Kinetics and Thermodynamics of Hemicellulose Adsorption onto Nanofibril Cellulose Surfaces by QCM-D. *ACS Omega* **2021**, *6*, 30618–30626. [CrossRef] [PubMed]

42. Linder, Å.; Bergman, R.; Bodin, A.; Gatenholm, P. Mechanism of Assembly of Xylan onto Cellulose Surfaces. *Langmuir* **2003**, *19*, 5072–5077. [CrossRef]
43. Yook, S.; Park, H.; Park, H.; Lee, S.-Y.; Kwon, J.; Youn, H.J. Barrier coatings with various types of cellulose nanofibrils and their barrier properties. *Cellulose* **2020**, *27*, 4509–4523. [CrossRef]
44. An, M.F.; Zhang, Q.L.; Ye, K.; Lin, Y.F.; Wang, D.L.; Chen, W.; Yin, P.C.; Meng, L.P.; Li, L.B. Structural evolution of cellulose triacetate film during stretching deformation: An in-situ synchrotron radiation wide-angle X-Ray scattering study. *Polymer* **2019**, *182*, 121815. [CrossRef]
45. Sharma, A.; Thakre, S.; Kumaraswamy, G. Microstructural differences between Viscose and Lyocell revealed by in-situ studies of wet and dry fibers. *Cellulose* **2020**, *27*, 1195–1206. [CrossRef]
46. Kim, J.W.; Park, S.; Harper, D.P.; Rials, T.G. Structure and Thermomechanical Properties of Stretched Cellulose Films. *J. Appl. Polym. Sci.* **2013**, *128*, 181–187. [CrossRef]

MDPI
St. Alban-Anlage 66
4052 Basel
Switzerland
www.mdpi.com

Polymers Editorial Office
E-mail: polymers@mdpi.com
www.mdpi.com/journal/polymers

www.ingramcontent.com/pod-product-compliance
Lightning Source LLC
LaVergne TN
LVHW070642100526
838202LV00013B/862
* 9 7 8 3 0 3 6 5 9 4 9 8 9 *